T0326369

Genomic and Precision Medicine

Genomic and Precision Medicine

Primary Care

Third Edition

Edited by
Sean P. David
Stanford University School of Medicine, Stanford, CA, United States

Series Editors
Geoffrey S. Ginsburg
Duke University, Durham, NC, United States

Huntington F. Willard
Marine Biological Laboratory, Woods Hole, MA, United States;
University of Chicago, Chicago, IL, United States

ACADEMIC PRESS

An imprint of Elsevier
elsevier.com

British Library Cataloguing-in-Publication Data
A catalogue record for this book is available from the British Library

Library of Congress Cataloging-in-Publication Data
A catalog record for this book is available from the Library of Congress

ISBN: 978-0-12-800685-6

For Information on all Academic Press publications
visit our website at https://www.elsevier.com/books-and-journals

Working together
to grow libraries in
developing countries

www.elsevier.com • www.bookaid.org

Publisher: Mica Haley
Acquisition Editor: Peter Linsley
Editorial Project Manager: Lisa Eppich
Production Project Manager: Edward Taylor
Designer: Matthew Limbert

Typeset by MPS Limited, Chennai, India

Contents

4. Genetic Testing for Rare and Undiagnosed Diseases

Thomas Morgan

5. Health Risk Assessments, Family Health History, and Predictive Genetic/Pharmacogenetic Testing

Maria Esperanza Bregendahl, Lori A. Orlando and Latha Palaniappan

6. Pharmacogenetics and Pharmacogenomics

J. Kevin Hicks and Howard L. McLeod

7. Hypertension

Patricia B. Munroe and Helen R. Warren

8. Coronary Artery Disease and Myocardial Infarction

Themistocles L. Assimes

9. Lung Cancer

Yaron B. Gesthalter, Ehab Billatos and Hasmeena Kathuria

10. Breast Cancer

Paul K. Marcom

11. Colorectal Cancer

Roland P. Kuiper, Robbert D.A. Weren and Ad Geurts van Kessel

12. Prostate Cancer

Wennuan Liu, Rong Na, Carly Conran and Jianfeng Xu

13. Asthma

Michael J. McGeachie, Damien C. Croteau-Chonka and Scott T. Weiss

14. Diabetes

Miriam S. Udler and Jose C. Florez

15. Metabolic Syndrome

Matthew B. Lanktree and Robert A. Hegele

List of Contributors

Themistocles L. Assimes Stanford University School of Medicine, Stanford, CA, United States

Ehab Billatos Boston University School of Medicine, Boston, MA, United States

Maria Esperanza Bregendahl Stanford University School of Medicine, Stanford, CA, United States

Maria Chahrour University of Texas Southwestern Medical Center, Dallas, TX, United States

Carly Conran NorthShore University HealthSystem, Evanston, IL, United States

Damien C. Croteau-Chonka Harvard Medical School, Boston, MA, United States

Jose C. Florez Massachusetts General Hospital, Boston, MA, United States

Yaron B. Gesthalter Boston University School of Medicine, Boston, MA, United States

Susanne B. Haga Duke University School of Medicine, Durham, NC, United States

Robert A. Hegele University of Western Ontario, London, ON, Canada

J. Kevin Hicks DeBartolo Family Personalized Medicine Institute, Moffitt Cancer Center, Tampa, FL, United States

Jean Jenkins National Institutes of Health, National Human Genome Research Institute, Bethesda, MD, United States

Samuel G. Johnson Virginia Commonwealth University School of Pharmacy, Washington, DC, United States

Hasmeena Kathuria Boston University School of Medicine, Boston, MA, United States

Roland P. Kuiper Radboud University Medical Centre, Nijmegen, The Netherlands, Princess Máxima Center for Pediatric Onolocy, Nijmegen, The Netherlands

Matthew B. Lanktree McMaster University, Hamilton, ON, Canada

Wennuan Liu NorthShore University HealthSystem, Evanston, IL, United States

Paul K. Marcom Duke University School of Medicine, Durham, NC, United States

Michael J. McGeachie Harvard Medical School, Boston, MA, United States

Howard L. McLeod DeBartolo Family Personalized Medicine Institute, Moffitt Cancer Center, Tampa, FL, United States

Thomas Morgan Novartis Institutes for Biomedical Research, Cambridge, MA, United States

Patricia B. Munroe Queen Mary University of London, London, United Kingdom

Rong Na NorthShore University HealthSystem, Evanston, IL, United States

Lori A. Orlando Duke University, Durham, NC, United States

Latha Palaniappan Stanford University School of Medicine, Stanford, CA, United States

Keyur Patel Toronto General Hospital, Toronto, ON, Canada; Toronto University, Toronto, ON, Canada

Laura Lyman Rodriguez National Institutes of Health, National Human Genome Research Institute, Bethesda, MD, United States

Akanksha Saxena University of Texas Southwestern Medical Center, Dallas, TX, United States

Nicholas A. Shackel University of New South Wales, Sydney, NSW, Australia; Ingham Institute, Liverpool, NSW, Australia; Liverpool Hospital, Liverpool, NSW, Australia

Thomas Tu Universitätsklinikum Heidelberg, Heidelberg, Germany

Miriam S. Udler Massachusetts General Hospital, Boston, MA, United States

Ad Geurts van Kessel Radboud University Medical Centre, Nijmegen, The Netherlands

Helen R. Warren Queen Mary University of London, London, United Kingdom

Scott T. Weiss Harvard Medical School, Boston, MA, United States

Robbert D.A. Weren Radboud University Medical Centre, Nijmegen, The Netherlands

Jianfeng Xu NorthShore University HealthSystem, Evanston, IL, United States

Preface

From the time of completion of the Second Edition of "Genomic and Personalized Medicine" until today, the broad field of genomic medicine has advanced from a period of rapid discovery from genome-wide association studies to—according to the National Human Genome Research Institute's Eric Green, M.D.—enhancing the understanding of the biology of many diseases and has entered into a phase of advancing the science of medicine with an ultimate endpoint of improving the effectiveness of health care. Although many evidence gaps remain, such as the need to demonstrate clinical validity and clinical utility for most disease associated genetic variants and pharmacogenomics, the potential of "precision medicine" to vastly improve the efficacy of treatments for cancer, neurological diseases, preventive medicine, and reducing health disparities was deemed substantial enough for President Obama to launch the "Precision Medicine Initiative" in his 2015 State-of-the-Union Address. However, as the field of medicine stands poised to advance genomic medicine on many fronts, ensuring that the entire population of patients can benefit from innovations in an evidence-based fashion will rely on partnerships between primary care physicians and clinical genetics professionals. The present text addresses a continuum of domains of genomic medicine that are germane to the primary care of patients and the scope of primary care grounded in family history taking and appropriate referrals, such as the continuing education of all health professionals, what genetic and genomic testing and precision treatments are presently available, and what is on the horizon for a wide range of conditions and population health challenges. It is our hope that primary care providers will, as a result of using this text, develop transdisciplinary thinking and begin to share a common language and sense of partnerships with clinical genetics professionals as we continue to forge this new frontier together for the ultimate goal of improved health, healthcare and more evidence-based, personalized, and patient-centered medicine for the 21st century.

The present, "Primary Care" volume is one of a series of texts tailored to clinicians from a range of medical specialties and academic disciplines. As genomic medicine transitions from a research aspiration to an integral component of personalized health care, the role of primary care physicians and allied health professionals is becoming paramount in the goal of leveraging genomic knowledge to better care for diverse populations. We, therefore, have sought to cover a sample of topics that are essential for building a foundation of general knowledge that we hope will guide primary care clinicians, educators, and healthcare institutions in advancing translation into practice. The range of topics are not comprehensive, but do provide entry-level content for a range of major health concerns that are equally useful for students, residents, attending physicians, other primary care health professionals, healthcare organizations, and policy makers.

The preface to the previous volume asserted that "We stand at the dawn of a profound change in science and medicine's predictive nature and in our understanding of the biological underpinnings of health and disease" but noted "grand challenges" to implementation of precision medicine from the potential to exaggerate health disparities to the need for educating the healthcare workforce and developing frameworks for aligning appropriate delivery models with good evidence and appropriate economic incentives. We have attempted to address many of these grand challenges, which align with the National Human Genome Research Institute's Grand Challenges II (genomics to health) and III (genomics to society) in forward-thinking chapters spanning multiple domains including:

- The role of primary care clinicians in genomic medicine and frameworks for integration with primary care redesign and clinical implementation science
- Genetic screening and diagnostic testing for rare diseases from preconception to neonates and throughout the life span
- Family history and its application to health risk assessment and predictive genetic testing
- Educational strategies for genomic medicine in primary care
- Policy, ethical, and societal considerations
- Current precision medicine treatments and future directions in research for common diseases (cancer, cardiovascular disease, hypertension, diabetes and metabolic syndrome, and autism spectrum disorder)

These topics represent only a fraction of the many diseases and thousands of type of genetic tests and clinical scenarios that are rapidly expanding in number. As Francis Collins envisioned in 2003, with increasing knowledge about the role of genetics in disease risk prediction, "many primary care physicians will become practitioners of genomic medicine, having to explain complex statistical risk information to healthy individuals who are seeking to enhance their chances of staying well. This will require substantial advances in the understanding of genetics by a wide range of clinicians." This prediction was prescient given the burgeoning research output of genetic studies and the availability of direct-to-consumer genetic testing and diminishing costs of next generation sequencing. We hope this text provides utility to all of us who practice primary care and prevention as vital stakeholders poised to learn together to build a more patient-centered, evidence-based, and personalized healthcare experience for all patients.

Sean P. David,
Huntington F. Willard
and Geoffrey S. Ginsburg

Chapter 1

Genomic Medicine in Primary Care

Samuel G. Johnson
Virginia Commonwealth University School of Pharmacy, Washington, DC, United States

Chapter Outline

The last two decades has seen unprecedented genomic discovery and major clinical advances in the management of common diseases including cardiovascular disease and cancer [1]. In the fast-paced practice environment with limited time to stay current with new medical literature, many primary care physicians may not highly prioritize continuing education about genomic medicine in particular. Moreover, many primary care physicians express little confidence in their ability to make clinical decisions when genetic or genomic information is involved. This challenge is not unique to primary care; as new medical discoveries often outpace an individual specialist practitioner's ability to master it as well. Even so, primary care physicians now have better resources to help them incorporate new medical knowledge, including genomic medicine, into practice [1]. Emphasis on generalist practice principles, especially the value of maintaining a broad knowledge base, spurs many to stay current with new literature while prioritizing what is most important to their patients' health. This may manifest as reliance on clinical practice guidelines in addition to cultivation of networks of trusted colleagues (through both informal "curbside" and formal consultations). With the rise of genomic medicine, these networks will increasingly include geneticists, genetic counselors, informaticists, and pharmacists.

Knowledge learned through such patient-centered interactions contributes to innovation in systems design that streamlines management of complex medical information. Such clinical decision support systems are increasingly being introduced into electronic health records across North America, and provide just-in-time alerts to front-line clinicians for many issues; including, adverse drug interactions or overdue health maintenance interventions. Efforts are already underway within leading health systems to incorporate genomic data into the electronic records in a similar fashion, in order to create systems to help manage large quantities of genomic information [2]. Since it is also important to prepare the future primary care physician workforce for this innovation, primary care residencies may also include more focus on genomics education and training [3].

The potential benefits of genomic medicine are many and include improved disease-risk assessment as well as precise selection of drug therapy. Potential detriments include provider and patient anxiety, the unnecessary and expensive tests and procedures that might follow from a genomic result, and many scenarios where current scientific understanding fails to ascertain actionable results [4]. Further, despite rapid advances in understanding the genetic architecture of many diseases, translational research that demonstrates outcome improvement from this knowledge has lagged. The full risk–benefit ratio is thus unknown for almost all genomic tests, particularly for long-term clinical outcomes. Since primary care practice fosters a culture of evidence-based medicine that seeks to maximize health benefits and minimize unnecessary harms to patients, primary care physicians may be reluctant to integrate genomics into clinical practice. While certain genomic tests have been better studied than others—for example, variants in the *BRCA1/2* genes which have proven implications for the risk assessment and management of hereditary breast and ovarian cancer, and pharmacogenetic considerations for efficacy and safety on the labels of more than 130 medications including clopidogrel, warfarin, and citalopram [5]—developing an evidence base for most genomic tests comparable to what is known about *BRCA* testing, for example, will require decades of research in large populations. Pending such research; however, primary care physicians may still make clinical decisions to benefit individual patients despite an underdeveloped evidence base.

Upon the completion of the Human Genome Project in 2003 and on the 50th anniversary of Watson and Crick's landmark discovery of the double-helical nature of DNA [6], then National Human Genome Research Institute Director Francis Collins, M.D., Ph.D. (now Director for the National Institutes of Health) envisioned a blueprint for research in the genomic era and a series of Grand Challenges (Table 1.1) to guide the translation of genomic knowledge to enhance understanding of biological mechanisms of disease (Grand Challenge I), genomics to health (Grand Challenge II), and genomics to society (Grand Challenge III). Rapid advances in Grand Challenge I rapidly ensued over the following decade. From 2005 to 2012, the number of genome-wide association studies (GWAS) increased exponentially with 1,350 publications in 2012 [7]. More than 150 GWAS markers have been associated with common diseases including cancer, type 2 diabetes mellitus, dyslipidemia, multiple sclerosis, nicotine dependence, and psoriasis, there are also dozens of GWAS hits associated

TABLE 1.1 National Human Genome Research Institute Genomics to Health Grand Challenges [1]

Grand Challenge II-1	*Develop robust strategies for identifying the genetic contributions to disease and drug response*
Grand Challenge II-2	*Develop strategies to identify gene variants that contribute to good health and resistance to disease*
Grand Challenge II-3	*Develop genome-based approaches to prediction of disease susceptibility and drug response, early detection of illness, and molecular taxonomy of disease states*
Grand Challenge II-4	*Use new understanding of genes and pathways to develop powerful new therapeutic approaches to disease*
Grand Challenge II-5	*Investigate how genetic risk information is conveyed in clinical settings, how that information influences health strategies and behaviors, and how these affect health outcomes and costs*
Grand Challenge II-6	*Develop genome-based tools that improve the health of all*

with altered drug response [8]. The dawn of the genomic age has also catalyzed drug discovery and development—resulting in new and more effective therapeutic agents for diseases once considered untreatable; including cystic fibrosis and advanced lung cancer and the potential for many more effective treatments to come through drug repurposing and repositioning [8–10]. In parallel with genomic research advances, the health-care field has grown increasingly complex with evolving roles for physicians in a rapidly learning health system [11,12], electronic health records, mobile devices, and the addition of patient-centered medical homes [13,14].

These and other advances have enabled rapid progress toward Grand Challenge II, pertaining to the practice of "personalized medicine." According to Collins, "Personalized medicine uses and individual's genetic profile and individual information about environmental exposures to guide decisions made in regard to the prevention, diagnosis, and treatment of disease" and "knowledge of a patient's genetic profile can help health-care providers select the proper medication or therapy and administer it using the proper dose or regimen" [10]. Prior to the completion of the Human Genome Project, genetic testing in clinical settings was mainly confined to diagnosing highly penetrable Mendelian disorders and resided largely in the realm of clinical geneticists and genetic counselors or medical specialists such as oncologists. The stakes of inappropriate genetic testing and communication of diagnoses or disease risks and the main role of primary care providers (including primary care physicians, physician assistants, and nurse practitioners) was conditioned to be one of gathering family histories and referring patients at high risk to providers with specialized genetics knowledge. National educational programs were proposed and implemented by the National Human Genome Research Institute to educate the primary care physician workforce on basic genetic principles and promote partnership models with other health-care providers and based on advancing knowledge in genomics research across five domains: understanding genome structure, genome biology, disease biology, advancing the science of medicine, and improving health-care effectiveness [15,16]. Historically, much of the leadership and success in genetics education is attributable to the nursing profession, which has implemented core competencies across 50 organizations, and primary care physician and physician assistant residency programs which have implemented core competencies in 20 family residency programs since 2001 [17,18]. Genetic diagnostic, predictive, prenatal, preimplantation, newborn screening, or carrier status testing for germline or somatic mutations will continue to be appropriate only for rigorously trained genetic or other medical subspecialty professionals to manage either as consultants or partners. That said, future primary care practice models will likely include more genetic or genomic information, making it imperative for primary care physicians to build and maintain core competencies in genomic medicine. To effectively realize this, much work remains to raise awareness among primary care physicians currently in clinical practice as well as those in training. Recent surveys of patients and primary care physicians continue to demonstrate limited knowledge of genetics and that most patients lack confidence in genetic clinical skills of their physicians, particularly with classical clinical genetics [i.e., diagnosis, prediction, and prenatal decision making for heritable cancers, neurodevelopmental, and other disorders (e.g., cystic fibrosis)] [19,20]. Fortunately, resources presently exist to assist frontline clinicians in improving core competencies in genomic medicine; including, the American Academy of Pediatrics Genetics in Primary Care Institute (www.geneticsinprimarycare.org), and the NIH-NHGRI-sponsored Genetics/Genomics Competency Center (www.g-2-c-2.org), which all provide free online education targeted for all health professionals (e.g., MDs, RNs, PAs, GCs, and PharmDs).

However, the pace of progress in genomic science has broadened the scope of application beyond what was imagined when many of the core competencies were originally proposed, especially for core competencies for pharmacogenomic testing in clinical practice. Remarkable progress has been achieved in *pharmacogenetics* (the study of inherited variation in a single gene and associated effects on drug disposition, metabolism, toxicity, and response) and *pharmacogenomics* (the study of inherited variation across many different genes that determine drug disposition, metabolism, toxicity, and response). To date, there are more than 130 Food and Drug Administration (FDA) pharmacogenomic biomarker drug labels (http://www.fda.gov/Drugs/ ScienceResearch/ResearchAreas/Pharmacogenetics/ucm083378.htm) recommending or requiring genetic testing to inform prescribers regarding drug exposure and clinical response variability, risk for adverse events, genotype-specific dosing, mechanisms of drug action, and polymorphic drug pharmacodynamics and pharmacokinetics [20]. For a relatively small subset of these FDA-approved tests, there are evidence-based dosing algorithms, clinical implementation guidelines, and clinical annotations available to the medical community; however, available education guidelines have not kept pace with these developments. Table 1.2 presents select FDA drug labels containing pharmacogenetic biomarker information with checkmarks for those that also have corresponding guidance from at least two of the following sources: (1) Pharmacogenomics Knowledge Base (PharmGKB; www.pharmgkb.org) high-level-of-evidence clinical annotations; (2) Clinical Pharmacogenetics Implementation Consortium (CPIC) guidelines; or (3) Evaluation of Genomic Applications in Practice and Prevention (EGAPP) guidelines. Many of these drugs are not commonly prescribed by primary care physicians; however, some, such as statins, antidepressants, and antithrombotics are [21,22,23].

Of the FDA-approved pharmacogenetic tests and companion diagnostics, more than 40 have published evidence-based guidelines or clinical annotations by CPIC and/or PharmGKB, but only one recommendation in favor of testing to guide treatment for colorectal cancer (*KRAS* for antiepidermal growth factor receptor treatments)—which is unfortunately not directly relevant to primary care. It is important to highlight differences in approach to the development of the CPIC and EGAPP guidelines, and how these compare to other available guidelines produced by the National Comprehensive Cancer Network (NCCN) and the American Society of Clinical Oncology (ASCO). The principal difference between the growing number of CPIC guidelines and EGAPP recommendations is the focus of the guidelines. The EGAPP recommendations, for example, were originally derived from the USPSTF model and recommend for or against using a pharmacogenetic test for a specific application. Traditionally, a rigorous evidence threshold is required for analytic validity and clinical utility as well as public health impact, which illustrates the time-consuming process inherent to EGAPP reports [19]. In contrast, CPIC guidelines are principally focused on recommendations for tailoring pharmacotherapy based on the assumption that the primary care physician already has pharmacogenetic test results available (as would be the case in a preemptive genotyping or "just-in-case" models) [24]. Table 1.2 provides a detailed comparison of available clinical practice guidelines for specific pharmacogenetic tests.

Pharmacogenomic testing has become a limited part of the standard of care for medical subspecialists such as clinical geneticists, pulmonologists, or oncologists for advanced treatment of fatal, heritable disorders [25]. While the current role of the primary care physician as a purveyor of pharmacogenomic information is unclear and concerns remain about core

TABLE 1.2 Select FDA Pharmacogenomic Drug Labels and Evidence-Based Clinical Resources

Drug	Therapeutic Area	Biomarker	Label Sections	PharmGKB Annotation	CPIC Guideline	EGAPP Guideline
Abacavir	Antivirals	HLA-B*5701	Boxed warning, contraindications, warnings and precautions, and patient counseling information	√	√	
Atomoxetine	Psychiatry	CYP2D6	Dosage and administration, warnings and precautions, drug interactions, and clinical pharmacology		√	√
Azathioprine	Rheumatology	TPMT	Dosage and administration, warnings and precautions, drug interactions, adverse reactions, and clinical pharmacology	√	√	
Capecitabine	Oncology	DPD	Contraindications, precautions, and patient information	√	√	
Carbamazepine	Neurology	HLA-B*1502	Boxed warning, warnings and precautions	√	√	
Codeine	Pain	CYP2D6	Dosage and administration, warnings and precautions, drug interactions, and clinical pharmacology	√	√	
Clomipramine	Psychiatry	CYP2D6	Drug interactions		√	

Drug	Therapeutic area	Biomarker	Label sections		
Clopidogrel	Cardiovascular	CYP2C19	Boxed warning, dosage and administration, warnings and precautions, drug interactions, and clinical pharmacology	✓	✓
Fluoxetine	Psychiatry	CYP2D6	Warnings, precautions, and clinical pharmacology	✓	✓
Irinotecan	Oncology	UGT1A1	Dosage and administration, warnings, and clinical pharmacology	✓	✓
Ivacaftor	Pulmonary	CFTR (G551D)	Indications and usage, adverse reactions, use in specific populations, clinical pharmacology, and clinical studies	✓	
Paroxetine	Psychiatry	CYP2D6	Clinical pharmacology and drug interactions	✓	✓
Peginterferon alfa-2b	Hepatitis C	IL28B	Clinical pharmacology	✓	
Phenytoin	Neurology	HLA-B*1502	Warnings	✓	
Warfarin	Anticoagulation	CYP2C9, VKORC1	Dosage and administration, precautions, and clinical pharmacology	✓	

competencies and appropriate use of genomic testing technology; nevertheless, there is tremendous opportunity for primary care physicians to incorporate genomic principles into future clinical practice models [26]. The National Human Genome Research Institute Inter-society Coordinating Committee (ISCC) along with the National Academy of Medicine's Roundtable on Genomics and Precision Health have encouraged a continuing medical education approach to instill core competencies appropriate for all health professionals. Moreover, ISCC contributed a framework for developing physician competencies, articulating entrustable professional activities, which constitute critical elements that operationally define what a competent professional may do without direct supervision (Table 1.3) [27]. A limited number of academic medical centers have instituted protocols for pharmacogenetic testing to guide clopidogrel and warfarin therapy for selected patients [28]. However, the translation of genomic knowledge toward the delivery of personalized healthcare by primary care physicians has advanced at a cautious pace. The implications of genomics for society—and not just healthcare—are substantial and far-reaching. The rise of direct-to-consumer genomic testing has major ethical, legal, and social implications—particularly regarding discrimination and loss of patient confidentiality—despite existing federal and state legislation to ostensibly protect consumers and patients [29,30].

Grand Challenge II (Genomics to Health) is most germane to primary care and established objectives to guide translating genome-based knowledge into health benefits. Tremendous progress has been achieved in the ensuing decade in Grand Challenges II-1–4, but research and health-care communities are

TABLE 1.3 Entrustable Professional Activities for Physicians in Genomic Medicine [27]

Family History	Elicit, document, and act on relevant family history pertinent to the patient's clinical status
Genomic Testing	Use genomic testing to guide patient management
Patient Treatment Based on Genomic Test Results	Use genomic information to make treatment decisions
Somatic Genomics	Use genomic information to guide the diagnosis and management of cancer and other disorders involving somatic genetic changes
Microbial Genomic Information	Use genomic tests that identify microbial contributors to human health and disease, as well as genomic tests that guide therapeutics in infectious diseases

only just beginning to understand the impact, constraints, and ecology of Grand Challenges II-5–6, which have profound ethical, legal, and social implications and require careful implementation and diffusion across a multiprofessional workforce through appropriate training and education (Table 1.1) [31]. Two remaining barriers to rapid implementation of personalized medicine are the need to determine that testing improves clinical outcomes (clinical utility) and is cost-effective for health systems [32].

Therefore, given the proliferation of knowledge and complexity of its translation to personalized medicine over the last decade, it is critical to refresh the vision of the role of the primary care physician and distill out the most important aspects of precision medicine that every primary care physician should incorporate in their clinical practice (Table 1.4 lists key takeaways for broad consumption).

All patients should have formal evaluation to develop a detailed family history. Most common chronic illnesses and variability in drug response are influenced by genetic, environmental, and behavioral factors. Family history captures aggregate genetic and other familial influences and can be used to aid in health promotion and disease prevention. Every patient should have a family history that queries major medical disorders and age of onset in first-, second-, and third-degree relatives, and while this may be cumbersome in certain clinical scenarios, there are emerging data that patient self-reported family histories are increasingly useful in practice [31,32]. Genetic testing or genomic sequencing is not an adequate substitution for a detailed family history because of the complexity of gene–environment interactions and epigenetic interactions, for which insufficient evidence exists to capture genomic influences on disease risk and drug response from genotype alone.

Though patients may have concerns about privacy, it is imperative to note that personal genetic information is protected in part by federal law. Genetic information poses risk for stigmatization and discrimination. The Genetic Information Nondiscrimination Act of 2008 (GINA) provides some protection against job discrimination and loss of health insurance coverage, but concerns persist regarding the potential for misuse of genomic information by the public and primary care providers who may not have adequate training to interpret genetic risk and test appropriately for medical indications outside their areas of expertise [33]. GINA is enforced by federal agencies with violations including corrective action including fines. GINA prohibits health insurance companies from using individuals' genetic information (including family history) to determine eligibility or premiums or for requesting or requiring individuals to undergo a genetic test. As well GINA prevents employers from using genetic test information to make employment decisions including hiring, firing, assignments, or dictate employment terms. However, GINA does not prohibit health professionals from recommending genetic tests for patients or life, disability or long-term care coverage insurers from requesting or requiring genetic tests or setting premiums based upon genetic test results. GINA also does not prohibit

TABLE 1.4 Top Precision Medicine Takeaways for Primary Care Providers

All patients should have a detailed family history	• Most common, chronic illnesses and variability in drug response are under the influence of genetic and environmental factors • Every patient should have a three generational family history that queries major medical disorders and age of onset in first-, second-, and third-degree relatives	• Surgeon General's Family Health History Initiative (https://familyhistory.hhs.gov/fhh-web/home.action)
Personal genetic information is protected in part by federal law	• GINA *prohibits* health insurance companies from using individuals' genetic information to determine eligibility or premiums, requesting or requiring individuals to undergo a genetic test or employers from using genetic test information to make employment decisions including hiring, firing, assignments, or terms of employment • GINA *does not prohibit* health professionals from recommending genetic tests for patients, life, disability, or long-term care coverage insurers from requesting or requiring genetic tests or setting premiums based upon genetic test results or medical underwriting based on current health status and does not mandate coverage for genetic tests or treatments • GINA does not apply to the military • GINA is enforced by federal agencies with violations including corrective action including fines	• The Genetic Information Nondiscrimination Act (GINA) of 2008 (http://www.genome.gov/24519851)
PCPs and genetics professionals can and should learn from each other through partnerships models to ensure that genetic tests are used prudently	• Genetics professionals should participate in the education of PCPs at every level of training (undergraduate, graduate, and continuing medical education) within a learning health system • Each health organization should clarify the roles of PCPs as either one of identification of high risk patients to refer or more active participants in screening • PCPs should proactively seek guidance from genetic counselors regarding the types of referrals they do and do not consider appropriate	• National Human Genome Research Institute FAQs about genetic testing (http://www.genome.gov/19516567)

When to refer a patient and family for genetic counseling, a clinical geneticist or both; and ensure that development and implementation of patient-centered medical homes integrates genomic medicine principles	• Patients with high-risk family histories for hereditary disorders should be referred to genetic counselors and clinical geneticists	• When to refer to genetic professionals (http://www.geneticsinprimarycare.org/YourPractice/When-to-Refer/Pages/When-to-Refer.aspx)
Be aware of the appropriate scope of care for PCPs regarding genomic testing and when it is not appropriate to order genomic tests	• Genetic tests for highly penetrable genetic disorders should not be ordered by PCPs • Examples of inappropriate tests for PCPs to order (unless orders are requested by consulting genetic counselors or clinical geneticists) include BRCA1/BRCA2 (breast cancer) or HNPCC (colon cancer) • Patients with high-risk personal or family histories for genetic disorders should be referred to genetic counselors • Genetic risk tests that combine additive genetic information from multiple polymorphisms should not be ordered unless and until the validity and clinical utility of such tests have been established and supported by evidence-based clinical guidelines	• National Coalition for Health Professional Education in Genetics (NCHPEG) • American College Genetics and Genomics • National Comprehensive Cancer Network guidelines (http://www.nccn.org) • US Preventative Services Task Force (http://www.uspreventiveservicestaskforce.org)
PCPs should be aware of evidence-based guidelines and clinical annotations for pharmacogenomics pertinent to the scope of primary care practice	• Some genetic tests may be appropriate for PCPs to order to guide drug prescribing but only for PCPs trained in or possessing expertise if there is strong evidence of clinical utility • Evidence-based guidelines are available from trusted sources including the CDC/Evaluation of Genomic Applications in Practice and Prevention (EGAPP), the Clinical Pharmacogenomics Implementation Consortium (CPIC), and the Pharmacogenomics Knowledge Base (PharmGKB)	• Evaluation of Genomic Applications in Practice & Public Health (EGAPP) (http://www.egappreviews.org) • PharmGKB (http://www.pharmgkb.org/search/clinicalAnnotationList.action?levelOfEvidence=top) • CPIC (http://www.pharmgkb.org/page/cpic)

(Continued)

TABLE 1.4 Top Precision Medicine Takeaways for Primary Care Providers (Continued)

Only laboratory tests performed in CLIA-approved laboratories for FDA-approved indications should be used to guide care by health professionals trained in their use	• Genetic test results from research studies or commercial genetic testing services that are not conducted in CLIA-approved laboratories are not reliable and should not be used to tailor treatment in clinical practice • Patients who present with results from direct-to-consumer genomic testing companies should be advised to interpret results with caution	• Institute of Medicine recommendations (http://www.iom.edu/Reports/2012/Evolution-of-Translational-Omics.aspx)
PCPs should be aware of and literate in the core competencies for genetics in primary care and all health professionals	• National Coalition for Health Professional Education in Genetics (NCHPEG) *Core Competencies for All Health Professionals* have been included in many physician and nursing training programs and are available for continuing medical education	• NCHPEG core competencies (http://www.nchpeg.org/index.php?option=com_content&view=article&id=237&Itemid=84) • Genetics in Primary Care Institute (www.geneticsinprimarycare.org)
Personalized medicine training resources are available in a growing number of learning health-care systems	• Genetic tests are currently available for more than 1200 diseases and more than 120 drug indications • Academic and community hospitals are learning how to integrate genetic testing for common, chronic illnesses into patient care through multiprofessional teams, electronic health records, and continuing medical education initiatives	• eMERGE Network (http://emerge.mc.vanderbilt.edu) • Duke Personalized Medicine (http://www.dukepersonalizedmedicine.org) • Geisinger Genomic Medicine Institute (http://www.geisinger.org/research/centers_departments/genomics/index.html) • Partners Center for Personalized Genetic Medicine (http://pcpgm.partners.org)
Knowledge of personalized medicine enables PCPs to be advocates for patients who may in limited cases receive inferior care without appropriate use of genetic testing	• Genetic sequencing is an evidence-based standard of care for some medical disorders (e.g., advanced lung cancer) that are not cared for by PCPs • Patients of PCPs receiving care for these disorders may not necessarily receive the genomic testing and stratified treatment that is indicated • Informed PCPs have a potential advocacy role to ensure that their patients are referred to genetics health professionals and pertinent medical subspecialists	• Genetic Alliance (http://www.geneticalliance.org)

medical underwriting based on current health status and GINA does not mandate coverage for genetic tests or treatments. In addition, individual states have legislation that either mirrors, or in some cases (e.g., Vermont) is more stringent than GINA [33, 34].

Collaboration and communication between primary care physicians and genetics specialists is necessary to ensure prudent use of genetic tests. Primary care physicians should proactively seek guidance from genetic counselors regarding appropriate referrals. Other providers with specialized genetics knowledge—including medical geneticists, pharmacogeneticists, and genetic nurse specialists—should participate in the education of primary care physicians at every level of training (undergraduate, graduate, and continuing medical education) within a learning health system. Each health organization should clarify roles of primary care physicians to improve identification and referral of high-risk patients and to actively participate in broad screening efforts.

Knowing when (and to whom) to refer a patient and family for genetic counseling, clinical geneticist care, or both, is critical. Patients with high-risk family histories for hereditary disorders should be referred to genetic counselors and clinical geneticists. The determination of high risk depends upon the disease in question and can often be complex, requiring synthesis of guidelines from more than one source. For example, the NCCN updates guidelines annually for breast, ovarian, and colorectal cancer. High-risk status, requiring referral to genetic counseling for Lynch Syndrome (i.e., hereditary nonpolyposis colorectal cancer) requires having either a family history of Lynch Syndrome or family history of polyposis syndromes considered to be in the high-risk category; but this requires knowledge of all high-risk polyposis syndromes (i.e., classical familial adenomatous polyposis, attenuated familial adenomatous polyposis, MUTYH-associated polyposis, Peutz–Jeghers Syndrome, juvenile polyposis syndrome, and serrated polyposis syndrome). Similar complexity exists for determination of high-risk family histories of breast and ovarian cancer (i.e., *BRCA1/2* carrier, Cowden Syndrome, and Li–Fraumeni Syndrome) and determination of risk is based upon more than family history alone, which is determined by the number of first- and second-degree relatives (Claus Model) or age, age at menarche, number of breast biopsies, age at first live birth, and number of first-degree relatives with breast cancer (Gail Model) [35,36]. Criteria for referral and testing are equally as complex for several other genetic screening tests. Therefore, it is crucially important that primary care physicians work within their institution's established protocols for screening in partnership with genetic health professionals so that a goal of universal literacy is achieved regarding referral criteria and if in doubt, consult guidelines from the US Preventative Services Task Force and EGAPP.

With renewed emphasis on development of innovative care delivery models from the Affordable Care Act, ensuring that development and implementation of patient-centered medical homes includes attention to genomic medicine principles as well as clinical pathways for integration. The emergence of patient-centered medical homes presents another opportunity for integration of genomic medicine into the primary care space with more efficiency, such as provider-patient entry of pedigree information, genetic test results reporting, clinical decision support tools, patient education materials, and communication with genetic specialists.

Clearly define the scope of care for primary care physicians regarding genomic testing including scenarios where such testing is inappropriate. Genetic tests for highly penetrable genetic disorders should not routinely be ordered by primary care physicians. Illustrative examples include *BRCA1/BRCA2* (breast cancer) or *HNPCC* (colon cancer). Patients with high-risk personal or family histories for genetic disorders should be referred to genetic counselors. Genetic risk tests that combine additive genetic information from multiple polymorphisms should not be ordered unless and until the clinical validity and utility of such tests have been established and supported by evidence-based clinical guidelines. The clinical utility and value to society of commercially available genomic risk testing for common, multigenetic disorders have not been compellingly demonstrated [37].

As data for clinical utility continue to emerge regarding clinical applications for pharmacogenetic testing, primary care physicians should be aware and make use of available resources to guide changes in therapy for optimal outcomes. Evidence-based guidelines are presently available from trusted sources including EGAPP, CPIC, and PharmGKB. As more health systems adopt and integrate electronic health records into care delivery, these guidelines establish a solid foundation for informatics tools to deliver "just-in-time" clinical decision support to frontline clinicians. The rapid translation of genomic knowledge to evidence-based clinical guidelines and clinical implementation within leading academic medical centers and some community hospitals is paving the way for the larger systems of care across the United States and potentially reducing the activation energy required for broader implementation of pharmacogenomic testing.

Only genetic test results from CLIA-certified laboratories should be used to guide care decisions by health professionals trained in their use. Despite exponential increases in genetic test results from research studies or commercial genetic testing services that are not conducted in CLIA-approved laboratories, it is critical for primary care physicians to understand that such results do not meet the threshold required to change clinical management and should not be used to tailor treatment in clinical practice [13]. Patients who present with results from direct-to-consumer genomic testing companies should be advised to interpret results with caution and if unsure about the interpretation of direct-to-consumer test results, it is best to err on the side of caution and refer to genetic counselors or medical consultants proficient in screening tests for the disease in question. The IOM recommends that genomic tests used for clinical interventions conducted in research studies or routine clinical use should only be performed in CLIA-approved clinical laboratories [38].

Primary care physicians should be aware of and literate in the core competencies for genetics established for their profession or those that have been developed by the NHGRI-sponsored Genetics and Genomics Competency Center (G2C2). Core competencies for all health professionals have been included in many physician and nursing training programs and are available for continuing medical education. Links to the core competencies and educational resources in genomics for primary care physicians are in Table 1.4.

Genomics educational resources are available in a growing number of health care systems. Genetic tests are currently available for more than 1200 diseases and more than 120 drug indications. Academic and community hospitals are learning how to integrate genetic testing for common, chronic illnesses into patient care through multiprofessional teams, electronic health records, and continuing medical education initiatives. Leading academic medical centers are shepherding programs to merge research on clinical implementation of genomic science into the clinical space and health care organizations are learning effective means of delivering patient centered genomic medicine [28]. Patients, primary care physicians (primary care physicians and nurses), genetics professionals, and health-care organizations are interdependent stakeholders in genomic medicine and should proactively consult each other in the process of clinical implementation to ensure accountability and high standards for clinical utility for genomic personalized medicine.

Knowledge of genomics enables primary care physicians to advocate for patients who may otherwise receive inferior care without appropriate use of genetic testing. Genetic sequencing is an evidence-based standard of care for some disorders (e.g., advanced lung cancer) that are not typically managed by primary care physicians. Patients receiving care for these disorders may not necessarily receive the genomic testing and targeted treatment that is indicated. The role of the primary care physician as advocate for patients who are appropriately referred to subspecialist consultants proficient in treating complex and often lethal diseases such as cystic fibrosis or advanced cancers is primarily a holistic approach ensuring that other measures (e.g., preventive care and behavioral health) are also available. In addition, informed primary care physicians must also advocate for patients to ensure that they are appropriately referred to genetics health professionals and pertinent medical subspecialists under special circumstances when there is clear opportunity to optimize treatment. Evolving future clinical practice will incentivize primary care physicians who are more knowledgeable and savvy with respect to application of evidence-based guidelines for stratified medicine, cementing their value in the increasingly complex health-care system.

Despite existing evidence gaps or lack of physician knowledge in genomic medicine, some patients will seek genomic testing. Even without demonstrated clinical utility, patients in certain parts of the world may still request whole genome sequencing or seek direct-to-consumer genetic testing out of a belief that it will equip them to improve their health, resulting in misalignment of patient, and physician values. But while primary care physicians may believe that the clinical utility of such genetic testing does not merit its incorporation into routine practice, patients' perceptions of its personal utility might. This concept incorporates a respect for each patient's health attitudes and preferences and is a central tenet of primary care. If providers do not meaningfully engage with patients in the health information they value, they may risk undermining the therapeutic relationship. Many primary care physicians use these episodes of patient engagement with their health as teachable moments to discuss medical evidence and explore the underlying health beliefs and values and personal preferences of their patients. These conversations can use existing models of shared decision-making that present options, discuss their risks and benefits, and elicit patient preferences [1]. In the case of genomic medicine, primary care physicians should educate patients about limits and the potential harms of uncertain and unwanted information, while listening to patients' beliefs and preferences for evaluating their genome. Primary care providers have an opportunity to lead the discussion on the role of genomic medicine in patient care. The concepts of shared decision-making and patient-centered care comprise the bedrock of primary care, and these ideals have grown out of respect for patient autonomy as well as the pragmatic recognition that optimal health can only be achieved if the patient is a partner in the care delivery process. By listening to their patients' values regarding genomic medicine, primary care physicians have the opportunity to harness those values for health.

In closing, 2 years prior to the completion of the Human Genome Project, Francis Collins, M.D., Ph.D., forecast: "Predictive genetic tests will become applicable first in situations where individuals have a strong family history of a particular condition; indeed, such testing is already available for several conditions, such as breast and colon cancers. But with increasing genetic information about common illnesses, this kind of risk assessment will become more generally available, and many primary care physicians will become practitioners of genomic medicine, having to explain complex statistical risk information to healthy individuals who are seeking to enhance their chances of staying well. This will require substantial advances in the understanding of genetics by a wide range of clinicians [39]." While Collins' statements were accurate, the requisite cultural changes for primary care physicians, educators, and health systems have not been fully realized. By establishing a reasonable consensus view of the most important principles to guide this enterprise, the focus is on mutual understanding and enhancing collective efforts to promote genomic literacy in the 21st century health system and beyond. This is imperative in order to achieve the full and equitable delivery of genomic services to all patients who have something to gain from it [40].

REFERENCES

[1] Vassy JL, Green RC, Lehmann LS. Genomic medicine in primary care: barriers and assets. Postgrad Med J 2013;89(1057):615–6.
[2] Vest JR, Gamm LD. Health information exchange: persistent challenges and new strategies. J Am Med Inform Assoc 2010;17(3):288–94.

[3]   Collins S, Piper KB, Owens G. The opportunity for health plans to improve quality and reduce costs by embracing primary care medical homes. Am Health Drug Benefits 2013;6(1):30–8.

[4]   Alzubi A, Zhou L, Watzlaf V. Personal genomic information management and personalized medicine: challenges, current solutions, and roles of HIM professionals. Perspect Health Inf Manag 2014;11(Spring):1c.

[5]   Conrado DJ, Rogers HL, Zineh I, Pacanowski MA. Consistency of drug–drug and gene–drug interaction information in US FDA-approved drug labels. Pharmacogenomics. 2013;14(2):215–23.

[6]   Collins FS, Green ED, Guttmacher AE, Guyer MS. A vision for the future of genomics research. Nature 2003;422(6934):835–47.

[7]   Begum F, Ghosh D, Tseng GC, et al. Comprehensive literature review and statistical considerations for GWAS meta-analysis. Nucleic Acids Res 2012;40(9):3777–84.

[8]   Collins FS. Reengineering translational science: the time is right. Sci Transl Med 2011;3(90):90cm17.

[9]   Collins FS. Mining for therapeutic gold. Nat Rev Drug Discov 2011;10(6):397.

[10]   Institute-of-Medicine. Genome-based therapeutics: targeted drug discovery and development: workshop summary. Washington, DC: Institute of Medicine of the National Academies; 2012.

[11]   Institute-of-Medicine. The learning healthcare system: workshop summary. Washington, DC: Institute of Medicine of the National Academies; 2007.

[12]   Etheredge LM. A rapid-learning health system. Health Aff (Millwood) 2007;26(2):w107–18.

[13]   Institute-of-Medicine. The Learning Health System and its Innovation Collaboratives: Update Report. Washington, DC: Institute of Medicine of the National Academies; 2012.

[14]   David SP, Johnson SG, Berger AC, et al. Making personalized health care even more personalized: insights from activities of the IOM genomics roundtable. Ann Fam Med 2015;13(4):373–80.

[15]   Burke W, Emery J. Genetics education for primary-care providers. Nat Rev Genet 2002;3(7):561–6.

[16]   Green ED, Guyer MS, National Human Genome Research Institute. Charting a course for genomic medicine from base pairs to bedside. Nature 2011;470(7333):204–13.

[17]   Jenkins J, Calzone KA. Establishing the essential nursing competencies for genetics and genomics. J Nurs Scholarsh 2007;39(1):10–16.

[18]   Laberge AM, Fryer-Edwards K, Kyler P, Lloyd-Puryear MA, Burke W. Long-term outcomes of the "Genetics in Primary Care" faculty development initiative. Fam Med. 2009;41(4):266–70.

[19]   Marshall E. Human genome 10th anniversary. Waiting for the revolution. Science. 2011;331(6017):526–9.

[20]   Feero WG, Green ED. Genomics education for health care professionals in the 21st century. JAMA 2011;306(9):989–90.

[21]   Food-and-Drug-Administration. Table of pharmacogenomic biomarkers in drug labels. 2013; <http://www.fda.gov/Drugs/ScienceResearch/ResearchAreas/Pharmacogenetics/ucm083378. htm>. Accessed 03/26/13, 2013.

[22]   Relling MV, Klein TE. CPIC: Clinical Pharmacogenetics Implementation Consortium of the Pharmacogenomics Research Network. Clin Pharmacol Ther 2011;89(3):464–7.

[23]   Teutsch SM, Bradley LA, Palomaki GE, et al. The Evaluation of Genomic Applications in Practice and Prevention (EGAPP) initiative: methods of the EGAPP Working Group. Genet Med 2009;11(1):3–14.

[24]   Altman RB. Pharmacogenomics: "noninferiority" is sufficient for initial implementation. Clin Pharmacol Ther 2011;89(3):348–50.

[25] Feero WG, Guttmacher AE, Collins FS. Genomic medicine—an updated primer. N Engl J Med 2010;362(21):2001–11.

[26] Collins FS. Preparing health professionals for the genetic revolution. JAMA 1997;278(15):1285–6.

[27] Korf BR, Berry AB, Limson M, et al. Framework for development of physician competencies in genomic medicine: report of the Competencies Working Group of the Inter-Society Coordinating Committee for Physician Education in Genomics. Genet Med 2014;16(11):804–9.

[28] Ginsburg GS, Staples J, Abernethy AP. Academic medical centers: ripe for rapid-learning personalized health care. Sci Transl Med 2011;3(101):101cm127.

[29] De Francesco L. Genetic profiteering. Nat Biotechnol 2006;24(8):888–90.

[30] Allison M. Direct-to-consumer genomics reinvents itself. Nat Biotechnol 2012;30(11):1027–9.

[31] Emery JD, Reid G, Prevost AT, Ravine D, Walter FM. Development and validation of a family history screening questionnaire in Australian primary care. Ann Fam Med 2014;12(3):241–9.

[32] Walter FM, Prevost AT, Birt L, et al. Development and evaluation of a brief self-completed family history screening tool for common chronic disease prevention in primary care. Br J Gen Pract 2013;63(611):e393–400.

[33] Hudson KL, Holohan MK, Collins FS. Keeping pace with the times—the Genetic Information Nondiscrimination Act of 2008. N Engl J Med 2008;358(25):2661–3.

[34] Vermont-Statute, The Vermont statutes online: Chapter 217: genetic testing <http://www.leg.state.vt.us/statutes/fullchapter.cfm?Title=18&Chapter=217>. In: Legislature VS, editor. 1999.

[35] Claus EB, Risch N, Thompson WD. Autosomal dominant inheritance of early-onset breast cancer. Implications for risk prediction. Cancer 1994;73(3):643–51.

[36] Gail MH, Brinton LA, Byar DP, et al. Projecting individualized probabilities of developing breast cancer for white females who are being examined annually. J Natl Cancer Inst 1989;81(24):1879–86.

[37] Institute-of-Medicine. The economics of genomic medicine: workshop summary. Washington, DC: Institute of Medicine of the National Academies; 2013.

[38] Institute-of-Medicine. Evolution of translational omics: lessons learned and the path forward. Washington, DC: Institute of Medicine of the National Academies; 2012.

[39] Collins FS, McKusick VA. Implications of the Human Genome Project for medical science. JAMA. 2001;285(5):540–4.

[40] Fears R, Weatherall D, Poste G. The impact of genetics on medical education and training. Br Med Bull 1999;55(2):460–70.

Chapter 2

Overview of Policy, Ethical, and Social Considerations in Genomic and Personalized Medicine

Susanne B. Haga

Duke University School of Medicine, Durham, NC, United States

Chapter Outline

ABBREVIATIONS

CLIA Clinical Laboratory Improvement Amendments
DTC direct to consumer
FDA U.S. Food and Drug Administration
GINA Genetic Information Nondiscrimination Act
GWAS Genome wide association studies

Genomic and Precision Medicine. DOI: http://dx.doi.org/10.1016/B978-0-12-800685-6.00005-9
19

INTRODUCTION

The translation of genome sciences into clinical practice is a step-wise process as with any other medical innovation. While some exciting genomic applications have begun to emerge, particularly those using next-generation sequencing technologies, a range of social and ethical issues continue to be debated, impacting both the research of genome discovery and validation as well as clinical utility, provider education, and delivery models. This chapter will provide an overview of the major policy, ethical, and social issues pertaining to genomics research, such as the return of research results, development and translation of genomic medicine applications, and issues related to the clinical use of genomic applications, including incidental/secondary findings, transition from single-gene tests to gene panels; oversight, coverage, and reimbursement of new genomics applications; and professional education. Subsequent chapters in this section will address other important issues as well as an in-depth analysis of several of the issues introduced here.

ISSUES IN GENETICS AND GENOMICS RESEARCH

The completion of the Human Genome Project marked the beginning of a new era, rather than the end of one [1,2]. In addition to the generation of a reference sequence of the human genome, the parallel technology development has enabled sequencing of huge numbers of human genomes (and nonhuman genomes) from around the world [3–8]. Many of the sequences are publicly available (i.e., 1000 Genomes Project). Analysis of individual genomes provides further insight into to development of diseases such as cancer and identification of causes of rare diseases [9–16].

While new technologies such as whole genome or exome sequencing has enabled comprehensive analysis of human genetic variation and identification of causes of disease, it has also raised a number of ethical issues in both the research and clinical setting, some carried over from family-based human genetics research, that warrant careful attention [17]. In particular, genomic research continues to raise concerns about the study of vulnerable populations, management of incidental or secondary findings, familial implications, and return of results.

Under-Represented and Vulnerable Populations

In contrast to human genetics research, which typically centers on the family, genomics research is frequently based on populations and/or large case-control cohorts. Population-based genomics research has spurred debate regarding the validity and use of racial categories as a variable in biological research [18–21]. Although studies have shown that individuals from populations around the world can be clustered genetically into six major groups [22], evidence has rapidly accumulated about the racial admixture and gene migration patterns in populations [23–25]. Ancestry informative markers (AIMs) [26], admixture maps [27–29] are used to assess admixture and better control for population stratification in genetic association studies [20,30].

Differing prevalence of gene variants is commonly observed between populations and is an important consideration in the development of clinical tests and estimating predictive value. In addition, the impact of a genetic variation on clinical outcome may vary between populations [31–35], implying a role for additional genetic or environmental factors and the need to study a diverse patient population to validate findings [36]. Other clinical applications based on genotype, such as warfarin dosing algorithms, have reported inaccurate dosing recommendations for African–Americans linked to failure to incorporate rare genetic variants observed in African–Americans [37–39]. Thus, extrapolating effects observed in individuals from one group to an entire race can lead to erroneous conclusions and potentially, incorrect clinical interventions. Some have speculated that the temptation to attribute health disparities to genetic causes may distract from other types of social and biomedical research that may lead to greater reduction in these observed differences [40].

Underserved populations have been under-represented in genomics research [41,42], which can result in limited benefits of clinical applications to these patient populations. In addition, genomics research has raised concerns about the study of vulnerable populations (e.g., Native Americans) and potential abuse of these populations [43]. What are the causes behind the lower participation of underserved populations in genomics research? Some commonly cited reasons include distrust of researchers and privacy concerns. However, others have reported that willingness to participate does not differ between groups [44], and that with additional effort and development of culturally appropriate recruitment materials, enrollment of a diverse population can be achieved [45]. Past abuses such of African–Americans in the Tuskegee syphilis study and political and social injustices endured by Native Americans underlie much of the distrust of medical researchers. The unsuccessful launch of the Human Genome Diversity Project (HGDP) in the late 1990s highlighted the importance of ongoing community engagement and consultation from the conception of a study to its completion [46,47]. Consultation with community representatives is essential to ensure that both researchers and prospective participants develop a clear understanding of the proposed research as well as the interest and concerns of the community, particularly if the research potentially involves areas of study that may conflict with community beliefs, traditions, or values [48–50]. More recent examples such as the failure to obtain consent for specific uses of Havusupai tribal member's donated DNA samples continues to engender feelings of distrust and abuse in these groups [51].

Biobanking

Several years prior to the completion of the Human Genome Project, a number of countries contemplated setting up a national research resource or "biobank" that would contain clinical specimens and a range of data for the purposes of large-scale genetic epidemiology studies. With the movement toward genome-wide association studies (GWAS), a large number of DNA samples were needed from individuals of various phenotypes for discovery and validation, particularly of findings with relatively low relative disease risk and multiple confounders that could lead to spurious findings. At present, several biobanks have been established across Europe and Asia (e.g., United Kingdom, Sweden, Estonia, Japan, Singapore, and Taiwan) that are owned or operated by universities, governmental agencies, and nonprofit groups [52]. Among the many issues that have been debated about biobanks are the scientific need and merit of a national biobank, the cost and feasibility to successfully establishing, and operating a biobank, the required infrastructure, accessibility to samples and data, informed consent, intellectual property, privacy and confidentiality of data, and disclosure of research results to participants. While current policies may be applicable to some of these issues as they are not unique to biobanks, new policies may be required to address some of these issues and to help reassure the public. As the United States begins to launch the Precision Medicine Initiative, which AIMs to collect one million samples and accompanying data, many of these issues have been at the forefront of discussion during the planning process.

One critical issue is public engagement since a national biobank depends on the public's support and participation [53–55]. Reports suggest that the public is very supportive of the establishment and goals of biobanks, despite their limited awareness about them [56–61]. However, some studies indicate that potential participants may decline donating to biobanks due to lack of perceived personal benefit and mistrust [62,63]. Different approaches have been used to communicate with the public about regional and national biobanks [64,65], most commonly involving a communication approach or a partnership approach [66]. The different approaches utilized are reflective of different cultures, attitudes toward scientific research, history, and government and healthcare systems. In Iceland, the communication approach included a brief public consultation took place in the form of radio and television programs, town hall meetings, and public surveys. In contrast, the United Kingdom used a partnership approach, embarking on a massive public 2-year consultation campaign about its proposed national databank. The national dialog included town hall meetings, focus groups, interactive workshops, and published reports. The exchange of information helped inform policy-making to ensure that public concerns were addressed while also raising public awareness about the project [66].

Consent of biobank donors is another key ethical concern. Due the prospective nature of biobanks, explicit detail regarding the risks and benefits of biobanked samples cannot be provided at the time of tissue collection. Thus, a variety of approaches have been suggested to obtain consent from donors and fulfill ethical obligations to donors. In general, consent for biobanks can be obtained through a tiered consent process (e.g., participants can consent to different categories of research such as disease types), an open (general or blanket) consent (often linked to study approval by an established ethics board with public representatives), or an initial consent for sample collection and subsequent consent for each study requesting use of the biobanked samples. Several studies have assessed public attitudes and preferences for biobank consents. Some studies report preference for an opt-in, general consent process, concluding it unnecessary to obtain consent for every single project using their biobanked samples [58,61,67]. Other studies report that reconsent is preferred for projects that significantly diverge from the original study or for each new project [60,68]. Yet others have advocated for a dynamic consent process [69], an approach that may be more feasible with implementation of electronic consents (e-consents) [70,71]. The appropriate type of consent may vary depending on public values, biobank administrators, and intended uses (e.g., cancer research only).

Returning Research Results

Translational research in genomics naturally overlaps with clinical care when assessing the validity and utility of new applications for various healthy and patient populations [72,73]. Thus, it is not surprising that the issue of returning research results to participants and the potential clinical benefit has been the subject of intense debate [74–76]. Briefly, research results can be divided into three categories: (1) individual research results or those anticipated to be generated from the study (i.e., related to the disease being studied); (2) incidental (or secondary) findings or findings not related to the study's disease focus; and (3) aggregate results (a summary of research findings with no individual information). If the samples are collected anonymously or pooled, aggregate reports would be the only way to inform participants about study outcomes. However, aggregate reports may dilute the value of results to individuals as participants will be left with uncertainty about the significance of the findings for themselves (e.g., whether they were found to be at high risk vs low risk) [77].

Researcher obligations and institutional policies appear to vary depending on the type of research results to be returned in question. The ongoing debate about what results, if any, should be returned to research participants has been considered by members of institutional review boards [78], ethicists, legal scholars [74,79–81], and prospective participants and the public [82–85] as well as informed by early practices [86]. Several studies report that research participants have expressed preference for access to individual research findings over general research summaries [82,87,88], even for conditions that currently are untreatable such as Alzheimer's disease [89]. Likewise, IRB members have been supportive of offering to return results [78], disclosure of research results, even those of a serious nature or negative results, have not been shown to cause adverse psychological impacts in the majority of research participants studied [90–94]. Furthermore, the type of research result will inform selection of the appropriate method of communication to participants [95].

Although several policies and recommendations on data disclosure have been proposed, no consensus has been reached about when to return research results [95,96]. A study of Canadian research institutions found that only 20% of decisions regarding return of results were governed by a policy or guideline at their institution [78]. For the numerous biobanks that have been established globally, only a minority has developed policies about return of results [97]. Of the published recommendations specific to genetic and genomic research, the importance of clinical relevance and utility is emphasized [98]. For example, the guidelines developed by a working group convened by the U.S. National Heart, Lung, and Blood Institute identified the availability of a clinical intervention (treatment or preventative) as a key criterion if results were to be returned to study participants [99,100].

It has been suggested that investigators have an ethical responsibility to offer participants access to research results [101]. Providing an option to disclose research data shows respect to the research participant [77]. Despite this belief, the practice does appear to be common [102–104]. Researchers may choose not to provide access to results for several reasons including time constraints, cost, and risk of harm due to the uncertain nature of some results. But one study reported that almost 80% of health professionals involved in cancer research were inclined to offer research results to participants based on the belief that most participants desire to be informed of study results [105]. Heaney et al. [103] found that more than half (54%) of respondents of a survey of authors of genetics/genomics papers had considered the issue of returning research results to participants, 28% offered to return individual research results, and 24% actually returned individual research results. The most common factors considered were clinically utility (18%) and respect for participants (13%).

However, given concerns about the potential harms of unvalidated or inconclusive data, it has been suggested that the option for access to research results should be provided but not necessarily encouraged and that the uncertain nature of the results may be emphasized [77]. In addition, the legal liability of researchers returning results has also been raised [79]. Because of the potential risks, it has also been suggested that a second consent should be obtained prior to disclosure of research results [106,107]. Different models of consent specifically for incidental findings have been proposed [108], the choice of an appropriate consent will likely be context specific [109,110]. For minors participating in genetic and genomics research, the issue of offering to return research results becomes further complicated—should only results for childhood disorders be returned for the child's benefit or should adult onset diseases be returned that could potentially benefit the parents as well as the child's to maintain the health of the parent(s) [111]?

The process to develop and implement a policy on return of results and procedures on consent, communication, and recontact of participants is resource intensive and requires continuous attention throughout the study [112]. For genome sequencing studies, the proportion of genome variants detected that meet the clinical criterion to offer return to participants was estimated to range from 4 to 16% [113]. This number is expected to increase in the coming years with greater understanding of the clinical impact of genetic variants. Thus, the amount of time required to offer participants research results can become increasingly burdensome, and it is not clear how researchers (or institutions and biobanks) can recover their effort [114,115].

Familial Implications

As genetics and genomics research can reveal information about biological family members, the issue of consent and return of results for family members has been the subject of some debate.

While the issue of disclosure of results is well known in clinical testing situations [116], the issue of returning results to family members in the research context has been relatively understudied until recently [117–122]. In clinical situations, the privacy of the patient is the primary factor in decisions regarding disclosure of results to family members. It is recommended that providers encourage patients to discuss testing with family members and/or share their results [123–126]. However, some familial situations are not conducive to sharing of results [127], and therefore, a provider may feel obligated to breach patient confidentiality and disclose the test results to a family member who may benefit from an available intervention to prevent disease onset or improve disease outcomes. With the availability of more patient resources, even for estranged families, disclosure is possible through a variety of ways and a provider's ethical obligation or duty to warn can be fulfilled by emphasizing to patients the importance of disclosure to family members and the dissemination of the resources available [128].

In contrast, in the research setting, the researcher does not have as clear an obligation or ethical responsibility to the clinical well-being of the participant as does a provider in a clinical setting. Thus, it is unclear what criteria should be used to establish policies regarding offering to return of research findings to relatives, especially the results of deceased participants. At present, relatives may access a deceased participant's genetic research result through a number of ways, which may only be partially limited by the participant's preference *not* to have the results shared with relatives [129]. However, it is recommended that researchers ascertain and document the participant's preferences regarding sharing of research results with legally authorized representatives at the time of enrollment or consent [129]. In the absence of knowledge of a participant's preferences, it is recommended that researchers do not actively seek to offer return of results to relatives unless deemed highly clinical actionable [129].

ISSUES RELATED TO INTEGRATION OF GENOMIC MEDICINE APPLICATIONS IN HEALTHCARE

The number and type of genetic and now genomic tests clinically available has substantially expanded over recent years. Genomic medicine utilizes a range of technologies to interrogate the many types of structural and sequence-based changes in the genome such as microarrays and sequencing, as well as RNA expression profiles. The expanded scope of clinical genomic applications necessitates consideration of the adequacy of current policies to ensure the safe, effective, and appropriate use and delivery of these new tests. In particular, revisions may be needed to the oversight of genomic tests, health professional education and clinical decision supports, and patient resources.

Oversight

The debate about the oversight of genetic testing has been ongoing for the past 20 years. During that time, several major U.S. reports have issued recommendations on the topic, concurring that additional oversight is needed, though varying with respect to the level or type [130–133]. Most of the recommendations for enhanced oversight in the Unites States have been directed at two federal agencies that currently have jurisdiction over clinical testing, including genetic testing: The U.S. Food and Drug Administration (FDA) and the Centers for Medicare and Medicaid Services (CMS), which administers the Clinical Laboratory Improvement Amendment (CLIA) program. Regular CLIA inspections are required for certification; inspections focus on the analytical validity of the test and laboratory environment, including personnel qualifications, documented quality assurance, and quality control protocols. In addition, several bills have been introduced in Congress to establish an oversight system for genetic testing [134–138].

However, despite the tremendous amount of time and effort that has been devoted to this issue with each succeeding committee or advisory group, little has changed. The FDA has continued to monitor and developed several guidances that impact genetic and genomic tests over the past several years, including guidance documents for drug metabolizing enzymes genotyping system [139], multiplex technology [140], analyte specific reagents [141,142], algorithm-based genetic tests [143,144], genetic toxicology studies [145], and pharmacogenetic tests [146]. In 2010 and 2011, the FDA convened public meetings to gather input from stakeholders specifically about oversight of Laboratory Developed Tests (LDTs); a range of views on this complex issue were provided, highlighting the challenges of developing oversight of LDTs that would achieve the goal of insuring the development of effective tests without adversely impacting innovation. In 2014, the FDA releases a draft guidance on LDTs, proposing a risk-based framework of oversight [147] followed by another public meeting in January 2015. To date, no final guidance has been issued. At present, the FDA and CMS/CLIA are working to harmonize regulatory requirements of clinical laboratories and remove duplicative oversight.

Secondary (Incidental) Findings

With advances in genotyping and sequencing technologies, the ability to test multiple genes or the whole genome or exome has become quite feasible and an accepted practice for some clinical circumstances. With these broader analyses, it is quite possible for a patient to learn about conditions unrelated to the indication for testing, though it can certainly happen with a single gene test, such as *ApoE* to predict statin response but also learn of one's risk for Alzheimers. As discussed in the section on returning incidental or secondary findings, this similarly has become a controversial topic in clinical genomic medicine. The main issue is whether to return these secondary test results to patients (or their parents since many the patients being sequenced are children) in addition to the main findings related to their condition. In 2013, the American College of Medical Genetics and Genomics released a position statement recommending that test results for 56 genes be returned to patients undergoing clinical sequencing [148]. These 56 genes, associated with 27 conditions, were selected because of high penetrance and clinical utility of the results. Following that initial report, a vigorous debate ensued amongst scholars and clinicians in the genetics and genomics community [149–152]. The ACMG issued additional statements clarifying the results for these 56 genes should be offered to patients if they desire to learn of them [153,154]. All clinical sequencing labs currently provide the option of a secondary report, and some labs have expanded the list to include others deemed clinical useful.

Direct-to-Consumer (DTC) Genetic Tests

Following the wave of GWAS discoveries and ongoing developments in genotyping and sequencing, several companies were established to provide consumers with direct access to information about their genetic risks, causing much debate and angst amongst health professionals and policy-makers [155–158]. In 2010, it was estimated that 30 companies provided testing services directly to consumers for about 400 tests without the involvement of a health professional [159]. Companies offered a range of testing for ancestry, physical traits (e.g., eye color), and disease risks based on the presence/absence of specific genome variants. In addition, DNA-tailored products such as personalized cosmetics and nutritional supplements were available. For tests directly purchased by consumers, the involvement of a health care professional is not required.

Several concerns about direct-to-consumer (DTC) testing were raised, including uncertain clinical validity and utility of testing, misleading claims, harms to consumers, and unnecessary follow-up testing [160–164]. The new model of delivery of genetic tests, in contrast to clinical testing typically ordered by physicians, has raised concerns about consumers' ability to make an informed decision about the test they are purchasing as well as correctly interpret their test results [157]. In 2004, the American College of Medical Genetics issued a statement recommending that genetic tests should only be ordered by an "appropriately qualified health care professional" to prevent potential harms such as test misinterpretation and inappropriate follow-up [163]. In 2007, the American Medical Association's House of Delegates passed Resolution 522 which requested further study of the practice of DTC advertising of genetic tests and, in particular, at the existing oversight of this field [165]. In their position statement on DTC testing, the European Society of Human Genetics recommended that clinical utility be a required factor for tests offered to the public [166].

In response to the concerns raised by DTC testing, states and federal authorities began to increase oversight of these companies. In 2008, the state of California sent several DTC companies cease-and-desist letters citing violations of multiple state business and laboratory codes. In 2010–11, the FDA sent letters to more than 20 DTC companies informing them that one or more of their tests were believed to be medical devices as defined under the Federal Food, Drug and Cosmetic Act. As such, these tests would require premarket clearance or approval as stipulated under the Medical Device Amendments of 1976. In July 2010, findings of an investigation by the Government Accounting Office (GAO) were presented at a Congressional hearing alongside testimony by executives of some DTC companies, physicians, and ethicists [167]. Eerily similar to a GAO investigation on DTC companies conducted in 2006 [168], the recent report concluded that the test results from five DTC companies investigated were "misleading and of little or no practical use to consumers" [167]. In the United Kingdom, the Human Genetics Commission developed a set of best practices for DTC companies to ensure consumer confidence in these services and establish quality standards for these companies [169].

However, several studies have been published on consumer interest and experiences with DTC testing. Consumers have expressed interest in these tests for a range of reasons including health, genealogy, curiosity, and recreation [170]. Of interest, these studies report a lack of demonstrable harm to consumers following receipt of their test results and minimal impact on health behaviors [171–174]. In addition, some groups may benefit from access to a personal genome report to fill in gaps of knowledge about their family history [175]. The future of DTC testing remains to be seen until the regulatory landscape is clarified, but evidence indicates that the anticipated harms of testing have not been realized.

Access/Reimbursement

Coverage of genomic applications is intimately linked to uptake of a new test and integration into clinical practice. The demonstration of the validity and utility of a new genomic application as well as cost-effectiveness is essential to securing reimbursement. More than 150 economic analyses of genetic and tests have been conducted, many for adult conditions such as cancer [176–178]. An analysis of published cost-effectiveness studies of genetic tests concluded that 17% of analyses found that testing was cost effective [179]. In addition, several insurance groups and Medicare contractors have issued technology assessments for genetic and genome tests, which serve as the basis for their coverage decisions. Given the rapid pace of development, many of these assessments have been updated one or more times to reflect the current evidence. For example, in 2005, the Blue Cross Blue Shield Association's Technology Evaluation Center (TEC) assessed four gene expression profiles for use in the management of breast cancer. Based on the available evidence at the time, the report concluded that none of the profiles met the TEC criteria [180]. The TEC criteria include approval by federal regulatory bodies and evidence demonstrating an improved outcome and comparable benefits to existing alternative treatments or applications. However, in 2007, a second review of the Oncotype DX test concluded that the test now met the TEC criteria and was considered to be useful specifically regarding adjuvant chemotherapy for women with estrogen receptor-positive, node-negative, tamoxifen-treated breast cancer [181]. The use of Oncotype DX for selecting adjuvant chemotherapy in patients with lymph-node-positive breast cancer, however, does not meet the TEC criteria [182].

While many tests have been shown to be cost effective, others still have questionable utility despite demonstrated clinical validity. For example, a commercially available molecular profiling test known as Allomap has been shown to be cost-effective to identify heart transplant patients at risk for rejection compared to endomyocardial biopsy to detect rejection, mainly due to the cost-savings gained from not having to perform a biopsy [183]. But several tests developed to guide therapy decisions are considered by some to be of questionable value [184]. For other tests such as HLA-B*5801 testing for adverse responses to the medication allopurinol, data have been conflicting about the cost-effectiveness of a test, potentially influenced by differing health care systems, predictive value of test, test and treatment cost, and values [185,186]. In addition to evaluating both clinical validity and utility, pharmacogenetic testing policies must also consider coverage of the drugs identified as being most likely to perform well or least likely to cause adverse side effects. For tests that are codeveloped with drugs, the cost of the testing and drug could be determined by a single-policy decision. But in the event that a test indicates a different drug, payors may be forced to revise their coverage policies for drugs that are not listed on their formularies.

Genetic Discrimination

The risk of genetic discrimination has existed prior to the start of the Human Genome Project and continues to be a perceived risk by patients and the general public [187–189], despite existing federal and state protections. The federal law, the Genetic Information Nondiscrimination Act (GINA); (PL-110-223), specifically prohibits discrimination on the basis of genetic information (which includes family history) by health insurers (Title I; 74 Fed. Reg. 51664, 10/7/09) and employers (Title II; 75 Fed. Reg. 68912, 11/9/10). The law defines genetic information as an individual's or family member's genetic test results as well as family medical history. Employers are prohibited from using genetic information in employment decisions and requesting, requiring, or purchasing genetic information. In addition to employers, the law also pertains to employment agencies, labor organizations and joint labor–management training, and apprenticeship programs. Similarly, health insurers are prohibited from using genetic information to inform decisions regarding eligibility, coverage, underwriting, or premium setting. Health insurers may not request or require genetic testing of individuals or their family members or disclosure of test results or family medical history.

However, equally important to what the law does prohibit is what it does not prohibit. The law does not apply to other types of insurers such as life insurers or long-term care insurers, adoption services, egg and sperm banks, educational admissions programs, and the military. Therefore, these groups may use genetic information to make decisions regarding insurability, premiums, donor suitability, or personnel decisions. In addition, there are several exceptions in which a request for genetic testing is permissible, such as in the context of research, so long as the results are not used for health insurance underwriting or employment decisions.

For genetic and genomic applications, the risk of genetic discrimination is recommended to be disclosed to research participants in the informed consent documents in the research setting [190]. In the research setting, participants have cited concerns about discrimination as the primary reason to enroll in a genetics study [191]. Likewise, in the clinical setting, the risk of discrimination is recommended to be discussed with patients considering genetic testing [192,193]. However, studies have demonstrated providers' and patients' limited awareness about GINA and the scope of the protections, raising concerns that patients are not appropriately informed of the potential risks for discrimination [194,195]. In some cases, learning about GINA may increase concerns rather than allay fears regarding discrimination [191]. Resources have been developed to provide more information about each part of GINA that can be as a reference (www.ginahelp.org and https://www.genome.gov/10002328).

Outside of the Unites States, various protections have been considered or enacted in the United Kingdom, Australia, and Canada to protect against genetic discrimination by life insurers and employers [196–198]. Similar to the Unites States, a patchwork of provincial legislative protections in Canada exist, though national legislation is being considered [199]. Austria, Belgium, France, the Netherlands, Luxembourg, Greece, and Italy have enacted legislation prohibiting access of genetic information without consent. In the United Kingdom, a moratorium is in effect banning the use of genetic testing information for insurance underwriting purposes [200].

Health Professional Education

The education of health professionals will be an important factor in the appropriate use of genetic and genomic tests. A number of surveys have documented the variable level of provider curricula/training and knowledge of genetics and personalized medicine [201–203]. The uptake and use of new applications will be stalled until health professionals gain some understanding about the appropriate use of these tests and the interpretation and application of test results.

With the broader use of genetic and genomic applications beyond the traditional specialties of medical genetics, pediatrics, and obstetrics and the relatively small number of the trained genetic specialists, enhanced training in genetics and genomic for all providers is warranted. Some nongenetics providers are being trained to provide genetic services, particularly nurses [204]. For example, the cancer family nurse specialist is trained to identify patients with a family history of cancer which would place them at increased risk and indicate testing [204]. Studies have demonstrated the equivalence of care provided by nurses trained in genetics compared to board-certified genetic counselors [205].

To better prepare providers to incorporate testing into their practice, the continuum of health education from graduate or professional school to continuing education should aim to increase awareness and understanding of these new tools and therapies across virtually all medical specialties [206–208]. In response to new genomic advancements and applications, a range of new educational initiatives or curricula have been developed [209–213]. In addition, clinical decision support tools have gained more attention with the widespread implementation of electronic medical records [214,215]. For practicing providers, continuing education will be particularly important on keeping providers' up-to-date; one study reported that participation in continuing education was associated with higher knowledge of genetic screening guidelines [216]. Professional genetic competencies have been developed for several clinical professions [217–221].

CONCLUSION

While the advancements stemming from the genome sciences is certain to significantly impact the practice of medicine and our understanding of human development, biology, biochemistry, and physiology, the safe and appropriate use of this new information and its associated technologies will warrant new policies and an improved level of understanding by patients, consumers, and providers. The ethical and social aspects of genome research and medical applications will substantially guide and influence policy-making related to the development and use of these new tools. New policies to address ethical and social issues should not obstruct or prevent scientific advancements from moving forward but rather should AIM to facilitate the expansion of the field in a manner that addresses concerns about the direction, pace, and application of the science. Given the wide range of stakeholders involved in the translation of genomic medicine applications from bench to bedside, it will be critical to gather their perspectives as new policies are being considered and developed.

REFERENCES

[1] Collins, F.S. (May 2003). Testimony Before the Subcommittee on Health Committee on Energy and Commerce, United States House of Representatives. Available at <http://www.genome.gov/11007447>.

[2] Butler D. Human genome at ten: science after the sequence. Nature 2010;465:1000–1.

[3] Levy S, Sutton G, Ng PC, Feuk L, Halpern AL, Walenz BP, et al. The diploid genome sequence of an individual human. PLoS Biol 2007;5(10):e254.

[4] Wheeler DA, Srinivasan M, Egholm M, Shen Y, Chen L, McGuire A, et al. The complete genome of an individual by massively parallel DNA sequencing. Nature 2008;452:872–6.

[5] Ashley EA, Butte AJ, Wheeler MT, Chen R, Klein TE, Dewey FE, et al. Clinical assessment incorporating a personal genome. Lancet 2010;375:1525–35.

[6] Wang J, Wang W, Li R, Li Y, Tian G, Goodman L, et al. The diploid genome sequence of an Asian individual. Nature 2008;456:60–5.

[7] Schuster SC, Miller W, Ratan A, Tomsho LP, Giardine B, Kasson LR, et al. Complete Khoisan and Bantu genomes from southern Africa. Nature 2010;463(7283):943–7.

[8] Ahn SM, Kim TH, Lee S, Kim D, Ghang H, Kim DS, et al. The first Korean genome sequence and analysis: full genome sequencing for a socio-ethnic group. Genome Res 2009;19:1622–9.

[9] O'Roak BJ, Deriziotis P, Lee C, Vives L, Schwartz JJ, Girirajan S, et al. Exome sequencing in sporadic autism spectrum disorders identifies severe de novo mutations. Nat Genet 2011; 43(6):585–9.

[10] Alazami AM, Patel N, Shamseldin HE, Anazi S, Al-Dosari MS, Alzahrani F, et al. Accelerating novel candidate gene discovery in neurogenetic disorders via whole-exome sequencing of prescreened multiplex consanguineous families. Cell Rep 2015;10(2):148–61.

[11] Mackenroth L, Fischer-Zirnsak B, Egerer J, Hecht J, Kallinich T, Stenzel W, et al. An overlapping phenotype of Osteogenesis imperfecta and Ehlers–Danlos syndrome due to a heterozygous mutation in COL1A1 and biallelic missense variants in TNXB identified by whole exome sequencing. Am J Med Genet A 2016;170A(4):1080–5. [Epub ahead of print].

[12] Mardis ER, Ding L, Dooling DJ, Larson DE, McLellan MD, Chen K, et al. Recurring mutations found by sequencing an acute myeloid leukemia genome. N Engl J Med 2009;361(11):1058–66.

[13] Lupski JR, Reid JG, Gonzaga-Jauregui C, Rio Deiros D, Chen DC, Nazareth L, et al. Whole-genome sequencing in a patient with Charcot–Marie–Tooth neuropathy. N Eng J Med 2010;362:1181–91.

[14] Chapman MA, Lawrence MS, Keats JJ, Cibulskis K, Sougnez C, Schinzel AC, et al. Initial genome sequencing and analysis of multiple myeloma. Nature 2011;471:467–72.

[15] Pelak K, Shianna KV, Ge D, Maia JM, Zhu M, Smith JP, et al. The characterization of twenty sequenced human genomes. PLoS Genet 2010;6(9):e1001111. Available from: http://dx.doi.org/10.1371/journal.pgen.1001111.

[16] Wadman M. Fifty genome sequences reveal breast cancer's complexity. Nat News 2011 Available at <http://www.nature.com/news/2011/110402/full/news.2011.203.html>.

[17] Kaye J, Boddington P, de Vries J, Hawkins N, Melham K. Ethical implications of the use of whole genome methods in medical research. Eur J Hum Genet 2010;18(4):398–403.

[18] Cooper RS, Kaufman JS, Ward R. Race and genomics. N Engl J Med 2003;348:1166.

[19] Burchard EG, Ziv E, Coyle N, Gomez SL, Tang H, Karter AJ, et al. The importance of race and ethnic background in biomedical research and clinical practice. N Engl J Med 2003;348:1170.

[20] Via M, Ziv E, Burchard EG. Recent advances of genetic ancestry testing in biomedical research and direct to consumer testing. Clin Genet 2009;76:225–35.

[21] Yudell M, Roberts D, DeSalle R, Tishkoff S. Taking race out of human genetics. Science 2016;351(6273):564–5.

[22] Rosenberg NA, Pritchard JK, Weber JL, Cann HM, Kidd KK, Zhivotovsky LA, et al. Genetic structure of human populations. Science 2002;298:2381–5.

[23] Narang A, Jha P, Rawat V, Mukhopadhyay A, Dash D, Indian Genome Variation Consortium Recent admixture in an Indian population of African ancestry. Am J Hum Genet 2011;89:111–20.

[24] Moorjani P, Patterson N, Hirschhorn JN, Keinan A, Hao L, Atzmon G, et al. The history of African gene flow into Southern Europeans, Levantines, and Jews. PLoS Genet 2011;7(4):e1001373.

[25] Hatin WI, Nur-Shafawati AR, Zahri MK, Xu S, Jin L, Tan SG, et al. Population genetic structure of peninsular Malaysia Malay sub-ethnic groups. PLoS One 2011;6(4):e18312.

[26] Mao X, Bigham AW, Mei R, Gutierrez G, Weiss KM, Brutsaert TD, et al. A genomewide admixture mapping panel for Hispanic/Latino populations. Am J Hum Genet 2007;80(6):1171–8.

[27] Smith MW, Patterson N, Lautenberger JA, Truelove AL, McDonald GJ, Waliszewska A, et al. A high-density admixture map for disease gene discovery in African Americans. Am J Hum Genet 2004;74(5):1001–13.

[28] Pasaniuc B, Zaitlen N, Lettre G, Chen GK, Tandon A, Kao WH, et al. Enhanced statistical tests for GWAS in admixed populations: assessment using African Americans from CARe and a Breast Cancer Consortium. PLoS Genet 2011;7(4):e1001371.

[29] Seldin MF, Pasaniuc B, Price AL. New approaches to disease mapping in admixed populations. Nat Rev Genet 2011;12(8):523–8.

[30] Klimentidis YC, Divers J, Casazza K, Beasley TM, Allison DB, Fernandez JR. Ancestry-informative markers on chromosomes 2, 8 and 15 are associated with insulin-related traits in a racially diverse sample of children. Hum Genomics 2011;5:79–89.

[31] Limdi NA, Wadelius M, Cavallari L. Warfarin pharmacogenetics: a single VKORC1 polymorphism is predictive of dose across 3 racial groups. Blood 2010;115:3827–34.

[32] Limdi NA, Arnett DK, Goldstein JA. Influence of CYP2C9 and VKORC1 on warfarin dose, anticoagulation attainment and maintenance among European–Americans and African–Americans. Pharmacogenomics 2008;9:511–26.

[33] Klein TE, Altman RB, Eriksson N. Estimation of the warfarin dose with clinical and pharmacogenetic data. N Engl J Med 2009;360:753–64.

[34] Helgadottir A, Manolescu A, Helgason A, Thorleifsson G, Thorsteinsdottir U, Gudbjartsson DF, et al. A variant of the gene encoding leukotriene A4 hydrolase confers ethnicity-specific risk of myocardial infarction. Nat Genet 2006;38:68–74.

[35] Corvol H, De Giacomo A, Eng C, Seibold M, Ziv E, Chapela R, et al. Genetic ancestry modifies pharmacogenetic gene–gene interaction for asthma. Pharmacogenet Genomics 2009;19:489–96.

[36] Bustamante CD, Burchard EG, De la Vega FM. Genomics for the world. Nature 2011;475:163–5.

[37] Perera MA, Cavallari LH, Limdi NA, Gamazon ER, Konkashbaev A, Daneshjou R, et al. Genetic variants associated with warfarin dose in African–American individuals: a genome-wide association study. Lancet 2013;382(9894):790–6.

[38] Drozda K, Wong S, Patel SR, Bress AP, Nutescu EA, Kittles RA, et al. Poor warfarin dose prediction with pharmacogenetic algorithms that exclude genotypes important for African Americans. Pharmacogenet Genomics 2015;25(2):73–81.

[39] Nagai R, Ohara M, Cavallari LH, Drozda K, Patel SR, Nutescu EA, et al. Factors influencing pharmacokinetics of warfarin in African–Americans: implications for pharmacogenetic dosing algorithms. Pharmacogenomics 2015;16(3):217–25.

[40] Sankar P, Cho MK, Condit CM, Hunt LM, Koenig B, Marshall P, et al. Genetic research and health disparities. JAMA 2004;291:2985–9.

[41] Need AC, Goldstein DB. Next generation disparities in human genomics: concerns and remedies. Trends Genet 2009;25(11):489–94.

[42] Haga SB. Impact of limited population diversity of genome-wide association studies. Genet Med 2010;12(2):81–4.

[43] Royal CD, Novembre J, Fullerton SM, Goldstein DB, Long JC, Bamshad MJ, et al. Inferring genetic ancestry: opportunities, challenges, and implications. Am J Hum Genet 2010;86:661–73.

[44] Wendler D, Kington R, Madans J, Van Wye G, Christ-Schmidt H, Pratt LA, et al. Are racial and ethnic minorities less willing to participate in health research? PLoS Med 2006;3(2):e19.

[45] Ochs-Balcom HM, Jandorf L, Wang Y, Johnson D, Meadows Ray V, Willis MJ, et al. "It takes a village": multilevel approaches to recruit African Americans and their families for genetic research. J Community Genet 2015;6(1):39–45.

[46] Dukepoo FC. It's more than the Human Genome Diversity Project. Politics Life Sciences 1999:293–7.

[47] Weiss KM, Cavalli-Sforza LL, Dunston GM, Feldman M, Greely HT, Kidd KK, et al. Proposed model ethical protocol for collecting DNA samples. Houst Law Rev 1997;33:1431–73.

[48] Santos L. Genetic research in native communities. Prog Community Health Partnersh 2008;2(4):321–7.

[49] Jacobs B, Roffenbender J, Collmann J, Cherry K, Bitsói LL, Bassett K, et al. Bridging the divide between genomic science and indigenous peoples. J Law Med Ethics 2010;38(3):684–96.

[50] Goering S, Holland S, Fryer-Edwards K. Transforming genetic research practices with marginalized communities: a case for responsive justice. Hastings Cent Rep 2008;38:45–53.

[51] Mello MM, Wolf LE. The Havasupai Indian tribe case—lessons for research involving stored biologic samples. N Engl J Med 2010;363(3):204–7.

[52] Zika E, Paci D, den Bäumen TS, Braun A, RijKers-Defrasne S, Deschênes M, et al. Biobanks in Europe: prospects for harmonization and networking. : Joint Research Centre, European Commission; 2010. Available at <http://ftp.jrc.es/EURdoc/JRC57831.pdf>.

[53] Cambon-Thomsen A. The social and ethical issues of post-genomic human biobanks. Nat Rev Genetics 2004;5(11):866–73.

[54] Cambon-Thomsen A, Ducournau P, Gourraud PA, Pontille D. Biobanks for genomics and genomics for biobanks. Comp Funct Genomics 2003;4(6):628–34.

[55] Mitchell R, Waldby C. National biobanks: clinical labor, risk production, and the creation of biovalue. Sci Technol Hum Values 2010;35(3):330–55.

[56] Hoeyer K. Science is really needed—that's all I know. Informed consent and the non-verbal practices of collecting blood for genetic research in Sweden. New Genet Soc 2003;22(3):224–9.

[57] Hoeyer K, Olofsson BO, Mjörndal T, Lynöe N. Informed consent and biobanks: a population-based study of attitudes towards tissue donation for genetic research. Scand J Public Health 2004;32(3):224–9.

[58]   Kettis-Lindblad A, Ring L, Viberth E, Hansson MG. Perceptions of potential donors in the Swedish public towards information and consent procedures in relation to use of human tissue samples in biobanks: a population-based study. Scand J Public Health 2007;35(2):148–56.

[59]   Kaufman D, Murphy J, Erby L, Hudson K, Scott J. Veterans' attitudes regarding a database for genomic research. Genet Med 2009;11(5):329–37.

[60]   Tupasela A, Sihvo S, Snell K, Jallinoja P, Aro AR, Hemminki E. Attitudes towards biomedical use of tissue sample collections, consent, and biobanks among Finns. Scand J Public Health 2010;38(1):46–52.

[61]   Simon CM, L'heureux J, Murray JC, Winokur P, Weiner G, Newbury E, et al. Active choice but not too active: public perspectives on biobank consent models. Genet Med 2011;13(9):821–31.

[62]   Melas PA, Sjöholm LK, Forsner T, Edhborg M, Juth N, Forsell Y, et al. Examining the public refusal to consent to DNA biobanking: empirical data from a Swedish population-based study. J Med Ethics 2010;36(2):93–8.

[63]   McDonald JA, Vadaparampil S, Bowen D, Magwood G, Obeid JS, Jefferson M, et al. Intentions to donate to a biobank in a national sample of African Americans. Public Health Genomics 2014;17(3):173–82.

[64]   O'Doherty KC, Burgess MM. Engaging the public on biobanks: outcomes of the BC biobank deliberation. Public Health Genomics 2009;12(4):203–15.

[65]   Lemke AA, Wu JT, Waudby C, Pulley J, Somkin CP, Trinidad SB. Community engagement in biobanking: Experiences from the eMERGE Network. Genomics Soc Policy 2010;6(3):35–52.

[66]   Godard B, Marshall J, Laberge C, Knoppers BM. Strategies for consulting with the community: the cases of four large-scale genetic databases. Sci Eng Ethics 2004;10:457–77.

[67]   Hansson MG, Dillner J, Bartram CR, Carlson JA, Helgesson G. Should donors be allowed to give broad consent to future biobank research? Lancet Oncol 2006;7(3):266–9.

[68]   Murphy J, Scott J, Kaufman D, Geller G, LeRoy L, Hudson K. Public perspectives on informed consent for biobanking. Am J Public Health 2009;99(12):2128–34.

[69]   Stein DT, Terry SF. Reforming biobank consent policy: a necessary move away from broad consent toward dynamic consent. Genet Test Mol Biomarkers 2013;17:855–6.

[70]   Simon CM, Klein DW, Schartz HA. Traditional and electronic informed consent for biobanking: a survey of U.S. biobanks. Biopreserv Biobank 2014;12(6):423–9.

[71]   Simon CM, Klein DW, Schartz HA. Interactive multimedia consent for biobanking: a randomized trial. Genet Med 2016;18(1):57–64.

[72]   Smith ME, Aufox S. Biobanking: the melding of research with clinical care. Curr Genet Med Rep 2013;1(2):122–8.

[73]   Wolf SM, Burke W, Koenig BA. Mapping the ethics of translational genomics: situating return of results and navigating the research-clinical divide. J Law Med Ethics 2015;43(3):486–501.

[74]   Wolf SM. The role of law in the debate over return of research results and incidental findings: the challenge of developing law for translational science. Minn J Law Sci Technol 2012;13(2).

[75]   Christenhusz GM, Devriendt K, Dierickx K. Disclosing incidental findings in genetics contexts: a review of the empirical ethical research. Eur J Med Genet 2013;56(10):529–40.

[76]   Kocarnik JM, Fullerton SM. Returning pleiotropic results from genetic testing to patients and research participants. JAMA 2014;311(8):795–6.

[77]   Shalowitz DI, Miller FG. Disclosing individual results of clinical research: implications of respect for participants. JAMA 2005;294:737–40.

[78]   MacNeil SD, Fernandez CV. Attitudes of research ethics board chairs towards disclosure of research results to participants: results of a national survey. J Med Ethics 2007;33(9):549–53.

[79] Clayton EW, McGuire AL. The legal risks of returning results of genomics research. Genet Med 2012;14(4):473–7.

[80] Pike ER, Rothenberg KH, Berkman BE. Finding fault? Exploring legal duties to return incidental findings in genomic research. Georgetown Law J 2014;102:795–843.

[81] McGuire AL, Knoppers BM, Zawati MH, Clayton EW. Can I be sued for that? Liability risk and the disclosure of clinically significant genetic research findings. Genome Res 2014;24(5):719–23.

[82] Murphy J, Scott J, Kaufman D, Geller G, LeRoy L, Hudson K. Public expectations for return of results from large-cohort genetic research. Am J Bioeth 2008;8(11):36–43.

[83] Beskow LM, Smolek SJ. Prospective biorepository participants' perspectives on access to research results. J Empir Res Hum Res Ethics 2009;4(3):99–111.

[84] Couzin-Frankel J. What would you do? Science 2011;331:662–5.

[85] Trinidad SB, Fullerton SM, Ludman EJ, Jarvik GP, Larson EB, Burke W. Research ethics. Research practice and participant preferences: the growing gulf. Science 2011;331(6015):287–8.

[86] Fullerton SM, Wolf WA, Brothers KB, Clayton EW, Crawford DC, Denny JC, et al. Return of individual research results from genome-wide association studies: experience of the Electronic Medical Records and Genomics (eMERGE) network. Genet Med 2012;14(4):424–31.

[87] O'Daniel J, Haga S. Public perspectives on returning genetics and genomics research results. Public Health Genomics 2011;14(6):346–55.

[88] Dixon-Woods M, Jackson C, Windridge KC, Kenyon S. Receiving a summary of the results of a trial: qualitative study of participants' views. BMJ 2006;332(7535):206–10.

[89] Gooblar J, Roe CM, Selsor NJ, Gabel MJ, Morris JC. Attitudes of research participants and the general public regarding disclosure of Alzheimer disease research results. JAMA Neurol 2015;72(12):1484–90.

[90] Snowdon C, Garcia J, Elbourne D. Reactions of participants to the results of a randomised controlled trial: exploratory study. BMJ 1998;317(7150):21–6.

[91] Schulz CJ, Riddle MP, Valdimirsdottir HB, Abramson DH, Sklar CA. Impact on survivors of retinoblastoma when informed of study results on risk of second cancers. Med Pediatr Oncol 2003;41(1):36–43.

[92] Partridge AH, Wong JS, Knudsen K, Gelman R, Sampson E, Gadd M, et al. Offering participants results of a clinical trial: sharing results of a negative study. Lancet 2005;365(9463):963–4.

[93] Bunin GR, Kazak AE, Mitelman O. Informing subjects of epidemiologic study results. Children's Cancer Group. Pediatrics 1996;97(4):486–91.

[94] Green RC, Roberts JS, Cupples LA, Relkin NR, Whitehouse PJ, Brown T, et al. Disclosure of APOE genotype for risk of Alzheimer's disease. N Engl J Med 2009;361:245–54.

[95] Kollek R, Petersen I. Disclosure of individual research results in clinico-genomic trials: challenges, classification and criteria for decision-making. J Med Ethics 2011;37(5):271–5.

[96] Jarvik GP, Amendola LM, Berg JS, Brothers K, Clayton EW, Chung W, et al. Return of genomic results to research participants: the floor, the ceiling, and the choices in between. Am J Hum Genet 2014;94(6):818–26.

[97] Johnson G, Lawrenz F, Thao M. An empirical examination of the management of return of individual research results and incidental findings in genomic biobanks. Genet Med 2012;14(4):444–50.

[98] Wolf SM, Crock BN, Van Ness B, Lawrenz F, Kahn JP, Beskow LM, et al. Managing incidental findings and research results in genomic research involving biobanks and archived data sets. Genet Med 2012;14(4):361–84.

[99] Bookman EB, Langehorne AA, Eckfeldt JH, Glass KC, Jarvik GP, Klag M, et al. Reporting genetic results in research studies: summary and recommendations of an NHLBI working group. Am J Med Genet A 2006;140A:1033–40.

[100] Fabsitz RR, McGuire A, Sharp RR, Puggal M, Beskow LM, Biesecker LG, et al. Ethical and practical guidelines for reporting genetic research results to study participants: updated guidelines from a National Heart, Lung, and Blood Institute working group. Circ Cardiovasc Genet 2010;3(6):574–80.

[101] MacNeil SD, Fernandez CV. Offering results to research participants. BMJ 2006;332:188–9.

[102] Fernandez CV, Kodish E, Taweel S, Shurin S, Weijer C. Disclosure of the right of research participants to receive research results: an analysis of consent forms in the Children's Oncology Group. Cancer 2003;97:2904–9.

[103] Heaney C, Tindall G, Lucas J, Haga SB. Researcher practices on returning genetic research results. Genet Test Mol Biomarkers 2010;14(6):821–7.

[104] Ramoni RB, McGuire AL, Robinson JO, Morley DS, Plon SE, Joffe S. Experiences and attitudes of genome investigators regarding return of individual genetic test results. Genet Med 2013;15(11):882–7.

[105] Partridge AH, Hackett N, Blood E, Gelman R, Joffe S, Bauer-Wu S, et al. Oncology physician and nurse practices and attitudes regarding offering clinical trial results to study participants. J Natl Cancer Inst 2004;96(8):629–32.

[106] Fernandez CV, Kodish E, Weijer C. Informing study participants of research results: an ethical imperative. IRB 2003;25(3):12–19.

[107] Fernandez CV, Kodish E, Weijer C. Importance of informed consent in offering to return research results to research participants. Med Pediatr Oncol 2003;41:592–3.

[108] Appelbaum PS, Parens E, Waldman CR, Klitzman R, Fyer A, Martinez J, et al. Models of consent to return of incidental findings in genomic research. Hastings Cent Rep 2014;44(4):22–32.

[109] Beskow LM, Burke W. Offering individual genetic research results: context matters. Sci Transl Med 2010;2(38):38cm20.

[110] Bredenoord AL, Onland-Moret NC, Van Delden JJ. Feedback of individual genetic results to research participants: in favor of a qualified disclosure policy. Hum Mutat 2011;32(8):861–7.

[111] Abdul-Karim R, Berkman BE, Wendler D, Rid A, Khan J, Badgett T, et al. Disclosure of incidental findings from next-generation sequencing in pediatric genomic research. Pediatrics 2013;131(3):564–71.

[112] Roberts JS, Shalowitz DI, Christensen KD, et al. Returning individual research results: development of a cancer genetics education and risk communication protocol. J Empir Res Hum Res Ethics 2010;5(3):17–30.

[113] Cassa CA, Savage SK, Taylor PL, Green RC, McGuire AL, Mandl KD. Disclosing pathogenic genetic variants to research participants: quantifying an emerging ethical responsibility. Genome Res 2012;22(3):421–8.

[114] Black L, Avard D, Zawati MH, Knoppers BM, Hébert J, Sauvageau G, et al. Funding considerations for the disclosure of genetic incidental findings in biobank research. Clin Genet 2013;84(5):397–406.

[115] Bledsoe MJ, Clayton EW, McGuire AL, Grizzle WE, O'Rourke PP, Zeps N. Return of research results from genomic biobanks: cost matters. Genet Med 2013;15(2):103–5.

[116] Godard B, Hurlimann T, Letendre M, Egalité N, INHERIT BRCAs Guidelines for disclosing genetic information to family members: from development to use. Fam Cancer 2006;5(1):103–16.

[117] Tassé AM. The return of results of deceased research participants. J Law Med Ethics 2011;39(4):621–30.

[118] Rothstein MA. Disclosing decedents' research results to relatives violates the HIPAA privacy rule. Am J Bioeth 2012;12(10):16–17.

[119] Rothstein MA. Should researchers disclose results to descendants? Am J Bioeth 2013;13(10):64–5.

[120] Taylor HA, Wilfond BS. The ethics of contacting family members of a subject in a genetic research study to return results for an autosomal dominant syndrome. Am J Bioeth 2013;13(10):61.

[121] Milner LC, Liu EY, Garrison NA. Relationships matter: ethical considerations for returning results to family members of deceased subjects. Am J Bioeth 2013;13(10):66–7.

[122] Graves KD, Sinicrope PS, Esplen MJ, Peterson SK, Patten CA, Lowery J, et al. Communication of genetic test results to family and health-care providers following disclosure of research results. Genet Med 2014;16(4):294–301.

[123] American Society of Human Genetics Social Issues Subcommittee on Familial Disclosure Professional disclosure of familial genetic information. Am J Hum Genet 1998;62(2):474–83.

[124] Offit K, Groeger E, Turner S, Wadsworth EA, Weiser MA. The 'duty to warn' a patient's family members about hereditary disease risks. JAMA 2004;292(12):1469–73.

[125] Forrest LE, Delatycki MB, Skene L, Aitken M. Communicating genetic information in families: a review of guidelines and position papers. Eur J Hum Genet 2007;15(6):612–8.

[126] American Medical Association Code of Medical Ethics' Opinions on Genetic Testing AMA opinion 2.131. Disclosure of familial risk in genetic testing. Virtual Mentor 2009;11(9):683–5.

[127] Forrest K, Simpson SA, Wilson BJ, van Teijlingen ER, McKee L, Haites N, et al. To tell or not to tell: barriers and facilitators in family communication about genetic risk. Clin Gen 2003;64(4):317–26.

[128] Weaver M. The double helix: applying an ethic of care to the duty to warn genetic relatives of genetic information. Bioethics 2016;30(3):181–7.

[129] Wolf SM, Branum R, Koenig BA, Petersen GM, Berry SA, Beskow LM, et al. Returning a research participant's genomic results to relatives: analysis and recommendations. Am J Law Med Ethics 2015;43(3):440–63.

[130] National Institutes of Health and US Department of Energy, Working Group on Ethical Legal and Social Implications of Human Genome Research. Promoting safe and effective genetic testing in the United States. September 1997.

[131] Institute of Medicine Assessing genetic risks: implications for health and social policy. Washington, DC: National Academy Press; 1994.

[132] Secretary's Advisory Committee on Genetic Testing. (2000). Enhancing the oversight of genetic testing: recommendations of the SACGT. Available at <http://www4.od.nih.gov/oba/sacgt/reports/oversight_report.htm>.

[133] Secretary's Advisory Committee on Genetics, Health, and Society. (2008). U.S. System of Oversight of Genetic Testing: A Response to the change of the Secretary of Health and Human Services. Available at <http://oba.od.nih.gov/oba/SACGHS/reports/SACGHS_oversight_report.pdf>.

[134] US House of Representatives. (October 2011). Modernizing laboratory test standards for patients act of 2011. (H.R.3207).

[135] US House of Representatives. (May 2010). Genomics and personalized medicine act of 2010 (HR 5440).

[136] US Senate. (March 2007). Laboratory test improvement act (S. 736). Introduced by Senators Kennedy and Smith.

[137] US Senate. (March 2007).Genomics and personalized medicine act of 2008 (S. 976). Introduced by Senators Obama and Burr.

[138] US House of Representatives. (July 2008).Genomics and personalized medicine act of 2008 (H.R. 6498). Introduced by Representative Kennedy.

[139] Food and Drug Administration. (2005). Guidance for industry and FDA staff. Class II special controls guidance document: drug metabolizing enzyme genotyping system. Available at <http://www.fda.gov/cdrh/oivd/guidance/1551.pdf>.

[140] Food and Drug Administration. (2005). Guidance for industry and FDA staff. Class II special controls guidance document: instrumentation for clinical multiplex test systems. Available at <http://www.fda.gov/cdrh/oivd/guidance/1546.pdf>.

[141] Food and Drug Administration. (2006). Draft guidance for industry and FDA staff—commercially distributed analyte specific reagents (ASRs): frequently asked questions. Available at <http://www.fda.gov/cdrh/oivd/guidance/1590.pdf>.

[142] Food and Drug Administration, Center for Devices and Radiological Health, Center for Biologic Evaluation and Research. (2007). Guidance for industry and FDA staff. Commercially distributed analyte specific reagents (ASRs): frequently asked questions. September 14, 2007.

[143] Food and Drug Administration, Center for Devices and Radiological Health, Center for Biologic Evaluation and Research. (2007). Draft guidance for industry, clinical laboratories, and FDA staff: in vitro diagnostic multivariate index assays.

[144] Food and Drug Administration. (2006). Draft guidance for industry, clinical laboratories, and FDA staff—in vitro diagnostic multivariate index assays. Available at <http://www.fda.gov/cdrh/oivd/guidance/1610.pdf>.

[145] Food and Drug Administration. (June 2006). Guidance for industry and review staff. Recommended approaches to integration of genetic toxicology study results. Available at <http://www.fda.gov/cder/guidance/6848fnl.pdf>.

[146] Food and Drug Administration. (2007). Guidance for industry and FDA staff. Pharmacogenetic tests and genetic tests for heritable markers. Available at <http://www.fda.gov/cdrh/oivd/guidance/1549.pdf>.

[147] Food and Drug Administration. (October 2014). Draft guidance: framework for regulatory oversight of laboratory developed tests (LDTs). Available at <http://www.fda.gov/downloads/MedicalDevices/DeviceRegulationandGuidance/GuidanceDocuments/UCM416685.pdf>.

[148] Green RC, Berg JS, Grody WW, Kalia SS, Korf BR, Martin CL, et al. ACMG recommendations for reporting of incidental findings in clinical exome and genome sequencing. Genet Med 2013;15(7):565–74.

[149] Burke W, Antommaria AH, Bennett R, Botkin J, Clayton EW, Henderson GE, et al. Recommendations for returning genomic incidental findings? We need to talk!. Genet Med 2013;15(11):854–9.

[150] Holtzman NA. ACMG recommendations on incidental findings are flawed scientifically and ethically. Genet Med 2013;15(9):750–1.

[151] Wolf SM, Annas GJ, Elias S. Point-counterpoint. Patient autonomy and incidental findings in clinical genomics. Science 2013;340(6136):1049–50.

[152] Parsons DW, Roy A, Plon SE, Roychowdhury S, Chinnaiyan AM. Clinical tumor sequencing: an incidental casualty of the American College of Medical Genetics and Genomics recommendations for reporting of incidental findings. J Clin Oncol 2014;32(21):2203–5.

[153] American College of Medical Genetics and Genomics Incidental findings in clinical genomics: a clarification. Genet Med 2013;15(8):664–6.

[154] ACMG Board of Directors ACMG policy statement: updated recommendations regarding analysis and reporting of secondary findings in clinical genome-scale sequencing. Genet Med 2015;17(1):68–9.

[155] Evans JP, Green RC. Direct to consumer genetic testing: avoiding a culture war. Genet Med 2009;11(8):568–9.

[156] Hunter DJ, Khoury MJ, Drazen JM. Letting the genome out of the bottle—will we get our wish? N Engl J Med 2008;358(2):105–7.

[157] Altman RB. Direct-to-consumer genetic testing: failure is not an option. Clin Pharmacol Ther 2009;86(1):15–17.

[158] Schickedanz AD, Herdman RC. Direct-to-consumer genetic testing: the need to get retail genomics right. Clin Pharmacol Ther 2009;86(1):17–20.

[159] Javitt G. Which way for genetic-test regulation? Assign regulation appropriate to the level of risk. Nature 2010;466(7308):817–8.

[160] Hudson K, Javitt G, Burke W, Byers P. ASHG Statement on direct-to-consumer genetic testing in the United States. Obstet Gynecol 2007;110(6):1392–5.

[161] Wallace H. Most gene test sales are misleading. Nat Biotechnol 2008;26(11):1221.

[162] Farkas DH, Holland CA. Direct-to-consumer genetic testing: two sides of the coin. J Mol Diagn 2009;11(4):263–5.

[163] American College of Medical Genetics and Genomics ACMG statement on direct-to-consumer genetic testing. Genet Med 2004;6(1):60.

[164] Secretary's Advisory Committee on Genetics, Health, and Society. (2010). Direct-to-consumer genetic testing. Available at <http://oba.od.nih.gov/oba/sacghs/reports/SACGHS_DTC_Report_2010.pdf>.

[165] American Medical Association. (2007). Resolution 522: direct to consumer advertising and provision of genetic testing. Available at <http://www.ama-assn.org/ama1/pub/upload/mm/467/522.doc>.

[166] European Society of Human Genetics Statement of the ESHG on direct-to-consumer genetic testing for health-related purposes. Eur J Hum Genet 2010;18(12):1271–3.

[167] Government Accountability Office. (2010). Direct-to-consumer genetic tests: misleading test results are further complicated by deceptive marketing and other questionable practices (GAO-10-847T). Available at <http://www.gao.gov/products/GAO-10-847T>.

[168] Government Accountability Office. (2006). Nutrigenetic testing: tests purchased from four web sites mislead consumers (GAO-06-977T). Available at <http://www.gao.gov/new.items/d06977t.pdf>.

[169] UK Human Genetics Commission. (2010). A common framework of principles for direct-to-consumer genetic testing services. Available at <http://www.hgc.gov.uk/UploadDocs/DocPub/Document/HGC%20Principles%20for%20DTC%20genetic%20tests%20-%20final.pdf>.

[170] Su Y, Howard HC, Borry P. Users' motivations to purchase direct-to-consumer genome-wide testing: an exploratory study of personal stories. J Community Genet 2011;2(3):135–46.

[171] McBride CM, Wade CH, Kaphingst KA. Consumers' views of direct-to-consumer genetic information. Annu Rev Genomics Hum Genet 2010;11:427–46.

[172] James KM, Cowl CT, Tilburt JC, Sinicrope PS, Robinson ME, Frimannsdottir KR, et al. Impact of direct-to-consumer predictive genomic testing on risk perception and worry among patients receiving routine care in a preventive health clinic. Mayo Clin Proc 2011;86(10):933–40.

[173] Bloss CS, Schork NJ, Topol EJ. Effect of direct-to-consumer genomewide profiling to assess disease risk. N Eng J Med 2011;364(6):524–34.

[174] Carere DA, VanderWeele T, Moreno TA, Mountain JL, Roberts JS, Kraft P, et al. The impact of direct-to-consumer personal genomic testing on perceived risk of breast, prostate, colorectal, and lung cancer: findings from the PGen study. BMC Med Genomics 2015;8:63.

[175] Baptista NM, Christensen KD, Carere DA, Broadley SA, Roberts JS, Green RC. Adopting genetics: motivations and outcomes of personal genomic testing in adult adoptees. Genet Med 2016;18(9):924–32.

[176] Carlson JJ, Henrikson NB, Veenstra DL, Ramsey SD. Economic analyses of human genetics services: a systematic review. Genet Med 2005;7:519–23.

[177] Phillips KA, Van Bebber SL. A systematic review of cost-effectiveness analyses of pharmacogenomic interventions. Pharmacogenomics 2004;5:1139–49.

[178] Paulden M, Franek J, Pham B, Bedard PL, Trudeau M, Krahn M. Cost-effectiveness of the 21-gene assay for guiding adjuvant chemotherapy decisions in early breast cancer. Value Health 2013;16(5):729–39.

[179] D'Andrea E, Marzuillo C, Pelone F, De Vito C, Villari P. Genetic testing and economic evaluations: a systematic review of the literature. Epidemiol Prev 2015;39(4 Suppl 1):45–50.

[180] Blue Cross Blue Shield Association Technology Evaluation Center Gene expression profiling for managing breast cancer treatment. Asses Program 2005;20(3):1–5.

[181] Blue Cross Blue Shield Association's Technology Evaluation Center. (2007). Actions taken by the medical advisory panel. Available at <http://www.bcbs.com/betterknowledge/tec/press/actions-taken-by-the-medical-1.html>.

[182] Blue Cross Blue Shield Association Technology Evaluation Center Gene expression profiling in women with lymph-node-positive breast cancer to select adjuvant chemotherapy treatment. Asses Program 2010;25(1). Available at <http://www.bcbs.com/cce/vols/25/25_01.pdf>.

[183] Evans RW, Williams GE, Baron HM, Deng MC, Eisen HJ, Hunt SA, et al. The economic implications of noninvasive molecular testing for cardiac allograft rejection. Am J Transplant 2005;5:1553–8.

[184] Matchar DB, Thakur ME, Grossman I, McCrory DC, Orlando LA, Steffens DC, et al. Testing for cytochrome P450 polymorphisms in adults with non-psychotic depression treated with selective serotonin reuptake inhibitors (SSRIs). Evidence report/technology assessment no. 146. (Prepared by the duke evidence-based practice center under contract no. 290-02-0025.) AHRQ Publication No. 07-E002. Rockville, MD: Agency for Healthcare Research and Quality; 2006.

[185] Saokaew S, Tassaneeyakul W, Maenthaisong R, Chaiyakunapruk N. Cost-effectiveness analysis of HLA-B*5801 testing in preventing allopurinol-induced SJS/TEN in Thai population. PLoS One 2014;9(4):e94294.

[186] Dong D, Tan-Koi WC, Teng GG, Finkelstein E, Sung C. Cost-effectiveness analysis of genotyping for HLA-B*5801 and an enhanced safety program in gout patients starting allopurinol in Singapore. Pharmacogenomics 2015;16(16):1781–93.

[187] Allain DC, Friedman S, Senter L. Consumer awareness and attitudes about insurance discrimination post enactment of the Genetic Information Nondiscrimination Act. Fam Cancer 2012;11(4):637–44.

[188] Parkman AA, Foland J, Anderson B, Duquette D, Sobotka H, Lynn M, et al. Public awareness of genetic nondiscrimination laws in four states and perceived importance of life insurance protections. J Genet Couns 2015;24(3):512–21.

[189] Wauters A, Van Hoyweghen I. Global trends on fears and concerns of genetic discrimination: a systematic literature review. J Hum Genet 2016;61(4):275–82.

[190] US Department of Health and Human Services, Office of Human Research Protections. Genetic Information Non-Discrimination Act Guidance, 2009. Available at <http://www.hhs.gov/ohrp/policy/gina.html>.

[191] Green RC, Lautenbach D, McGuire AL. GINA, genetic discrimination, and genomic medicine. N Engl J Med 2015;372(5):397–9.

[192] Berliner JL, Fay AM, Cummings SA, Burnett B, Tillmanns T. NSGC practice guideline: risk assessment and genetic counseling for hereditary breast and ovarian cancer. J Genet Couns 2013;22(2):155–63.

[193] Prince AER, Roche MI. Genetic information, non-discrimination, and privacy protections in genetic counseling practice. J Genet Couns 2014;23(6):891–902.

[194] Laedtke AL, O'Neill SM, Rubinstein WS, Vogel KJ. Family physicians' awareness and knowledge of the Genetic Information Non-Discrimination Act (GINA). J Genet Couns 2012;21(2):345–52.

[195] Dorsey ER, Darwin KC, Nichols PE, Kwok JH, Bennet C, Rosenthal LS, et al. Knowledge of the Genetic Information Nondiscrimination act among individuals affected by Huntington disease. Clin Genet 2013;84(3):251–7.

[196] Van Hoyweghen I, Horstman K. European practices of genetic information and insurance: lessons for the Genetic Information Nondiscrimination Act. JAMA 2008;300:326–7.

[197] Rothstein MA, Joly Y. The Handbook of Genetics and Society: Mapping the New Genomic Era Atkinson P, Glasner P, Lock M, editors. Genetic information and insurance underwriting: contemporary issues and approaches in the global economy insurance. London: Routledge; 2009. p. 127–44.

[198] Otlowski M, Taylor S, Bombard Y. Genetic discrimination: international perspectives. Annu Rev Genomics Hum Genet 2012;13:433–54.

[199] Nicholls SG, Fafard P. Genetic discrimination legislation in Canada: moving from rhetoric to real debate. CMAJ 2016;188(11):788–9.

[200] Thomas R. Genetics and insurance in the United Kingdom 1995–2010: the rise and fall of "scientific" discrimination. New Genet Soc 2012;31:203–22.

[201] Metcalfe S, Hurworth R, Newstead J, Robins R. Needs assessment study of genetics education for general practitioners in Australia. Genet Med 2002;4:71–7.

[202] Marzuillo C, De Vito C, Boccia S, D'Addario M, D'Andrea E, Santini P, et al. Knowledge, attitudes and behavior of physicians regarding predictive genetic tests for breast and colorectal cancer. Prev Med 2013;57(5):477–82.

[203] Marzuillo C, De Vito C, D'Addario M, Santini P, D'Andrea E, Boccia A, et al. Are public health professionals prepared for public health genomics? A cross-sectional survey in Italy. BMC Health Serv Res 2014;14:239.

[204] Christensen KD, Vassy JL, Jamal L, Lehmann LS, Slashinski MJ, Perry DL, et al. Are physicians prepared for whole genome sequencing? A qualitative analysis. Clin Genet 2016;89(2):228–34.

[205] Bennett C, Burton H, Fardon P. Competences, education and support for new roles in cancer genetics services: outcomes from the cancer genetics pilot projects. Familial Cancer 2007;6:171–80.

[206] Torrance N, Mollison J, Wordsworth S, Gray J, Miedzybrodzka Z, Haites N, et al. Genetic nurse counsellors can be an acceptable and cost-effective alternative to clinical geneticists for breast cancer risk genetic counselling. Evidence from two parallel randomised controlled equivalence trials. Br J Cancer 2006;95(4):435–44.

[207] Challen K, Harris HJ, Julian-Reynier C, Ten Kate LP, Kristoffersson U, Nippert I, et al. Genetic education and non genetic health professionals: educational providers and curricula in Europe. Genet Med 2005;7(5):302–10.

[208] Gurwitz D, Weizman A, Rehavi M. Education: teaching pharmacogenomics to prepare future physicians and researchers for personalized medicine. Trends Pharmacol Sci 2003;24:122–5.

[209] Demmer LA, Waggoner DJ. Professional medical education and genomics. Annu Rev Genomics Hum Genet 2014;15:507–16.

[210] Waggoner DJ, Martin CL. Integration of internet-based genetic databases into the medical school pre-clinical and clinical curriculum. Genet Med 2006;8(6):379–82.

[211] Bean LJ, Fridovich-Keil J, Hegde M, Rudd MK, Garber KB. The virtual diagnostic laboratory: a new way of teaching undergraduate medical students about genetic testing. Genet Med 2011;13(11):973–7.

[212] Dhar SU, Alford RL, Nelson EA, Potocki L. Enhancing exposure to genetics and genomics through an innovative medical school curriculum. Genet Med 2012;14(1):163–7.

[213] Perry CG, Maloney KA, Beitelshees AL, Jeng LJ, Ambulos Jr NP, Shuldiner AR, et al. Educational Innovations in Clinical Pharmacogenomics. Clin Pharmacol Ther 2016;99(6):582–4.

[214] Shirts BH, Salama JS, Aronson SJ, Chung WK, Gray SW, Hindorff LA, et al. CSER and eMERGE: current and potential state of the display of genetic information in the electronic health record. J Am Med Inform Assoc 2015;22(6):1231–42.

[215] Nishimura AA, Shirts BH, Salama J, Smith JW, Devine B, Tarczy-Hornoch P. Physician perspectives of CYP2C19 and clopidogrel drug–gene interaction active clinical decision support alerts. Int J Med Inform 2016;86:117–25.

[216] Chan V, Blazey W, Tegay D, Harper B, Koehler S, Laurent B, et al. Impact of academic affiliation and training on knowledge of hereditary colorectal cancer. Public Health Genomics 2014;17(2):76–83.

[217] Jenkins J, Calzone KA. Establishing the essential nursing competencies for genetics and genomics. J Nurs Scholarsh 2007;39(1):10–16.

[218] Lewis JA, Calzone KM, Jenkins J. Essential nursing competencies and curricula guidelines for genetics and genomics. MCN Am J Matern Child Nurs 2006;31(3):146–55.

[219] National Coalition for Health Professional Education in Genetics. (2005). Core competencies in genetics essential for all health-care professionals. Available at <http://www.nchpeg.org/core/Corecomps2005.pdf>.

[220] National Health Service's National Genetics Education and Development Centre and the Skills for Health. (2007). Competences for genetics in clinical practice. Available at <http://www.geneticseducation.nhs.uk/develop/index.asp?id=44>.

[221] Korf BR, Berry AB, Limson M, Marian AJ, Murray MF, O'Rourke PP, et al. Framework for development of physician competencies in genomic medicine: report of the Competencies Working Group of the Inter-Society Coordinating Committee for Physician Education in Genomics. Genet Med 2014;16(11):804–9.

Chapter 3

Educational Issues and Strategies for Genomic Medicine

Jean Jenkins and Laura Lyman Rodriguez
National Institutes of Health, National Human Genome Research Institute, Bethesda, MD, United States

Chapter Outline

INTRODUCTION

Primary care providers (Pcps) such as physicians, nurse practitioners (NP), and physician assistants (PA) are often the first contact of individuals entering the healthcare system. Plus, PCP through their ongoing relationships with individuals are positioned to utilize diverse information (i.e., family history, medical history, genomics) over time to guide shared decision making that affects quality care [1]. Discoveries of genomic variation associated with health, disease, and treatment options when translated into practice can make a difference for a patient and his/her family. Currently, PCP are being urged to assess how the incorporation of genomic technologies and findings in routine clinical care can benefit their patients through personalizing clinical interventions [2]. All aspects of the healthcare continuum are influenced by genomic advances. A crucial

Genomic and Precision Medicine. DOI: http://dx.doi.org/10.1016/B978-0-12-800685-6.00006-0
45

starting point in determining the relevancy of genomic information is to explore what is known about the genomic influences of common issues encountered in the PCP practice. Examples of relevancy include risk-assessment indications, common chronic disease management, adverse treatment events, and targeted treatment decisions. Genomic information is relevant throughout a lifetime including: before birth (i.e., preimplantation testing, prenatal testing, and carrier testing); childhood (i.e., diagnostic testing); adolescent (i.e., diagnostic testing, treatment decisions, and pharmacogenomics); and adult (i.e., diagnostic testing, treatment decisions, pharmacogenomics, and direct to consumer/ancestry).

Based on their practice analysis, PCP who recognize the importance of adding genomic data with other known clinical predictors to make clinical recommendations may be motivated to begin a professional journey to become more educated about genomic influences of care. For example, a 40-year-old white male of Northern European ancestry on both the maternal and paternal lineage received a screening colonoscopy recommended by his primary care physician based on a single first degree relative diagnosed with colon cancer before the age of 50. The colonoscopy identified a 20-mm elevated, nodular lesion in the descending colon that was biopsied. No polyps were seen. Diagnostic evaluation revealed a Stage 1 adenocarcinoma. The PCP referred her patient to an oncologist for staging and treatment options [3].

Genomic technology will enable PCP to identify those at risk for genomic conditions with tools that include everything from family history to whole genome sequencing (WGS). PCP without specialized genetics training will be increasingly called upon to recognize those at risk for illness, order screening procedures, refer for genetic testing, and use the results in the care of their patients as described above [4]. Infrastructure to support the provision of predictive and preventive services is necessary to maximize the potential value of genomic information to personalizing clinical decisions. Identified challenges exist to the integration of genomic information in primary care such as the complexity and volume of the information, limited evidence of value, and difficulty of interpretation of test results for patients [1]. There are limitations to current understanding of the implications of genomic test results that need to be overcome in order to effectively apply results in practice [5]. Knowledgeable providers that can interpret test results for individuals and their family are sparse.

GAPS IN CURRENT HEALTHCARE PROFESSIONAL LITERACY

Although recommended educational preparation and practice of all healthcare practitioners (i.e., physicians, NP, and PA) are changing, current evidence indicates that healthcare providers and faculty have limited genomic competency [6]. The Secretary's Advisory Committee on Genetics, Health, and Society stated that education of healthcare professionals is a significant factor limiting the integration of genetics into clinical care [7]. Most healthcare professionals have not had genetics/genomics in academic preparation. In addition, keeping up with the emerging scope of genetic and genomic applications across all populations and clinical settings is difficult and time consuming.

Surveys of practicing physicians reveal limitations of overall knowledge of genomics. In a survey of 220 internists from academic medical centers, 74% rated their knowledge of genomics as somewhat or very poor [8]. Plus they reported knowledge of guidelines for genetic testing as somewhat or very poor (87%). These physicians widely endorsed the need for more training such as when to order tests (79%), how to counsel patients (82%), how to interpret test results (77%), and how to maintain privacy (81%). Similar knowledge deficits were reported by neurologists and psychiatrists and they too reported the need for more training [9]. A qualitative study of PCPs and cardiologists ($n = 20$) involved with a clinical study offering WGS to participants provided themes of factors associated with physician preparedness [10]. They shared reluctance to use such testing if they were unprepared to sufficiently interpret, explain, or respond to WGS results. These physicians also expressed interest in developing proficiency but recommended that time and support would be essential to develop genomic competencies.

Nurse and NP assessments reflect similar knowledge deficits as physicians [11–13]. Registered nurses employed at Magnet Hospitals were surveyed as part of an integration of genomics in practice study [11]. The majority of the 7798 respondents (57%) reported their genomic knowledge base to be poor. Despite that limitation, 65% felt that use of genomics could result in better treatment decisions. Knowledge gaps were found for all nurses regardless of highest level of nursing education (diploma through doctorate). Eighty-nine percent felt it was very or somewhat important for nurses to become more educated about the genetics of common diseases [11].

Despite the opportunities for the NP to contribute to advancing personalized healthcare that incorporates genomic information in their responsibilities, little has changed in the academic preparation of NPs from 2005 to 2010. A study was conducted to assess advance practice nursing faculty members' current knowledge of genetics/genomics and their integration of such content into academic curricula [13]. Although there was a reported increase of faculty comfort with teaching genetic/genomic concepts in that five year period (50% increased to 70%), participants rated their overall knowledge of genetic/genomic concepts as very low or low (66%). Such data indicate gaps remain in the educational preparation of NPs.

No surveys of knowledge or interest in genomic education have been reported for PAs. However, PAs published competency recommendations in 2007 [14] with updated recommendations recently published [15] indicating the importance of PAs acquiring competency in genomics to provide the best possible care for patients.

EDUCATIONAL OPPORTUNITIES AND RESOURCES TO ADDRESS GENOMIC LITERACY GAPS

PCPs have recommended competencies to guide development of educational opportunities and resources that address identified genomic literacy gaps [15–17]. Each of the PCP disciplines has published documents that indicate domains of professional competency specific to their scope of practice (available at: http://g-2-c-2.org/competency). Competency-based education has been identified as a new educational paradigm that will enable the medical education community to meet societal, patient, and learner needs [18]. Table 3.1 presents select competencies identified by PCP in six primary categories: basic genetic/genomic concepts; family history; genomic testing; treatment; referrals; and ethical, legal, social issues. These competencies illustrate the similarities of the foundational knowledge needed by all PCP. Failure of PCP to understand the relevancy of genomics for healthcare, to have a sufficient scientific foundation in genomics to comprehend the literature, and to have the capacity to teach this material is contributing to limited progress in the integration of genomics into clinical care.

WHAT DO HEALTHCARE PROVIDERS NEED TO UNDERSTAND TO IMPLEMENT GENOMIC HEALTHCARE?

There are opportunities across the healthcare journey where genomic information can be used to improve care of individuals. Table 3.2 offers scenarios that illustrate key areas of relevancy for the PCP with a sampling of what providers need to understand to implement genomic healthcare.

To begin with, the increase in direct-to-consumer access to genomic technology has shifted the dynamic in how questions about genomic information may be brought into clinical encounters. Many clients are intrigued by the ads seen on the television, internet, or in magazines about ancestry DNA testing, or those companies offering "recreational genomics" tests. Because the general public may have limited exposure to or understanding about genetic testing, they often bring questions to the PCP about the value of the tests that are available and/or how to interpret the results they receive. They may not recognize the limitations of such testing or how these results differ from other genetic testing results (e.g., WGS). This has implications for the PCP to learn about the basic genetic and genomic concepts that provide the foundational understanding of the types of genetic information and potential clinical ramifications of direct-to-consumer testing [19].

A second area of relevancy for the PCP is to understand the contribution of using risk-assessment tools to recognize individuals and/or families that could benefit from additional preventive screening. For example, more than 55 hereditary cancer syndromes have been identified with the most common cancer syndromes associated with breast, ovarian, and gastrointestinal cancers [20]. Many

TABLE 3.1 Select Common Competencies Across Disciplines

Competency	Physicians [16]	Physician Assistants [15]	Advanced Nurse [17]
Basic genetic and genomic concepts	Discern the potential clinical ramifications of genetic variation on risk stratification and individualized treatment	Define the role of genetic variation in health and disease	Integrate best genetic/genomic evidence into practice that incorporates values and clinical judgment
Family history	Elicit, document, and act on relevant family history pertinent to the patient's clinical status	Gather family history information and construct a multigenerational pedigree	Analyze a pedigree to identify potential inherited predisposition to disease
Genomic testing	Use genomic testing to guide patient management	Incorporate genetic tests into patient management	Select appropriate genetic/genomic tests and/or studies
Treatment	Use genomic information to make treatment decisions	Discuss the range of genetic and genomic-based approaches to the treatment of disease	Manage care of clients, incorporating genetic/genomic information and technology
Referrals	Make appropriate referrals for specialty evaluation based on results of family history	Identify patients who would benefit from referral to genetics professionals	Make appropriate referrals to genetic professionals or other health care resources
Ethical, legal, social issues	Explain to patient relevant social and legal risks related to family history as well as relevant legal protections	Discuss financial, ethical, legal, and social issues related to genetic testing and recording of genetic information	Implement effective strategies to resolve ethical, legal, and social issues related to genetics/genomics

at risk for these syndromes can be spotted through use of family history risk assessment. The interpretation of the red flags that indicate need for additional screening, referral, or even genomic testing is an additional PCP skill that will enhance the quality of care provided to those seen in the primary care setting [21]. Creating an environment that acknowledges the barriers as well as the

TABLE 3.2 Key Examples of Relevancy of Genomics for PCP

Competency Category	Representative Scenario
Basic genetic and genomic concepts	• Becca is being seen in your office. She brings up in conversation that she has been looking into her family's ancestry and is considering sending saliva in for genetic testing. She is excited because the company offering the testing also provides information about health risks and asks your opinion about the value of knowing such information for her care
Family history	• You recognize that Lauren is at risk for a hereditary breast cancer syndrome based on the family history you just collected. You know that guidelines recommend that she undergo genetic testing to determine preventive care options. She is asking you as her PCP for guidance so that she can make the best decision about next steps
Genomic testing	• Kong has come to your office complaining of diabetic neuropathy. You are considering ordering carbamazepine but recognize that his Chinese ethnicity may increase his risk for toxicity (i.e., severe skin rash). You are aware of a pharmacogenomics test that can be ordered to verify this risk so you explain the options
Treatment	• Shirley received gefitinib (Iressa) for the treatment of her metastatic nonsmall cell lung cancer (NSCLC). She has completed treatment and is now being seen by you as her PCP for a sinus infection. She mentions that she has been going to a support group and one of the attendees told her that she also has lung cancer but had not gotten that same drug. Shirley asks you if you know why they didn't receive the same treatment
Referrals	• As you review Kevin's family history, you note an extensive frequency of colon and endometrial cancer occurring at young ages, indicative of a hereditary syndrome in his family. You refer Kevin to a genetics specialist for education and counseling about genetic testing. You receive the genetic testing results that indicate that Kevin is at risk for hereditary colon cancer. As his PCP you recognize why he needs a referral for a colonoscopy now even though he's only 30
Ethical, legal, and social issues	• Stacey's sister, Peggy, is enrolled in a whole genome sequencing clinical study designed to determine why she experienced cardiovascular problems at a young age. Stacey is asking you as her PCP for guidance so that she can decide about also participating in the study as her sister has requested. Stacey shares concerns about the potential utilization of this information against her in the workplace. You're aware of the Genetic Information Nondiscrimination Act so you take time to discuss relevant privacy issues

benefits of health promotion and disease prevention for individuals and populations at large can facilitate adoption of family history in practice [22,23].

PCPs prescribe a substantial portion of drug treatments [24]. Many of the drugs are known to have increased adverse toxicities because of individual genomic variant(s). Options for drug selection may be stratified based on individual pharmacogenomic test results. The Kong scenario in Table 3.2 illustrates the role of the PCP in knowing what genomic tests are available [25] and guidelines for clinical pharmacogenetics implementation (see Table 3.3 for CPIC). There has also been a rapid ascent of the availability of targeted therapies improving care options for patients (i.e., cancer care) [26,27]. Most targeted therapies help treat cancer by interfering with specific proteins that help tumors grow and spread throughout the body. For example in treatment of nonsmall cell lung cancer, gefitinib (Iressa) is now available for those whose tumors have

TABLE 3.3 Recommended Resources

Clinical Pharmacogenetics Implementation Consortium (CPIC) Guidelines https://www.pharmgkb.org/page/cpic

ClinGen https://www.clinicalgenome.org/

ClinVar http://www.ncbi.nlm.nih.gov/clinvar/

Evaluation of Genomic Applications in Practice and Prevention (EGAPP) https://www.cdc.gov/egappreviews/

Genetic and Rare Diseases Information Center (GARD) https://rarediseases.info.nih.gov/gard

Genetic Counselors http://nsgc.org/p/cm/ld/fid=164

Genetic Testing Registry http://www.ncbi.nlm.nih.gov/GTR/

Genetics/Genomics Competency Center (G2C2) http://genomicseducation.net

Genomics and Health Impact Update https://www.cdc.gov/genomics/update/current.htm

Global Genetics and Genomics Community (G3C) http://g-3-c.org/

Inter-Society Coordinating Committee for Practitioner Education in Genomics (ISCC) http://www.genome.gov/27554614

Office of Rare Diseases Research https://rarediseases.info.nih.gov/resources/pages/25/how-to-find-a-disease-specialist

PharmGenEd http://pharmacogenomics.ucsd.edu

PharmGKB (Pharmacogenomics Knowledge Base) http://www.pharmgkb.org/

Precision Medicine Initiative https://www.nih.gov/research-training/allofus-research-program

Undiagnosed Diseases Network (UDN) http://commonfund.nih.gov/Diseases/index

specific mutations in the gene for the epidermal growth factor receptor (EGFR) [28]. The Food and Drug Administration has approved use of a companion diagnostic test, the *therascreen* EGFR roto-gene Q Mdx 5plex HRM Instrument (RGQ) polymerase chain reaction (PCR) Kit, to test tumor samples for *EGFR* mutations, and determine whether patients are candidates for treatment with gefitinib. This knowledge is important for the PCP to be able to explain to their patient who may not understand how treatments are now more personalized to their tumor's molecular profile or genetic make-up.

Unfortunately, there is a lack of provider awareness about the availability and value of genomic testing. In a study of 10,303 US physicians reported by Stanek, 98% of respondents agreed that genetic variation may influence drug response while only 10% felt adequately informed about pharmacogenomic (PGx) testing [29]. Participants reported that there had been no PGx education provided in medical school (85%); 77% had no PGx in postgrad training; and only 29% had received any PGx education at all. These results were found to influence clinical utilization of PGx testing decisions as physicians with prior PGx education were more likely to order PGx tests (odds ratio (OR) 1.63, 95% confidence interval (CI) 1.34–1.97, $P < 0.001$). Physicians who felt well informed about the availability and applications of PGx were more likely to order PGx testing (odds ratio (OR) 1.92, 95% confidence interval (CI) 1.51–2.45, $P < 0.001$) [29]. The utility of PGx testing will be optimized if PCP understand PGx test options and the potential impact of test results on both currently prescribed and future medications. Effective communication about PGx tests includes pretest information (i.e., risks/benefits, limitations, and alternative options) and posttest communication of test results [30]. Knowing that time with patients is often limited, referrals for additional education and counseling may be needed.

Referral to genetic specialists for screening, consultation, risk communication, and/or education may also be indicated when a risk-assessment reveals an increased lifetime risk of an illness that can be improved with preventive care. For example, Lynch syndrome is a cancer predisposition syndrome responsible for a significant proportion of cases of colorectal and endometrial cancers [31]. Identification of at risk individuals within a family with Lynch syndrome allows the application of targeted surveillance and preventive interventions. Recording a cancer family history is important to identify such families with predisposition to health problems and can be used to guide an efficient, cost-effective genomic testing strategy. In the case of Kevin (Table 3.2), a specific mutation was identified in the family, permitting cascade testing of at-risk family members restricted to the known mutation [32]. Kevin was then provided personalized screening recommendations based on published guidelines [33]. When the PCP appropriately communicates genomic information, refers, and then follows-up on testing and screening, the patient benefits [34].

Coming soon to clinical care is the opportunity for WGS [35]. The ability to sequence someone's entire genome for $1000 or less brings this technology into a cost range considered realistic for routine clinical application. Physicians express interest in using WGS results to complement other clinical data for decision making and motivating patient behavior. However, the technology to generate genomic information has outpaced our capacity to understand its contributions to disease and to translate that knowledge into practice. The current disconnect between the technological capacity and clinical understanding will be an on-going challenge for PCP as genomics continues to transition from the research laboratory to daily healthcare delivery.

The case of Peggy (see Table 3.2) highlights the ethical, legal, and social issues, such as privacy and confidentiality concerns, that must also be considered when integrating genomic information into clinical care. Due to the volume and nature of genomic information, its uncertain and evolving interpretation, and the fact that it pertains not only to the individual being seen, but also to family members, groups, and even populations, there are additional sensitivities to be cognizant of in discussions with patients. For instance, there are concerns about where genomic data are stored, who has access to the data and for what purpose, and how the information might be shared with others. These concerns have implications for the PCP when considering security of the electronic medical record (EMR), test reports, and communication methods for the return of genomic testing results. The general public has expressed concern about the possibility of insurance discrimination and the PCP can assist them with decisions by informing them of relevant legal protections such as the Genetic Information Nondiscrimination Act (http://www.genome.gov/10002077) and basic health privacy protections under the Health Insurance Portability and Accountability Act [36]. As discussed further below, genomic medicine brings many ethical, legal, and social issues to the forefront for PCP.

ETHICAL, POLICY, AND SOCIAL CONSIDERATIONS

The possibilities of genomic healthcare are simultaneously exciting and a bit disconcerting. PCP as a result of professional training already have a strong ethical foundation to guide them in conversations with patients, but even so, controversial issues remain to be addressed by the medical community and professional guidance on these topics often does not exist. The potential to improve care is real, but the knowledge and systems to support safe implementation for all are not yet developed [37]. Plus, not everyone can afford the cost of genetic testing or services that are recommended following genetic testing such as enhanced screening procedures. Reimbursement that covers such services is limited and this must be addressed to reduce costs encountered so that access to care that incorporates genomic information is equitably available to everyone.

One important ethical issue that may be grounded in the downstream research domain, but is crucial for the provision of quality care in the future, is the lack of diversity within study populations available for existing genomic studies providing the bases for new genomic medicine insights. It is imperative that ancestral diversity is represented within datasets used as engines for discovery and clinical interpretation development. PCP can help explore issues of concern with patients about participation in genomic or precision medicine research so that ongoing genomics research results provide pathways of quality care across all ethnic and racial groups.

Decisions about the return of research results for genomic testing have become increasingly complex as the technology has expanded beyond single gene tests, to multiplex testing, to WGS. The debate about if, when, and what to report back to the patient continues, but educational resources and recommendations are now available regarding communication of test results of adults and children [38,39]. Given the complexities of interpreting and communicating the clinical utility of test results, the PCP may benefit from consulting with and referring to geneticists, genetic counselors, or other experts in their area.

However, it is important to note that not everyone wishes to know about healthcare risks that may be identified through genomic information. Differences in cultural beliefs and values may influence how individuals interpret or feel about the use of genomic information in their clinical decision making. Awareness by the PCP of their patient's knowledge, attitudes, and beliefs can help them consider all options for care and incorporate sensitivities into personal communication and education strategies.

RESOURCES AVAILABLE

There are multiple opportunities for genomics education of the PCP during academic, preservice, and postgraduate training [40]. It is more challenging once in practice for the PCP to protect time to attend conferences or complete continuing education (CE) courses. Many professional organizations are offering CE that includes genomic content, but finding what is available can require substantial time given the distributed nature of the offerings. The Inter-Society Coordinating Committee on Practitioner Education in Genomics (ISCC, http:// www.genome.gov/27554614), is a partnership of more than 40 professional organizations representing generalists and specialists, such as the American Medical Association, the American Heart Association (AHA), and the American Society of Clinical Oncology (ASCO) [41]. ISCC members who meet annually have worked with the National Human Genome Research Institute to develop competencies for primary care physicians, as noted previously, and contribute curated educational resources to the Genetics/Genomics Competency Center for Education (G2C2, http://genomicseducation.net). G2C2 resources are searchable by genomic competencies, keywords, or for discipline specific content and can be a starting point for PCP looking for CE.

PCP organizations and other guideline setting organizations are instrumental in providing recommendations that can improve care outcomes. Recent documents published by organizations like the AHA and ASCO serve as resources providing education and recommendations for application of genetics and genomics to patient care [42,43]. Increased emphasis on translational genomic research that provides the evidence base for guideline development is recommended [44].

Fundamental changes are needed in the educational infrastructure and methods for lifelong learning about the foundational knowledge of genomic clinical applications. The Institute of Medicine held a workshop on Improving Genetics Education in Graduate and Continuing Health Professional Education [45]. A variety of approaches that could improve the teaching of genetics in the education of health professionals were reviewed as exemplars. These included online and interactive instruction, just-in-time approaches, the development of clinical decision-support tools, and the incorporation of genetics requirements into licensing and accreditation. Other methods for CE attainment include webinars, internet offerings, mobile devices, and simulated learning environments such as the Global Genetics and Genomics Community (G3C, http://g-3-c.org/). Evaluation of the most effective and efficient educational interventions for PCP education is advised.

One potential way for reaching all PCPs with essential genomic educational content is the Interprofessional Education (IPE) model. Although primarily designed for competency development within the academic environment, IPE promotes learning between individuals of two or more professions to improve collaboration and the quality of health care [46]. Because physicians, NP, and PA have identified similar genomic competencies, IPE education would promote efficient team-based learning and interactions. Bringing in additional healthcare disciplines for such training (i.e., nurses and pharmacists) could expand team-centered care [45]. Such discussion and sharing about roles and responsibilities could further define performance expectations as genomic medicine emerges.

Infrastructure that supports role delineation, clinical education, and processes for application of genomics in care must also be addressed. Individuals who are leaders within the practice environment are key components of the infrastructure needed to integrate genomics in care [11]. These champions are valuable members of the team determining what is relevant in their setting, assessing gaps in knowledge, informing others of the benefits and challenges, and determining institutional factors hindering clinical integration. For example, an investment in bioinformatics technology that supports family history documentation; enables EMR alerts about appropriate use of PGx tests; and administers point of care education as needed [47,48]. Clinical data support systems such as ClinVar, ClinGen, and Genetic Test Registry (Table 3.3) developed in the last few years have the potential to improve how genomic information will inform PCP clinical decision making [49].

CONCLUSION

The building of genomic healthcare requires continued research progress and evidence of value, healthcare systems that establish resources for learning and applying genomics in clinical care, and healthcare professionals who recognize the relevance and value of genomics for care. PCP who are motivated to evaluate their personal genomic competency and plan how to achieve competency is vital. Utilizing leadership skills as a champion in the healthcare environment and within professional organizations will accelerate the process and ultimately improve patient care. Each of these building blocks will contribute a strong foundation for genomic competency from which to prepare for integration of genomics in care by the PCP now and into the future.

GLOSSARY TERMS

Primary care provider
Genetics
Genomics
Whole genome sequencing

REFERENCES

[1] David SP, Johnson SG, Berger AC, Feero WG, Terry SF, Green LA, et al. Making personalized health care even more personalized: insights from activities of the IOM genomics roundtable. Ann Fam Med 2015;13(4):373–80.

[2] Manolio TA, Green ED. Leading the way to genomic medicine. Am J Med Genet C Semin Med Genet 2014;166C(1):1–7.

[3] Group ICSW. Utilizing family history to identify Lynch Syndrome Case Study. 2014.

[4] Vassey J, Korf B, Green R. How to know when physicians are ready for genomic medicine. Sci Transl Med 2015;7(287):1–3.

[5] Van Ness B. Applications and limitations in translating genomics to clinical practice. Transl Res 2015;168:1–5.

[6] Passamani E. Educational challenges in implementing genomic medicine. Clin Pharmacol Ther 2013;94(2):192–5.

[7] Secretary's Advisory Committee on Genetics Health, and Society. Genetics Education and Training; 2011.

[8] Klitzman R, Chung WD, Marder K, Shanmugham A, Chin LSJ, Stark M, et al. Attitudes and practices among internists concerning genetic testing. J Genet Couns 2013;22(1):90–100.

[9] Salm M, Abbate K, Appelbaum P, Ottman R, Chung W, Marder K, et al. Use of genetic tests among neurologists and psychiatrists: knowledge, attitudes, behaviors, and needs for training. J Genet Couns 2014;23(2):156–63.

[10] Christensen KD, Vassy JL, Jamal L, Lehmann LS, Slashinski MJ, Perry DL, et al. Are physicians prepared for whole genome sequencing? a qualitative analysis. Clin Genet 2016;89(2):228–34.

[11] Calzone KA, Jenkins J, Culp S, Caskey S, Badzek L. Introducing a new competency into nursing practice. J Nurs Regul 2014;5(1):40–7.

[12] Calzone KA, Jenkins J, Culp S, Bonham Jr. VL, Badzek L. National nursing workforce survey of nursing attitudes, knowledge and practice in genomics. Pers Med 2013;10(7):719–28.

[13] Maradiegue A, Edwards Q, Seibert D. 5-Years later – have faculty integrated medical genetics into nurse practitioner curriculum? Int J Nurs Educ Scholarsh 2013;10(1):245–54.

[14] Rackover M, Goldgar C, Wolpert C, Healy K, Feiger J, Jenkins J. Establishing essential physician assistant clinical competencies guidelines for genetics and genomics. J Phys Assist Educ 2007;18(2):47–8.

[15] Goldgar C, Michaud E, Park N, Jenkins J. Physician assistant genomic competencies. J Phys Assist Educ 2016;27(3):110–16.

[16] Korf B, Berry A, Limson M, Marian A, Murray MF, O'Rourke P, et al. Framework for development of physician competencies in genomic medicine: report of the Competencies Working Group of the Inter-Society Coordinating Committee for Physician Education in Genomics. Genet Med 2014;16(11):804–9.

[17] Greco K, Tinley S, Seibert D. Essential genetic and genomic competencies for nurses with graduate degrees. MD: Silver Spring; 2012.

[18] Caccia N, Nakajima A, Kent N. Competency-based medical education: the wave of the future. J Obstet Gynaecol Can 2015;37(4):349–53.

[19] McGowan ML, Fishman JR, Settersten Jr. RA, Lambrix MA, Juengst ET. Gatekeepers or intermediaries? The role of clinicians in commercial genomic testing. PLoS One 2014;9(9) e108484.

[20] Lindor NM, McMaster ML, Lindor CJ, Greene MH. National Cancer Institute DoCPCO, Prevention Trials Research G. Concise handbook of familial cancer susceptibility syndromes – second edition. J Natl Cancer Inst Monogr 2008;38(38):1–93.

[21] Institute NCI. Cancer Genetics Overview–for health professionals (PDQ®). Available at <https://www.cancer.gov/about-cancer/causes-prevention/genetics/overview-pdq>; 2015.

[22] Wu RR, Orlando LA. Implementation of health risk assessments with family health history: barriers and benefits. Postgrad Med J 2015;91(1079):508–13.

[23] Doerr M, Edelman E, Gabitzsc E, Eng C, Teng K. Formative evaluation of clinician experience with integrating family history-based clinical decision support into clinical practice. J Pers Med 2014;4:115–36.

[24] Haga SB, LaPointe NM, Cho A, Reed SD, Mills R, Moaddeb J, et al. Pilot study of pharmacist-assisted delivery of pharmacogenetic testing in a primary care setting. Pharmacogenomics 2014;15(13):1677–86.

[25] Relling MV, Evans WE. Pharmacogenomics in the clinic. Nature 2015;526(7573): 343–50.

[26] Masters GA, Krilov L, Bailey HH, Brose MS, Burstein H, Diller LR, et al. Clinical cancer advances 2015: annual report on progress against cancer from the American Society of Clinical Oncology. J Clin Oncol 2015;33(7):786–809.

[27] Institute NCI. Targeted Cancer Therapies. Available at <https://www.cancer.gov/about-cancer/treatment/types/targeted-therapies/targeted-therapies-fact-sheet>; 2015.

[28] Institute NCI. With FDA Approval, Gefitinib Returns to U.S. Market for Some Patients with Lung Cancer. Available at <https://www.cancer.gov/news-events/cancer-currents-blog/2015/fda-gefitinib>; 2015.

[29] Stanek EJ, Sanders CL, Taber KA, Khalid M, Patel A, Verbrugge RR, et al. Adoption of pharmacogenomic testing by US physicians: results of a nationwide survey. Clin Pharmacol Ther 2012;91(3):450–8.

[30] Mills R, Voora D, Peyser B, Haga SB. Delivering pharmacogenetic testing in a primary care setting. Pharmgenomics Pers Med 2013;6:105–12.

[31] Jasperson K, Burt R. The genetics of colorectal cancer. Surg Oncol Clin N Am 2015;24:683–703.

[32] Group EGAPP. Recommendations from the EGAPP Working Group: genetic testing strategies in newly diagnosed individuals with colorectal cancer aimed at reducing morbidity and mortality from Lynch syndrome in relatives. Genet Med 2009;11(1):35–41.

[33] Network NCC. Genetic/Familial High Risk Assessment: Colorectal. Available at <https://www.nccn.org/professionals/physician_gls/f_guidelines.asp#genetics_colon>; 2016.

[34] Clark D, Kowal S. Communicating genomic risk in primary health care: challenges and opportunities for providers. Med Care 2014;52(10):933–4.

[35] Vassey J, Christensen K, Slashinski MJ, Lautenbach D, Raghavan S, Robinson JO, et al. 'Someday it will be the norm': physician perspectives on the utility of genome sequencing for patient care in the MedSeq Project. Pers Med 2015;12(1):23–32.

[36] Hudson KL, Holohan MK, Collins FS. Keeping pace with the times – the Genetic Information Nondiscrimination Act of 2008. N Engl J Med 2008;358(25):2661–3.

[37] Rodriguez L, Galloway E. Bringing genomics to medicine: ethical, legal, and social considerations. GS Ginsburg and H Willard, (Eds.), Genomics and Personalized Medicine: Foundations, Translation, and Implementation, 3rd ed, 2017, Academic Press.

[38] Green RC, Berg JS, Grody WW, Kalia SS, Korf BR, Martin CL, et al. ACMG recommendations for reporting of incidental findings in clinical exome and genome sequencing. Genet Med 2013;15(7):565–74.

[39] Wilfond B, Fernandez C, Green R. Disclosing secondary findings from pediatric sequencing to families: considering the "Benefit to Families". J Law Med Ethics 2015;Fall:552–8.

[40] Lamb N, Gunter C. Educational issues and strategies for genomic medicine Ginsburg GS, Willard H, editors. Genomic and personalized medicine (2nd ed). San Diego, CA: Elsevier Inc.; 2013.

[41] Manolio T, Murray MF. Genomics ISCCPEG. The growing role of professional societies in educating clinicians in enomics. Genet Med 2014;16:571–2.

[42] Robson ME, Bradbury AR, Arun B, Domchek SM, Ford JM, Hampel HL, et al. American Society of Clinical Oncology policy statement update: genetic and genomic testing for cancer susceptibility. J Clin Oncol 2015;33:1–9.

[43] Musunuru K, Hickey K, Al-Khatib S, Delles C, Fornage M, Fox C, et al. Basic concepts and potential applications of genetics and genomics for cardiovascular and stroke clinicians: a scientific statement from the American Heart Association. Circ Cardiovasc Genet 2015;8:216–42.

[44] Clyne M, Schully SD, Dotson WD, Douglas MP, Gwinn M, Kolor K, et al. Horizon scanning for translational genomic research beyond bench to bedside. Genet Med 2014;16(7):535–8.

[45] Berger AC, Johnson SG, Beachy S, Olson S. Improving genetics education in graduate and continuing health professional education: workshop summary. Washington, DC: Institute of Medicine; 2015.

[46] Panel IECE. Core competencies for interprofessional collaborative practice: Report of an expert panel. Washington, DC; 2011.

[47] Larson EA, Wilke RA. Integration of genomics in primary care. Am J Med 2015;128(11):1251. e1–1251.e5.

[48] Castaneda C, Nalley K, Mannion C, Bhattacharyya P, Blake P, Pecora A, et al. Clinical decision support systems for improving diagnostic accuracy and achieving precision medicine. J Clin Bioinformatics 2015;5(4):1–16.

[49] Rehm HL, Berg JS, Brooks LD, Bustamante CD, Evans JP, Landrum MJ, et al. ClinGen – the clinical genome resource. N Engl J Med 2015;372(23):2235–42.

Chapter 4

Genetic Testing for Rare and Undiagnosed Diseases

Thomas Morgan
Novartis Institutes for Biomedical Research, Cambridge, MA, United States

Chapter Outline

INTRODUCTION

Genetic diagnosis is in the midst of an astounding revolution. The unprecedented gain in technical ability to assay DNA sequence to establish or confirm a rare disease diagnosis has been one of the most immediate and tangible benefits of genomic medicine. As of 2015, approximately 3000 Mendelian disease genes have been identified and cataloged in Online Mendelian Inheritance in Man [1]. As more disease genes are identified, there is a corresponding increase in clinically available genetic tests. The Genetic Testing Registry lists over 31,000 genotypic and soluble marker tests for 5800 conditions (complex as well as Mendelian) relating to 3900 genes [2]. It appears probable, however, that

Genomic and Precision Medicine. DOI: http://dx.doi.org/10.1016/B978-0-12-800685-6.00002-3

genome sequencing (GS) will eventually become so widespread and afford-able that genetic testing at the point of care could be done *in silico* using an electronic medical record system [3], though gene panels may continue to be attractive because they avoid generation of potentially unwanted genomic data that are irrelevant to the indication for testing.

The enormous strides in GS and computational genomics call attention to the lack of commensurate technical progress in phenotyping of individual patients [4]. Although a case can be made that innovative diagnostic imaging may nearly be keeping pace with genomic advances, and that GS is part of a broader revolution in molecular pathology, most diagnostic technologies, such as the stethoscope and sphygmomanometer, seem quaintly antiquated. Family history taking tends to be cursory in most clinical settings, and coding family data into electronic format is challenging, though new tools are being developed [5]. As GS becomes ubiquitous in clinical medicine, the burgeoning "island of knowledge" could create an ever-widening "shoreline of ignorance" about the mechanism of genotype–phenotype correlation. It appears inevitable that much phenotype information will be captured using "big data" approaches that utilize wearable sensors and repeated sampling of blood and other somatic tissues. It is envisioned that electronic medical record systems will evolve to accommodate such data and lead to the identification of the genetics, first of all Mendelian disorders, and ultimately, all human disease with substantial genetic underpin-ning. Once a critical mass of genomes has been sequenced, stored, and linked to accurate phenotypic information, then stable genotype–phenotype associations will naturally emerge. The first half of the 21st century is poised to become the epoch of clinical genomics, commencing with the first published human GS in 2001. Whether it also emerges unequivocally as the era of genetically informed personalized medicine is harder to foresee, and this promises to be an ongoing topic of debate in the near future.

In spite of phenomenal progress, the genetic basis remains unknown for approximately half of all known Mendelian conditions [1], and many genetic disorders are likely to have escaped clinical recognition and formal classifica-tion due to the absence of readily recognizable phenotypes. Moreover, as the number of patients with GS data increases from thousands to millions, the exist-ing corpus of disease-specific pathological mutation databases will require criti-cal reappraisal. Diagnosis of a rare genetic disorder must be based on a solid scientific foundation. It is not sufficient to be certain that a particular mutation is present in the patient; a reasoned judgment that the mutation is specifically pathological is required.

GENOTYPE–PHENOTYPE CAUSATION

Cause–effect relationships between particular mutations and Mendelian phe-notypes ethically may not be determined by direct human experimentation.

Rather, causation must be inferred via the classic epidemiological precepts expounded by Doll and Hill to demonstrate that smoking causes lung cancer [6]. Specifically in rare disease genetics, there must be statistically significant association between the suspected disease mutation (i.e., consistent coinheritance of mutations with disease within families or a higher overall burden of gene-specific mutations in affected versus unaffected individuals). Such association should not be confounded by an environmental variable (e.g., poor household diet causing the intrafamilial clustering of pellagra, a nutritional not genetic disease) or by the ethnic background of the patients (i.e., some "rare" mutations are common in certain ethnic groups). The impressiveness of the risk conferred by the mutation (i.e., penetrance) is another important consideration. Finally, the putative mechanism by which a mutation causes disease should ideally be known. If not already known, Koch's postulates of infectious pathogenicity can be adapted to the investigation of experimentally induced mutations in model organisms [7]. For genetic specialists, the American College of Medical Genetics has recently published guidelines for the interpretation of DNA sequence variants [8].

GENETIC TESTING THEORY

It is imperative that physicians be familiar with the philosophical basis of genetic medicine so that nuances of diagnostic interpretation can be appreciated. The application of genetic cause–effect criteria varies according to the clinical context. Well-founded suspicion prior to testing remains the cornerstone of genetic diagnosis, even in the postgenomic era. By Bayes' theorem, all diagnostic tests serve to increase or decrease the pretest probability that a patient has a suspected disease [9,10], though it should be remembered that pretest probability formally applies to a hypothetical set of clinical characteristics possessed by a patient, whose personal probability of disease is either 0 or 1. Application of a genetic test to persons who are at low risk of having the relevant disease means that positive tests will generally have poor predictive value, though some genetic testing is so extremely sensitive and specific that an unequivocally positive molecular test may be practically tantamount to a diagnosis (e.g., Huntington's disease testing in middle-aged adults). Conversely, a negative genetic test in the face of high suspicion for the disease may create doubt, but it doesn't necessarily negate a presumptive diagnosis. Negative predictive value of negative genetic tests is limited in many cases by the inability to identify all "noncoding" sequence variations that regulate the expression of a gene but do not code for the amino acid sequence of a polypeptide. Such sequence variations include, in addition to promoters and splice sites within a given gene, genetic elements that may be situated far from the actual gene, such as enhancer and silencer sequences that modulate the quantitative expression of the gene and hence the amount of protein produced.

IMPORTANCE OF INDIVIDUAL PATIENT CHARACTERISTICS IN GENETIC TEST INTERPRETATION

As discussed above, diagnostic confirmation of a strongly suspected genetic disorder in a patient with readily observable phenotypic features is different from the approach to an asymptomatic patient without such features. The former is focused diagnostic confirmation, whereas the latter constitutes exploratory genetic screening. Patients at the extremes of age likewise present different interpretive problems. A child or fetus may have a mutation that portends the possible development of a disease phenotype later in life, whereas the older adult who has the same mutation yet never developed signs of disease calls penetrance into question (meaning the conditional probability, given the presence of a mutation, that disease will develop). In addition, the sex of the patient may dramatically alter the interpretation of a mutation, as in the case of X-linked diseases. Even the sex of the parent from whom a patient inherited a mutation is important for disorders involving genes that are imprinted [11]. This is known as an "epigenetic effect" because imprinting refers not to genetic sequence *per se*, but to the covalent chemical bonding of methyl groups to DNA that alters the transcription of the gene, and therefore, its translation into protein.

CLINICAL RATIONALE FOR GENETIC TESTING

With respect to diagnosis of a known or suspected rare disease, the rationale for genetic testing is as follows: (1) the diagnosis may not be confidently established clinically and molecular testing is substantially confirmatory, (2) knowledge of the specific molecular basis of disease may guide medical management in a way that could improve the patient's outcome, and/or (3) for genetic counseling about mode of inheritance and/or testing of at-risk relatives.

In the case of an undiagnosed disease, absent any likely diagnostic hypothesis, the rationale for testing is relatively challenging to articulate. By definition, the clinician does not have a suspected diagnosis firmly in mind, and therefore, the utility of such diagnosis remains to be seen. However, generalizations may be made about the likelihood of "medically actionable" findings in such cases. The utility varies depending on the clinical indication and setting for sequencing, as discussed below for specific indications.

TYPES OF GENETIC TESTING (SUMMARIZED IN TABLE 4.1)

Genetic testing involves either determination of DNA sequence, chemical modification of nucleotides, or enumeration of genetic copy number, ranging in scope from a single nucleotide insertion/deletion site, to nucleotide repeats (e.g., trinucleotide, tetranucleotide repeats, etc.), to the number of copies of whole chromosomes, which are long, continuous DNA macromolecules, or even to ploidy (number of copies of the entire genome). The "gold standard"

TABLE 4.1 Summary of Selected Genetic/Genomic Tests

Genetic Test Type	Clinical Indications	Disease Example	Advantages	Disadvantages
Single gene/panel gene sequencing	Known/suspected genetic diagnosis	Hypertrophic cardiomyopathy	Cost; no off-target incidental findings	Less sensitive than genomic tests
Exome sequencing	Known/suspected genetic syndrome	Syndromes; multiple congenital anomalies	Lower cost than whole genome; interpretability	Gaps in exome coverage; no noncoding DNA
Genome sequencing	Known/suspected genetic syndrome	Syndromes; multiple congenital anomalies	Full coverage of DNA sequence	Cost, turnaround time, and analytical challenges
Oligonucleotide microarray	Known/suspected chromosomal imbalance	Syndromes or developmental delay/dysmorphism	High resolution, good copy number detection	Nondetection of balanced rearrangements
Comparative genomic hybridization	Known/suspected chromosomal imbalance	Syndromes or developmental delay/dysmorphism	Good signal: noise and copy number detection	Nondetection of balanced rearrangements
Karyotype	Known/suspected chromosomal imbalance	Syndromes or developmental delay/dysmorphism	Rapid <1 week; detects balanced rearrangements	Poor resolution for imbalance less than 5 megabases

for DNA sequence determination is Sanger sequencing. However, many other mutation detection methods exist and can be put to clinical use. Methylation analysis, often using a combination of methylation sensitive and insensitive restriction endonucleases (produced by various bacterial species), is employed in the diagnosis of imprinting disorders (e.g., Beckwith–Wiedemann syndrome). The karyotype permits direct visualization of stained chromosomes by light microscopy, and fluorescence in situ hybridization (FISH) couples a DNA-sequence-specific probe with a fluorescent molecule that can also be seen under a microscope. A molecular karyotype, which can be either a chromosomal microarray (based on single nucleotide polymorphisms) or comparative genomic hybridization (based on longer stretches of DNA derived from bacterial artificial chromosomes), combines the comprehensiveness of karyotyping with the high resolution of FISH and even surpasses FISH in the detection of very small segmental aneuploidies. Molecular karyotyping is, however, limited in that it can't detect balanced chromosomal translocations or differences in ploidy.

Finally, there is next-generation sequencing (NGS), a term that encompasses a range of technologies that determine DNA sequence in ultraminiaturized, massively parallel format (meaning multiple chemical reactions going on simultaneously, in parallel), and applications include targeted panels that capture a list of genes or other DNA sequence of interest, exome sequencing (capturing the 1.2% of the genome that codes for amino acids), or GS (which aims at comprehensiveness but still has small gaps in coverage). At this time in the rapidly evolving history of NGS, exome sequencing arguably provides a similar amount of interpretable genetic information as GS, because the clinical effects of mutations outside of the coding regions of genes are difficult to predict using existing methods. However, as costs of GS drop, it is becoming a more attractive option for technical reasons as well as enhanced coverage of the genome. GS is currently best suited for sequence determination rather than assay of copy number, but improving research methods for NGS analysis show potential for NGS to double as a molecular karyotype. For now, GS and molecular karyotyping are complementary tests that typically must be ordered separately. Methylation analysis and determination of nucleotide repeat number are specialized tests ordered for appropriate indications (e.g., Fragile X CGG repeat number/promoter methylation status).

POTENTIAL INDICATIONS FOR CLINICAL GENOME/EXOME SEQUENCING (CGES)

General Indications

According to the 2012 ACMG Policy Statement, "Points to Consider in the Clinical Application of Genomic Sequencing," clinical genome/exome sequencing (CGES) should be considered when (1) a genetic syndrome is strongly suspected but can't be diagnosed using a targeted method and (2) the disorder being tested can be caused by mutations in so many different genes that it is more practical to employ CGES [8]. At this time, CGES has been introduced into clinical care. A seminal paper in the New England Journal of Medicine containing a diverse set of patients reported a diagnostic yield of 25% for ES specifically [12]. The diagnostic yield and rationale for testing vary according to the clinical indication and setting.

THE ACUTELY ILL INFANT

Newborn and pediatric intensive care units serve a clinical population that suffers disproportionately from the effects of genetic syndromes. In addition, parents and physicians in the neonatal and pediatric intensive care (NICU/PICU) setting often must make emotionally wrenching decisions about treatment under relentless time pressure. Thus, an undiagnosed genetic disorder in an acutely ill infant is a particularly compelling indication for CGES. In a recent effort to address this critical unmet medical need, the StatSeq Project at Children's Mercy Hospital in Kansas City has demonstrated that 20 of 35 acutely ill infants (57%) could be diagnosed by rapid GS of parent–infant trios (completed within 50 h) [13]. Of those 20, 13 (65%) had a diagnosis deemed useful in acute medical management. Nine of the 20 diagnoses were unsuspected prior to GS. A limitation of this approach is that it requires availability of both biological parents for GS. However, generalization of the StatSeq approach to other clinical settings and scenarios is under active investigation [13].

MULTIPLE CONGENITAL ANOMALIES

Multiple congenital anomalies (MCA) affect approximately 1% of children, and by definition, include two or more major malformations or at least three minor or major malformations. Chromosomal microarray is the first-tier test for MCA [14], and MCA may have substantial overlap with the acutely ill infant as well as the child with a neurodevelopmental disorder. Genetic testing is not generally indicated for children with a single, isolated congenital anomaly unless the nature of the anomaly suggests a particular syndromic diagnosis (e.g., cardiac rhabdomyoma suggesting tuberous sclerosis). GS may be considered in particular cases, according to criteria discussed below in the section on the Undiagnosed Disease Network.

DEVELOPMENTAL DELAY/INTELLECTUAL DISABILITY/ AUTISTIC SPECTRUM DISORDERS

Intellectual disability (ID) affects ~3% of the population aged 5 years or older (when the cutoff is set at 2 standard deviations below mean performance on psychometric testing), and ID may be preceded by global developmental delay

(GDD), which also involves domains including gross or fine motor skills, speech/language, cognitive, social/personal, and activities of daily living [15]. Autistic spectrum disorders are diagnosed using standardized psychometric instruments and comprehensive developmental evaluation, and the prevalence of autistic spectrum disorders (ASD) by age 8 years is approximately 1 in 88 [16]. The large number of individuals with ASD and/or ID poses a challenge for the implementation of genomic diagnostics. However, there are published guidelines outlining the standard approach to genetic evaluation of GDD/ID as well as ASDs. Such evaluation involves, in cases that are undiagnosed following thorough clinical evaluation, chromosomal microarray testing for genomic imbalances (deletions/duplications/aneuploidies) and Fragile X testing [15,17]. Additional genetic testing is considered on a case-by-case basis. With the introduction of the chromosomal microarray/comparative genomic hybridization technology, the diagnostic yield of genetic evaluation has increased in the past decade from 6–10 to 30–40%, without including GS. However, GS is considered an emerging diagnostic technology not yet in routine clinical use for ID/GDD/ASD [15,17]. The presumed causative mutation detection rate in trios with one child with severe ID is 42% [18], the same diagnostic yield found in quartets comprised two autistic siblings and both parents [19]. Notably, affected siblings did not share the same presumed high-penetrance ASD mutation in two-third of cases, highlighting the genetic heterogeneity of this disorder. GS shows great promise for determining the etiology in many cases of ID/GDD/ASD. The immediate utility of molecular diagnosis is that it informs genetic counseling and provides diagnostic closure, but current therapies such as early intervention and applied behavior analysis are applied regardless of genotype. The discovery of the fundamental basis of developmental brain disorders, however, has led to intensive research investigation of pathways and holds potential for genetically guided therapeutics.

OTHER RARE DISORDERS AND SYNDROMES

Disorders of virtually any organ system may be sufficiently difficult to diagnose with precision that CGES may be considered, although diagnostic gene panels may be more efficient if they are comprehensive and yield a diagnosis in a substantial fraction of patients. Examples of disorders that may be efficiently diagnosed via a gene panel include Epileptic encephalopathies (often due to brain ion channel mutations) [20], neuromuscular disorders [21], ophthalmological disorders [22], cardiomyopathies (often due to mutations in sarcomeric proteins) [23], and hereditary cancer predisposition syndromes [24].

It is important to recognize that the choice of a gene panel versus CGES involves trade-offs of certain values in exchange for others. Accordingly, there may be some disagreement about best clinical practices, and choices should involve shared decision-making among patient, physician, genetic counselor, and other healthcare providers. Choices may be influenced (duly or unduly) by geographic and economic factors related to supply of testing options. The main advantages of gene panels are their lower cost, relative analytical simplicity, lesser potential for generation of unwanted incidental findings, and focused optimization of sensitivity and specificity for detection of particular genes or particularly important gene variants. The disadvantage of a panel is that it is biased in favor of particular candidate genes, and a negative test may necessitate broader sequencing, and it is theoretically possible that a variant of uncertain significance detected by a panel could promote premature closure of a diagnostic investigation that could have been solved more accurately and/or thoroughly via comprehensive GS.

PREIMPLANTATION AND PRENATAL RARE DISEASE DIAGNOSIS

Preimplantation genetic diagnosis (PGD) constitutes either genetic screening or focused testing for a particular rare disorder for which the conceptus is known or suspected to be at risk. The goal is the implantation into the uterus of an embryo that will not have a specific genetic disorder or a set of defined genetic disorders. In PGD, conception occurs in vitro, and then a blastocyst cell (trophoectoderm portion of eight-cell inner mass stage embryo) is biopsied, typically at day 5 postconception. Cellular DNA is then amplified by a high-fidelity phi 29 DNA polymerase to produce sufficient quantities of DNA to meet the technical specifications for genetic testing. In most cases, such testing is focused on a particular disorder that is a known heightened risk for a child of that couple, often due to the prior birth of an affected sibling or other relative, or because of parental carrier or genetic disease status.

In addition to focused testing for a particular genetic disorder, PGD-derived, whole genome amplified DNA can be routinely assayed for chromosomal aneuploidy, which is a generally elevated risk in the in vitro fertilization setting as well as a specific risk in the case of a parent with a Robertsonian chromosomal translocation (due to fusion of two acrocentric chromosomes—13, 14, 15, 21, or 22—into a doubled-up isochromosome, with live birth possible for trisomies 13 or 21). Partial aneuploidy (deletion/duplication) is a risk when a parent is a balanced reciprocal chromosomal translocation carrier, and PGD has been used successfully for this indication [25]. Diagnosis of many microdeletion syndromes is beyond the confident limit of resolution for current standard molecular karyotyping done on whole genome amplified DNA, though methods may continue to improve rapidly [26]. In addition, NGS is being actively evaluated for possible translation into the clinical practice of PGD, but currently is not in routine use [27,28].

Prenatal genetic testing for rare diseases has been revolutionized by the development of methods to retrieve and analyze fetal cell-free DNA from maternal serum, also known as noninvasive prenatal testing (NIPT). The initial indication for NIPT was as a screening option for women at high risk for fetal aneuploidy (age >35 years), but recent research has demonstrated the superiority of NIPT over standard screening methods in first trimester (10–14 weeks) pregnancies at low risk for aneuploidy as well [29]. There is currently no consensus on the application of NIPT to low- and intermediate-risk pregnancies, but this is an area of intense research that should be followed closely. Genetic counseling for NIPT is complex and should contain all of the key elements required for women to make informed decisions about various testing options [30]. As in the case of PGD, there are technical challenges and limited experience with molecular karyotyping for microdeletion syndromes using cell-free fetal DNA in NIPT [30], which also must undergo whole genome amplification prior to analysis, with concern for artifactual alteration of copy number of chromosome segments.

Fetal GS has become a technical reality, still challenging for whole genome amplified cell-free DNA obtained from maternal circulation [31,32], but relatively straightforward for DNA obtained by invasive methods (chorionic villus sampling at 10–14 weeks; amniocentesis at 15–18 weeks). The profound ethical and social ramifications of this new technology have been reviewed recently [33]. Despite all the highly thoughtful ethical analysis that has taken place, it is likely that we won't truly know how preimplantation and prenatal genetic testing for rare diseases will affect our society until we have being doing it for a substantial period of time. Hypothetical choices have been supplanted by real ones, and they're being made by real people right now in medicine. To what extent rare diseases will be prevented prenatally remains to be seen.

PROCESS OF GENETIC TESTING

Pretest and posttest counseling is the foremost consideration in the process of genetic testing, which should represent an informed choice by physician and patient or parent alike. The importance of careful history-taking; obtaining a thorough understanding of all relevant clinical data about the patient; cautious deliberation about diagnostic hypotheses; consulting with experienced colleagues as needed; and finally, a clear discussion about the potential risks, benefits, and alternatives of genetic testing can't be overemphasized as an ideal toward which genetic medicine should strive. Unfortunately, the time pressures on healthcare providers typically do not facilitate ideal practice, and the consequence of inappropriate genetic testing may result from omission of any of the cardinal principles involved in genetic counseling. It should be noted that informed consent should be part of all diagnostic and therapeutic intervention in medicine, and in that regard, there is nothing unique about genetics. Complex genetic counseling and testing, however, do require specialized knowledge and training. Providers should be certain that they have the personal knowledge and experience to perform such testing, or else consult or refer to a qualified genetic specialist. Given the expense and potential for harm from misinterpreted genetic tests, appropriate resource stewardship is essential so that genetic medicine will not suffer social setbacks. Diagnostic yield should be a consideration in the decision to test or not, and testing for improbable diagnoses should only be considered when there is an effective treatment. All criteria for analytical validity of the test should be known (sensitivity, specificity, quality control, and robustness), and clinical genetic testing in the United States must occur in a CLIA-certified laboratory (Clinical Laboratory Improvement Amendments of 1988) that undergoes rigorous proficiency checks. In addition, laboratory testing should comply with regulations of the U.S. Food and Drug Administration, wherever applicable, which has recently announced its intention to regulate all genomic and other laboratory-developed tests [34]. The rapid rise of genomic tests has created numerous systemic challenges that need to be addressed in order to fully realize the potential of genomic medicine.

One such challenge is what to do about incidental findings produced by GS. The American College of Medical Genetics provided a list containing 56 genes that when mutated would represent "clinically actionable" findings found in about 1% of patients, and most of which involve cancer predisposition or cardiovascular-related genes that present risk of sudden death from arrhythmia or aortic rupture [35]. While the clinical rationale behind the list itself was not particularly controversial, its publication led to deep, divided debate about whether reporting of such mutations should be optional or required for patients undergoing exome or GS [36]. The debate remains unsettled, and there are many ongoing sub-debates pertaining to special populations (e.g., children and pregnant women) and various clinical settings (research versus clinical care) about ethical obligations and best practices. At present, thorough pretest counseling to enable shared decision-making with the patient or parents is recommended, and in most cases, patients/parents agree to receive incidental findings. Some providers of genetic healthcare, however, may not agree to withhold potentially life-saving information from a patient, so if there is disagreement, artful negotiation is required.

In addition to the problem of incidental findings, a more serious threat to the clinical implementation of genetic testing is cost. Although technical costs have dropped dramatically for GS and other genotyping technology, actual charges in clinical practice remain high enough to act as a formidable barrier to access. In addition, insurance coverage for genetic testing in the United States varies so widely that it is hard to make any accurate generalization about what is covered, for whom, and in what clinical setting [37].

Determinations of coverage can be quite cumbersome and time-consuming to obtain, requiring the writing of medical necessity letters, lengthy follow-up discussions with insurance companies lacking in genetic expertise, and appeals of unwarranted denials of coverage [37]. In addition, insurance policies tend to be written with a narrow focus on the covered individual with signs or symptoms of a genetic disease, with no allowance made for testing of an apparently asymptomatic person determined to be at high risk because of an incidental genetic finding or cascade genetic testing for a mutation found in a family member. In addition, in order to screen family members for a mutation, it may be necessary first to test a person in the family who is already known to be affected by the disease, but coverage for such testing may be denied on the grounds that the person's diagnosis is already known and treatment of the individual will not be altered by the test result.

Moreover, in the case of complicated syndromic diagnoses and undiagnosed rare diseases, an attitude of diagnostic nihilism may prevail. "Why is it necessary to know a diagnosis if there's no cure?" may be asked, with complete emotional detachment. The intangible value of a diagnosis is hard to communicate when cost containment efforts conflict with the needs of patient, parent, and physician. Research has shown that explaining why disease occurred is intrinsically important to patients [38], but there is no monetary conversion factor to aid value-based healthcare purchasing. Often the most compelling argument is that genetic testing may bring an end, at last, to a "diagnostic odyssey," [39] or in some cases, the harm that may result if diagnosis is not made in a timely fashion, and another child with the same condition is born to parents or extended relatives without warning.

In summary, it is critically important for healthcare providers who order genetic tests for rare diseases to have a thorough understanding of the scientific, clinical, interpersonal, social, legal, and regulatory context in which such testing will occur. "Checking the box" to order the test triggers a whole series of potentially weighty responsibilities that should be appreciated and handled by healthcare providers who have the experience and capabilities to do so.

UNDIAGNOSED DISEASES NETWORK

A welcome new development in diagnosis of rare diseases is the Undiagnosed Diseases Network (UDN), which was set to begin patient recruitment in August 2015 [40]. The UDN is an expansion of a multidisciplinary model that was developed at the National Institutes of Health Clinical Center, also known as "America's Research Hospital," the largest to be devoted exclusively to research. In anticipation of widespread adoption of GS in clinical care, the purpose of the UDN is to support clinical translational research that will improve the care and outcomes of people with rare diseases and their families. In addition to the UDN centers, many individual medical centers are developing their own genomic medicine programs. At this critical juncture in the introduction of GS and other methods for rare disease diagnosis, it is important to pay equal attention to the comprehensiveness of phenotyping as of genotyping. In order to understand what mutations mean, we need standardized and specialized phenotypes to correlate with genotypes. Moreover, all of this information must be entered into harmonized databases (e.g., PhenomeCentral) so that it can be analyzed meaningfully [40,41]. Our ability to critically appraise the clinical utility of GS depends pivotally on the quality of the research that is undertaken at this historic moment. It will not be long until GS will become inexpensive enough that most people in relatively affluent countries will be able to afford it. This situation provides momentous scientific opportunities, but it also comes with a risk of grievous harms if not implemented well. Misdiagnosis, misguided therapy, psychic harms, social backlash, and serious setbacks in implementation of valuable technological advances into appropriate clinical use are ominous risks. However, never before has there been so much progress, promise, and hope for people with undiagnosed rare diseases. In addition, the scientific insights gleaned from rare diseases may, as they have done before, continue to elucidate fundamental molecular pathways that can be targeted by novel therapeutics. The public health benefits of improved rare disease diagnosis may well be the greatest for the vast numbers of people who suffer from common diseases that result from milder derangements of the same pathways involved in rare diseases.

CONCLUSION

We are in the midst of a revolution in the technical ability to assay DNA sequence and chromosomal structure. Such revolutionary advances present a singular opportunity to correlate the particular characteristics of individual genomes with personal phenotypes. We must proceed with great caution, but proceed we must. The potential to understand genetic influence on disease is here, and the potential to improve human health is extraordinary. It is hoped that the genomic revolution will provide inspiration for parallel advances in phenotyping technology and renewed emphasis on careful, systematic observation in clinical medicine, without which the human genome is without meaning [1–16].

REFERENCES

[1] Chong JX, Buckingham KJ, Jhangiani SN, Boehm C, Sobreira N, Smith JD, et al. The genetic basis of Mendelian phenotypes: discoveries, challenges, and opportunities. Am J Hum Genet 2015;97(2):199–215.

[2] Rubinstein WS, Maglott DR, Lee JM, Kattman BL, Malheiro AJ, Ovetsky M, et al. The NIH genetic testing registry: a new, centralized database of genetic tests to enable access to comprehensive information and improve transparency. Nucleic Acids Res 2013;41(Database issue):D925–35.

[3] Phimister EG, Feero WG, Guttmacher AE. Realizing genomic medicine. N Engl J Med. 2012;366(8):757–9.

[4] MacRae CA. A critical need for clinical context in the genomic era. Circulation. 2015;132(11):992–993 [Epub ahead of print].

[5] Welch BM, Dere W, Schiffman JD. Family health history: the case for better tools. JAMA. 2015;313(17):1711–2.

[6] Doll R. Proof of causality: deduction from epidemiological observation. Perspect Biol Med. 2002;45(4):499–515.

[7] Marian AJ. Causality in genetics: the gradient of genetic effects and back to Koch's postulates of causality. Circ Res. 2014;114(2):e18–21.

[8] Richards S, Aziz N, Bale S, Bick D, Das S, Gastier-Foster J, et al. Standards and guidelines for the interpretation of sequence variants: a joint consensus recommendation of the American College of Medical Genetics and Genomics and the Association for Molecular Pathology. Genet Med 2015;17(5):405–24.

[9] Smith JE, Winkler RL, Fryback DG. The first positive: computing positive predictive value at the extremes. Ann Intern Med 2000;132(10):804–9.

[10] Ogino S, Wilson RB. Bayesian analysis and risk assessment in genetic counseling and testing. J Mol Diagn 2004;6(1):1–9.

[11] Lawson HA, Cheverud JM, Wolf JB. Genomic imprinting and parent-of-origin effects on complex traits. Nat Rev Genet 2013;14(9):609–17.

[12] Yang Y, Muzny DM, Reid JG, Bainbridge MN, Willis A, Ward PA, et al. Clinical whole-exome sequencing for the diagnosis of Mendelian disorders. N Engl J Med 2013;369(16):1502–11.

[13] Willig LK, Petrikin JE, Smith LD, Saunders CJ, Thiffault I, Miller NA, et al. Whole-genome sequencing for identification of Mendelian disorders in critically ill infants: a retrospective analysis of diagnostic and clinical findings. Lancet Respir Med 2015;3(5):377–87.

[14] Miller DT, Adam MP, Aradhya S, Biesecker LG, Brothman AR, Carter NP, et al. Consensus statement: chromosomal microarray is a first-tier clinical diagnostic test for individuals with developmental disabilities or congenital anomalies. Am J Hum Genet 2010;86(5):749–64.

[15] Moeschler JB, Shevell M, Committee on Genetics Comprehensive evaluation of the child with intellectual disability or global developmental delays. Pediatrics 2014;134(3):e903–18.

[16] Autism and Developmental Disabilities Monitoring Network Surveillance Year 2008 Principal Investigators; Centers for Disease Control and Prevention. Prevalence of autism spectrum disorders—Autism and Developmental Disabilities Monitoring Network, 14 sites, United States, 2008. MMWR Surveill Summ. 2012;61(3):1–19.

[17] Schaefer GB, Mendelsohn NJ, Professional Practice and Guidelines Committee Clinical genetics evaluation in identifying the etiology of autism spectrum disorders: 2013 guideline revisions. Genet Med 2013;15(5):399–407.

[18] Gilissen C, Hehir-Kwa JY, Thung DT, van de Vorst M, van Bon BW, Willemsen MH, et al. Genome sequencing identifies major causes of severe intellectual disability. Nature 2014;511(7509):344–7.

[19] Yuen RK, Thiruvahindrapuram B, Merico D, Walker S, Tammimies K, Hoang N, et al. Whole-genome sequencing of quartet families with autism spectrum disorder. Nat Med. 2015;21(2):185–91.

[20] Mercimek-Mahmutoglu S, Patel J, Cordeiro D, Hewson S, Callen D, Donner EJ, et al. Diagnostic yield of genetic testing in epileptic encephalopathy in childhood. Epilepsia 2015;56(5):707–16.

[21] Ankala A, da Silva C, Gualandi F, Ferlini A, Bean LJ, Collins C, et al. A comprehensive genomic approach for neuromuscular diseases gives a high diagnostic yield. Ann Neurol 2015;77(2):206–14.

[22] Consugar MB, Navarro-Gomez D, Place EM, Bujakowska KM, Sousa ME, Fonseca-Kelly ZD, et al. Panel-based genetic diagnostic testing for inherited eye diseases is highly accurate and reproducible, and more sensitive for variant detection, than exome sequencing. Genet Med 2015;17(4):253–61.

[23] Alfares AA, Kelly MA, McDermott G, Funke BH, Lebo MS, Baxter SB, et al. Results of clinical genetic testing of 2,912 probands with hypertrophic cardiomyopathy: expanded panels offer limited additional sensitivity. Genet Med 2015;17(11):880–888 [Epub ahead of print].

[24] LaDuca H, Stuenkel AJ, Dolinsky JS, Keiles S, Tandy S, Pesaran T, et al. Utilization of multigene panels in hereditary cancer predisposition testing: analysis of more than 2,000 patients. Genet Med 2014;16(11):830–7.

[25] Li G, Jin H, Xin Z, Su Y, Brezina PR, Benner AT, et al. Increased IVF pregnancy rates after microarray preimplantation genetic diagnosis due to parental translocations. Syst Biol Reprod Med 2014;60(2):119–24.

[26] Stern HJ. Preimplantation genetic diagnosis: prenatal testing for embryos finally achieving its potential. J Clin Med 2014;3(1):280–309.

[27] Yang Z, Lin J, Zhang J, Fong WI, Li P, Zhao R, et al. Randomized comparison of next-generation sequencing and array comparative genomic hybridization for preimplantation genetic screening: a pilot study. BMC Med Genomics 2015;8:30.

[28] Kumar A, Ryan A, Kitzman JO, Wemmer N, Snyder MW, Sigurjonsson S, et al. Whole genome prediction for preimplantation genetic diagnosis. Genome Med 2015;7(1):35.

[29] Norton ME, Jacobsson B, Swamy GK, Laurent LC, Ranzini AC, Brar H, et al. Cell-free DNA analysis for noninvasive examination of trisomy. N Engl J Med 2015;372(17):1589–97.

[30] Sachs A, Blanchard L, Buchanan A, Bianchi DW. Recommended pre-test counseling points for noninvasive prenatal testing using cell-free DNA: a 2015 perspective. Prenat Diagn. 2015;35(10):968–971 [Epub ahead of print].

[31] Fan HC, Gu W, Wang J, Blumenfeld YJ, El-Sayed YY, Quake SR. Non-invasive prenatal measurement of the fetal genome. Nature 2012;487(7407):320–4.

[32] Kitzman JO, Snyder MW, Ventura M, Lewis AP, Qiu R, Simmons LE, et al. Noninvasive whole-genome sequencing of a human fetus. Sci Transl Med 2012;4(137):137ra76.

[33] Van den Veyver IB, Eng CM. Genome-wide sequencing for prenatal detection of fetal single-gene disorders. Cold Spring Harb Perspect Med 2015;5(10) pii: a023077. [Epub ahead of print].

[34] Evans BJ, Burke W, Jarvik GP. The FDA and genomic tests—getting regulation right. N Engl J Med 2015;372(23):2258–64.

[35] Green RC, Berg JS, Grody WW, Kalia SS, Korf BR, Martin CL, et al. ACMG recommendations for reporting of incidental findings in clinical exome and genome sequencing. Genet Med 2013;15(7):565–74.

[36] Blackburn HL, Schroeder B, Turner C, Shriver CD, Ellsworth DL, Ellsworth RE. Management of incidental findings in the era of next-generation sequencing. Curr Genomics. 2015;16(3):159–74.

[37] Deverka PA, Kaufman D, McGuire AL. Overcoming the reimbursement barriers for clinical sequencing. JAMA 2014;312(18):1857–8.

[38] Meisel SF, Carere DA, Wardle J, Kalia SS, Moreno TA, Mountain JL, et al. Explaining, not just predicting, drives interest in personal genomics. Genome Med 2015;7(1):74.

[39] Sawyer SL, Hartley T, Dyment DA, Beaulieu CL, Schwartzentruber J, Smith A, et al. Utility of whole-exome sequencing for those near the end of the diagnostic odyssey: time to address gaps in care. Clin Genet 2015 Aug 18:275–84. [Epub ahead of print].

[40] Brownstein CA, Holm IA, Ramoni R, Goldstein DB, Members of the Undiagnosed Diseases Network Data sharing in the Undiagnosed Diseases Network. Hum Mutat 2015;36(10):985–988 [Epub ahead of print].

[41] Buske OJ, Girdea M, Dumitriu S, Gallinger B, Hartley T, Trang H, et al. PhenomeCentral: a portal for phenotypic and genotypic matchmaking of patients with rare genetic diseases. Hum Mutat 2015;36(10):931–940 [Epub ahead of print].

Chapter 5

Health Risk Assessments, Family Health History, and Predictive Genetic/ Pharmacogenetic Testing

Maria Esperanza Bregendahl[1], Lori A. Orlando[2] and Latha Palaniappan[1]
[1]*Stanford University School of Medicine, Stanford, CA, United States*, [2]*Duke University, Durham, NC, United States*

Chapter Outline

INTRODUCTION

Most primary-care physicians spend a majority of their time treating chronic conditions [1]. Chronic disease such as cardiovascular diseases, obesity, and cancer are long-term medical conditions responsible for seven of every 10 deaths in the United States [2]. In the last decade, thousands of genetic variations have been discovered [3] that might explain differences in chronic disease prevalence and pharmacologic response in individuals, offering primary-care physicians the unprecedented opportunity to personalize disease predication and treatment. In addition, the costs for these tests have decreased rapidly, and in turn, they are becoming more mainstream [4]. In order for personalized medicine to be fully realized in the electronic health record (EHR) era, several tools to effectively tailor prediction and treatment, such as health risk assessment (HRA)

Genomic and Precision Medicine. DOI: http://dx.doi.org/10.1016/B978-0-12-800685-6.00003-5

(including family history), medication reconciliation, and physician decision support, should be incorporated into health systems and clinical workflows [5].

HRAs evaluate a healthy individual by predicting susceptibility to a disease, and in some cases it also provides a recommended plan of action for risk management [6]. An essential component of HRA is the family history, the strongest individual predictor of common chronic diseases such as diabetes [7], stroke [8], and cardiovascular diseases [9], as well as for other medical conditions such as breast cancer [10] and colon cancer [11]. Many chronic diseases have a genetic origin and identifying the underlying causes through genetic (targeting single genes) or genomic (looking at a person's entire gene set) testing will help provide unparalleled customization of patient care [12]. For the first time, population health guidelines, such as those published by the U.S. Preventive Services Task Force (USPSTF) and the Centers for Disease Control (CDC) recommend genetic testing for risk prediction of breast cancer and colon cancer, respectively, in healthy individuals in primary care [13,14]. Both of these guidelines require some knowledge and synthesis of the patient's family history information. Although precision medicine tools hold promise, widespread adoption in primary care is quite slow due to a number of limitations such as insufficient time and resources to collect and update family history and physicians' lack of knowledge and training to recognize genetic patterns related to syndromes [5].

The Joint Commission named medication reconciliation a National Patient Safety Goal in 2005 [15]. Although this safety goal was created to avoid potential drug–drug interactions and to ensure continuity of prescribed medications across the inpatient and outpatient settings, this initiative can also be harnessed to usher in the pharmacogenetics era in primary care.

Making sure medication lists are accurate and up to date in the primary care setting can enable physician decision support tools for evidence-based and reimbursable pharmacogenetic testing. For instance, Medicare and other insurers cover CYP2C19 testing in patients who are on clopidogrel, a drug with a black box warning advising physicians to identify potential poor metabolizers of this antiplatelet drug, often used in cardiac patients after stent placement. Clopidogrel is just one example of a commonly used medication with a pharmacogenomic testing recommendation. The list of potential medication–gene combinations is ever expanding and there are now more than 50 drugs with differential dosing guidelines recommended by the Clinical Pharmacogenetics Implementation Consortium (CPIC) [16]. Given the enormity of keeping track of each of these drug–gene combinations, physician decision support tools at the point of care are critical to alerting physicians to patients that could potentially benefit from testing, and suggest which test(s) to order.

Primary care, the first point of contact for patients, is in an ideal position to promote the integration of genomic medicine in wider healthcare use. As a result, several initiatives to address the objective of personalizing care—promoting the use of HRAs, improving collection of family history, and use of genetic and genomic testing for disease management have been implemented [6,17,18].

The goal of this chapter is to discuss the importance of how HRAs, particularly family history followed by appropriate genetic and genomic testing will impact primary care.

Health Risk Assessments and Family Health History

HRA is one of the basic components of personalized medicine that evaluates a person's overall health and the probability of developing a certain disease [6]. Given that primary-care physicians have a continuous partnership with patients, they would be the natural choice to partner in ensuring proper collection of HRA. Basic HRA data includes demographics, lifestyle, personal and family health history, and physiological data (such as blood pressure, weight, cholesterol, etc.) [5]. For example, the USPSTF recommends several breast cancer risk prediction tools that incorporate varying levels of family history information [13]. The inclusion of genetic information, can in some cases further refine risk calculations and potentially enhance prevention, prediction, and treatment. The case of *PALB2* is an excellent example of how combining genetic results with traditional risk information, in this case family history, can refine risk information [19]. As primary care's role is largely addressing the general well-being of the patient, collecting an accurate health risk assessment and family history will likely improve risk stratification and shared medical decisions with other medical providers. As a result, improving delivery of care throughout the healthcare system.

Collecting Family Health History with Electronic Health Record

Obtaining family health history is a time-consuming process. Studies have shown that very few physicians actually collect health history during healthcare-related visits [20] and many are not properly trained to collect such information. In addition, patient recall of personal and family health history can be very inaccurate and may drastically diminish the quality and effectiveness of the information that is collected [20].

It has been noted that younger physicians are more likely to collect family health history. However, the amount of time spent on the discussion and collection of family history is widely varied. Furthermore, the amount of time primary-care physicians spend on this is significantly less than the amount of time spent by medical geneticists. On average, providers spend less than 3 min during the collection of family health history for new patients, and less than 2 min for patients that are returning [21]. Unfortunately, many physicians also feel that they do not have adequate training and preparation to collect family history. This is especially true when it comes to diseases that they may not have had prior information or training about. This poses a problem particularly for genetic diseases, as this is one of the most important aspects in which family history is useful.

Certain tools are currently in place that facilitate the collection of family health history from patients. The CDC, U.S. Surgeon General, and the National Human Genome Research Institute of the National Institutes of Health collaborated to create My Family Portrait, a web-based family health history collection tool that is publicly available for free. However, most are not yet capable of transferring information entered by patients into health information technology (HIT) systems, such as EHRs [22], and little standardization exists between systems. The systems that are currently in place and ones that will be developed need to be uniform and interoperable—for example, nomenclature must be standardized, information security strengthened, data storage and transmission standards adopted, and others. In addition, systems that are patient-facing (i.e., require patients to independently input and/or collect their own data) should cater their levels of comfort and medical knowledge—in terms of language and education, for example.

A systematic review by de Hoog et al. identified six potential tools for collecting family history that is applicable in primary-care settings [23]. The tools are available on the web (My Family Health Portrait, Family Healthware, and Health Heritage), computer (MeTree^{TM}), and paper (family health questionnaire) format [23]. The My Family Health Portrait Tool, developed under the Surgeon General's Family Health History Initiative [24], allows individuals to input data regarding family health history and print it out so that they can take it to their doctor's appointment. Another tool, Family Healthware, identifies familial and hereditary risk for six conditions including breast, ovarian and colorectal cancer, diabetes, and stroke. It also collects data on health behaviors, screening tests, and health history among first and second-degree relatives [25]. MeTree^{TM}, a relatively new self-administered computerized tool, collects a three-generation family health history of 48 conditions [17]. This is the only tool that provides clinical decision support based on evidence-based guidelines in managing conditions such as breast cancer, ovarian cancer, colon cancer, thrombosis, and hereditary cancer syndrome [17]. Other common family history tools that apply to primary-care settings are described in detail in the same review [23].

In order to improve the quality and effectiveness of the collection of family health history, several recommendations have been made. One recommendation is that collected data is presented in the form of a special pedigree known as a genogram. A genogram takes into account multigenerational family history and incorporates biological, social, and psychosocial issues. For example, a genogram denotes which family members live together and even goes into detail regarding the nature of relatives' relationships, such as whether or not two people are on talking terms, or are estranged [21]. Another recommendation is to have patients complete their family history assessment on their own time. Previous studies have shown that patient-entered data contains higher quality family history data than ones taken during routine primary-care visits (e.g., <4% medical charts contained superior family history information for a single relative before MeTree^{TM} was implemented) [26]. Using MeTree^{TM}, high-quality information on at least one relative in the family history is documented (<4% at baseline to >99% with MeTree™) [27]. Patient-entered data can help improve risk stratification and engage patients in shared decision-making with their physicians regarding their health [20,23]. Ultimately, EHRs should be restructured to standardize the capture of family history and the documentation of relevant information that would facilitate adoption of risk stratification and genomic medicine in primary care. In a qualitative study by Scheuner et al., the majority of physicians believed that genomic medicine would drive EHR content within the next 5 years or so [28]. As it's clear that healthcare providers are currently unprepared for the rapid progress in genomic medicine and lack the training to implement genomic medicine, customizing EHR content will facilitate adoption of genomic medicine in clinics.

Collecting family history is not the end-goal, but a means for physicians to provide a concrete, actionable risk management plan for their patients. Clinical utility of family history information depends upon physicians correctly interpreting the collected data. However, a study by Baldwin et al. revealed that not only do physicians feel incompetent interpreting evidence-based guidelines for risk stratification but that when they did so they both over and underestimated their patients' risks [29]. Therefore, using a reliable family history tool that improves discrimination between at-risk individuals and identifies appropriate risk management strategies can both automate workflows and mitigate compromised care.

Challenges in Implementing Evidence-Based Guidelines

Adherence to and implementation of evidence-based guidelines, such as screening for breast or colorectal cancer, improves the quality of care delivered to patients through early detection and treatment [13,14], and reduces disease-specific mortality [30,31]. As many screening tests (such as mammograms and colonoscopies) require a referral, primary-care physicians play a pivotal role in implementing screening recommendations. However, there are a number of challenges that need to be addressed to successfully integrate these guidelines into clinical practice. Professional organizations such USPSTF, CDC, and Evaluation of Genomic Applications in Practice and Prevention (EGAPP) Working Group support physicians by offering recommendations on clinical services based on the combination of published empirical research and clinical experience [13,14,30]. See Figs. 5.1 and 5.2 for recommendations for breast cancer (USPSTF) and colon cancer (EGAPP), respectively; but despite this uptake is limited.

Some of the pervasive challenges include the following: (1) lack of a unified risk assessment tool that will identify the threshold for at-risk patients, (2) failure of current risk assessment tools to inter-operate with EHR, (3) inadequate time for physicians to perform risk assessment, (4) inadequate training of physicians on identification of genetic risks, and (5) lack of clear-cut referral

Risk assessment, genetic counseling, and genetic testing for BRCA-related cancer in women
Clinical summary of U.S. Preventive services task force recommendation.

Population	Women who have not been diagnosed with BRCA-related cancer and who have no signs or symptoms of the disease	
Recommendation	Screen women whose family history may be associated with an increased risk for potentially harmful BRCA mutations. Women with positive screening results should receive genetic counseling and, if indicated after counseling, BRCA testing	Do not routinely recommend genetic counseling or BRCA testing to women whose family history is not associated with an increased risk for potentially harmful BRCA mutations.
	Grade: B	**Grade: D**
Risk Assessment	Family history factors associated with increased likelihood of potentially harmful BRCA mutations include breast cancer diagnosis before age 50 years, bilateral breast cancer, family history of breast and ovarian cancer, presence of breast cancer in $\geq$ 1 male family member, multiple cases of breast cancer in the family, $\geq$ 1 family member with 2 primary types of BRCA-related cancer, Ashkenazi Jewish family ethnicity	
	Several familial risk stratification tools are available to determine the need for in-depth genetic counseling, such as the Ontario Family History Assessment tool, Manchester Scoring System, Referral Screening Tool, Pedigree Assessment Tool, and FHS-7.	
Screening Tests	Genetic risk assessment and BRCA mutation testing are generally multistep processes involving identification of women who may be at increased risk for potentially harmful mutations, followed by genetic counseling by suitably trained health care providers and genetic testing of selected high-risk women when indicated.	
	Tests for BRCA mutations are highly sensitive and specific for known mutations, but interpretation of results is complex and generally requires posttest counseling.	
Treatment	Interventions in treatment in women who are BRCA mutation carriers include earlier, more frequent or intensive cancer screening; risk reducing medications (e.g. tamoxifen or raloxifene); and risk reducing surgery (e.g. mastectomy or salpingo-oophorectomy)	
Balance of Benefits and Harms	In women whose family history is associated with an increased risk for potentially harmful BRCA mutations, the net benefit of genetic testing and early intervention is moderate.	In women whose family history is not associated with an increased risk for potentially harmful BRCA mutations, the net benefit of genetic testing and early intervention ranges from minimal to potentially harmful.
Other Relevant USPSTF Recommendations	The USPSTF has made recommendations on medications for the reduction of breast cancer risk and screening for ovarian cancer. These recommendations are available at www. uspreventiveservicestaskforce.org	

FIGURE 5.1 USPSTF Recommendation on Breast Cancer related risk assessment, Genetic counseling and Genetic testing. *Adapted from U.S Preventive Services Task Force. Final Recommendation Statement: BRCA-related Cancer: Risk Assessment, Genetic Counseling and Genetic Testing [Internet]. 2013 [cited 2016 Feb 1]. Available from: http://www.uspreventiveser vicestaskforce.org/Page/Document/RecommendationStatementFinal/brca-related-cancer-risk-assessment-genetic-counseling-and-genetic-testing.*

Lynch Syndrome EGAPP Recommendation

Background

Approximately 3% of people who develop colorectal cancer have an inherited condition known as Lynch syndrome. This condition is also referred to as hereditary nonpolyposis colorectal cancer or HNPCC. People with Lynch syndrome have a greatly increased chance of developing colorectal cancer, especially at a young age (under 50). Close biological relatives (parents, children, sisters, brothers) of people with Lynch syndrome have a 50% chance of inheriting this condition. Other close relatives, such as grandparents, aunts, uncles, cousins, nieces and nephews, are also at increased risk to have Lynch syndrome. If an individual is found to carry a genetic change (mutation) associated with Lynch syndrome, his/her relatives can be tested to determine if they also carry the mutation. Relatives found to have a Lynch syndrome gene mutation can lower their chance of developing or dying from colorectal cancer by having earlier and more frequent colonoscopies.

EGAPP™ Recommendation Statement

Summary of Findings on Genetic Testing for Lynch Syndrome

The independent Evaluation of Genomic Applications in Practice and Prevention (EGAPP™) Working Group 🖉 found good scientific evidence to recommend that *all* individuals with a new diagnosis of colorectal cancer (regardless of age or family history) be offered genetic testing for Lynch syndrome, in order to help prevent cancer in their close relatives.

There are several laboratory testing approaches available to implement genetic testing for Lynch syndrome in clinical practice. However, the EGAPP™ working group determined that there is not enough evidence to recommend the use of one specific approach.

For the General Public

Patient Summary : Genetic Testing for Lynch Syndrome

For Health Professionals

EGAPP™ Recommendation Statement 🔧 [PDF 219 KB]

"The Evaluation of Genomic Applications in Practice and Prevention (EGAPP™) Working Group found sufficient evidence to recommend offering genetic testing for Lynch syndrome to individuals with newly diagnosed colorectal cancer (CRC) to reduce morbidity and mortality in relatives. We found insufficient evidence to recommend a specific genetic testing strategy among the several examined."

FIGURE 5.2 EGAPP recommendation for Lynch Syndrome Adopted from Centers for Disease Control and Prevention. Lynch Syndrome EGAPP recommendation [Internet]. 2011 [cited 2016 Feb 1]. Available from: http://www.cdc.gov/genomics/gtesting/EGAPP/recommend/lynch.htm.

processes to promote teamwork within the medical community leading to disjointed and uncoordinated care—that is, lack of a clear-cut referral process and followup. Examples of how these challenges specifically apply to breast cancer and colorectal screening candidates are described in detail below.

Lack of a unified screening tool to identify the threshold for at-risk patients. The USPSTF recommends multiple-validated risk-assessment tools for primary-care physicians to use in evaluating patients' risk in breast cancer and to detect the possible presence of harmful variations in BReast CAncer or BRCA genes. The recommended tools are (1) Ontario family history assessment tool, (2) Manchester scoring system, (3) Referral screening tool, (4) Pedigree assessment tool, and (5) FHS-7 (a seven-questionnaire family history screening instrument) [13]. These tools help primary-care physicians identify female patients in need of further evaluation by genetic counselors [13]. Each tool has unique strengths and weaknesses. For instance, the FHS-7 has good sensitivity in identifying hereditary breast cancer phenotypes but is unable to correctly identify patients with Li-Fraumeni syndrome (a rare disorder that greatly increases the risk of developing several types of cancer including breast) [31]. On the other hand, the updated version of the Referral screening tool (Breast Cancer Genetics Referring Screening Tool or B-RST) is the quickest to administer and can be completed in less than 5 minutes [32]. The Pedigree assessment is a simple tool that has a higher predictive ability (100% sensitivity and 93% specificity) than the Gail model in identifying women with BRCA mutations [33]. In general, these tools are specifically used by nonspecialists to identify and refer at-risk women to genetic counselors. The limitations of these tools prevent experts from endorsing one tool over another, which in turn has limited widespread adoption in primary-care practices. In addition, none of these tools provide clear guidance or thresholds on when a primary-care physician should refer to a genetic counseling service. This is in contrast to the clear guidelines and thresholds that primary-care physicians are exposed to in other common diseases, such as high cholesterol [34] and high blood pressure [35]. Although a variety of options might seem ideal, the ability to recommend or create a single tool, with clear thresholds for referral, will help streamline the genetic counseling referral process and optimize the use of appropriate genetic testing in primary care.

Failure to integrate into EHR. Previous studies have shown the promise of EHRs to improve the quality of care and increase screening services [36]. For example, colorectal screening is a complicated process that often requires multiple steps involving other care team members. The multiple steps include dispensing, tracking, and following-up stool blood testing (fecal occult blood testing), which needs to be done annually based on current guidelines or scheduling colonoscopies, which require staff to train patients in a complex preparation procedure. We can learn from these examples of successful uptake of evidence-based guidelines using EHR clinical decision support, and apply them to genetic testing. For example, family members of patients with Lynch Syndrome should undergo screening for a Lynch Syndrome mutation (cascade screening); however, in our current system, this rarely occurs. An EHR could facilitate testing by alerting providers of family members when a relative has been diagnosed with Lynch, or any of the other monogenic highly penetrant syndromes [37]. In addition, primary-care physicians could receive electronic prompts to collect additional information from patients with either a family history concerning for a hereditary cancer syndrome or an inadequate family history that is not thorough enough to perform appropriate risk stratification [38]. This two-pronged EHR-based approach, both radiating from the proband, and providing additional screening and catchment from within usual primary care may offer higher rates of appropriate genetic testing for first-degree relatives.

Inadequate time for primary-care physicians to evaluate patient's risk. The physician's lack of time might compromise their ability to perform a risk assessment and thus the quality of preventive care delivered especially to at-risk patients. With the limited face-time physicians have during routine visits [39], prioritizing the collection of family history is next to impossible. As previously described, they have limited confidence in their skills in synthesizing family history data, performing risk calculations, identifying actionable recommendations, and even understanding the risk process outlined in many evidence-based guidelines— all resulting in over-estimating or under-estimating a patient's risk [29]. This mismatch can have serious clinical implications. For example, over-estimating colon cancer risk can harm patients by over-screening with test/procedures that themselves have inherent risks, increased anxiety, and unexpected health-care costs [29]. Risk assessment tools may help alleviate some of these challenges. For example, in a previous study of the efficacy of a family history tool in primary-care settings, physicians reported that it was easy to incorporate into the workflow, made their practice easier, and improved their understanding of family history [40]. These findings suggest that barriers such as time constraint and difficulty in synthesizing family history data into an actionable plan can be resolved by leveraging technology that are acceptable and valued by physicians.

Inadequate training for primary-care physicians. One of the barriers to integrating genomic medicine into practice is the lack of genetics training and education among primary-care physicians. Furthermore, the amount of genetics training greatly influences the decision to order genetic tests [41]. Previous studies have shown that primary-care physicians often feel inadequately versed in genomics and are hesitant to incorporate it into their practice [41,42]. The main goal of most primary-care physicians is to deliver to their patients' evidence-based medicine practices that increase health and reduce harm. As such, some of the primary-care physician's lack of confidence can be attributed to the limited evidence of a relationship between genomics and clinical outcomes. To be more specific, some genomic tests (such as testing for BRCA1 or BRCA2 mutations) are better studied than others and have established implications for risk assessment [43], resulting in evidence-based population-level clinical guidelines. Others, such as direct to consumer single nucleotide polymorphism testing and whole exome sequencing, have been less well studied and have limited data to suggest that they offer any health benefits. Early uptake of these tests may result in increased testing [44], without any health benefits, though they have not been found to increase harm in terms of psychological distress [45]. Training of primary-care physicians may lead to optimal utilization of genetic testing, which in turn can improve treatments and outcomes.

Lacking clear-cut referral process in clinics. To promote the integration of genomic medicine and improve the quality of care in clinics, a definitive referral process between primary-care physicians and genetic counselors'/specialists is necessary. EHRs offer a strong platform for assimilating referral guidelines into physician workflows at the time of care, and for primary-care providers to share key information with specialists at the time of the referral, thereby improving the coordination of care. Clear guidance on risk calculators, risk thresholds, and referral guidelines can improve the appropriate genetic testing process in primary care in several ways. Having a unified risk calculator/algorithm together with clinical decision support from within the EHR will facilitate implementation of evidence-based guidelines and referral process in clinics [46]. MeTree™ is the embodiment of the type of tool needed to improve the referral process in clinics. Although not yet publicly available, it includes a clinical decision support component which generates a patient and physician report providing actionable recommendations linked to the patient's risk level.

In pharmacogenetics, the patient's DNA is analyzed to determine drug metabolizing gene variants. The most studied are 50 liver enzymes that are part of the Cytochrome P450 system. These enzymes account for about 75% of the total number of different metabolic reactions, such as the 2C19 effect on Plavix (clopidogrel). The U.S. Food and Drug Administration or FDA has required a "black box warning" on Plavix that advises that genetic testing is available to identify poor metabolizers, that may be unable to respond to the drug, and places them at increased risk for heart attack and stroke. The U.S. Food and Drug Administration or FDA lists over 100 other drugs and combinations which could indicate ineffective treatment due to genetic profiles [47]. CPIC offers updated testing and dosing guidelines for over 50 drugs [16]. Although most electronic health systems have integrated drug–drug interactions systems and warnings, future decision support should be directed toward physician guidance in identifying potential drug–gene interactions and appropriate pharmacogenetic testing.

Since the Health Information Technology for Economic and Clinical Health Act (HITECH Act) was created in 2009 as part of the American Reinvestment and Recovery Act (ARRA) Bill, 80% of primary-care physicians now use EHRs [48]. Although HRA, family history, and medication reconciliation are traditionally considered time-consuming, we can now use newly available, highly penetrant, EHR platforms to more rapidly integrate these workflows into primary-care visits. One potential framework for this integration is the RE-AIM framework. The reach, effectiveness, adoption, implementation, and maintenance (RE-AIM) framework was proposed by the IOM Genomics Roundtable in 2015 as a pathway to implement genomics in primary care (Fig. 5.3) [49].

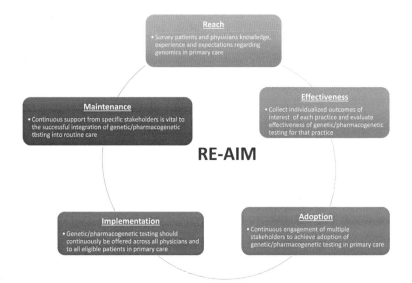

FIGURE 5.3 Using the RE-AIM framework to implement genetic/pharmacogenetic testing in primary care settings. Refer to [49] for more information regarding the framework.

The principles are as follows:

Reach—Prior to implementation, patient and physician surveys can be used to assess interest, knowledge, attitudes, barriers, and beliefs in each unique primary-care population.

Effectiveness—Each practice may have individualized outcomes of interest, and these should be collected to evaluate the effectiveness of genetic/pharmaco-genetic testing in practice. Possible outcomes may include decreased incidence of breast cancer or colon cancer or earlier detection. In the pharmacogenetic testing arena, fewer side effects or adverse drug reactions may be observed. EHRs may also facilitate this data collection and retrospective analysis.

Adoption—In addition to physicians, multiple members of the care team, information technology teams, finance and administration must be engaged to achieve adoption of genetic/pharmacogenetic testing in primary-care. These stakeholders need to continue to be involved and updated as genetic/pharmaco-genetic testing programs evolve and mature.

Implementation—Genetic and pharmacogenetic testing should be offered consistently, to all physicians and all eligible patients in a primary-care practice. Data on these process measures should be available transparently to all involved in the implementation process. Costs should be monitored, to assure return on investment for the various stakeholders. All of these metrics should be routinely reported and analyzed to allow for dynamic changes to the program.

Maintenance—Genetic/Pharmacogenetic testing is a rapidly progressing field. Maintenance of these programs will require integration into routine care, and continuous support from key stakeholders. Financial reimbursement is also a constantly changing landscape, and resources should be allocated to assure financial viability and compliance. These programs will likely grow over time as more genetic/pharmacogenetic tests become applicable to the primary-care patient population.

CONCLUSION

In order for personalized medicine to be fully realized, there is a need to improve systems in primary care. The EHR is the integrated system that binds these personalized medicine tools: HRAs, family history, and genetic information. Restructuring the EHR system to include the translation of genomics into primary care will mitigate barriers in clinical workflow: that is, inadequate time and resources, lack of clinical utility of family history information, lack of genetics content in EHR, and the lack of clinical decision support capabilities for physicians. In addition, allowing patients to enter their data in these family history tools interoperable with EHR will also improve their health literacy and motivate them to make shared decisions with their physicians. The future of personalized medicine relies heavily on the infrastructures that will support it.

REFERENCES

[1] Zamosky L. Chronic disease: a growing challenge for PCPs. Med Econ [Internet] 2013 [cited 2016 Feb 2]; Available from: <http://medicaleconomics.modernmedicine.com/medical-economics/content/tags/chronic-disease/chronic-disease-growing-challenge-pcps>.

[2] Centers for Disease Control and Prevention. Death and Mortality [Internet]. [cited 2016 Jan 29]. Available from: <http://www.cdc.gov/nchs/fastats/deaths.htm>.

[3] Hindorff LA, MacArthur J (European Bioinformatics Institute), Morales J (European Bioinformatics Institute), Junkins HA, Hall PN, Klemm AK and Manolio TA. A Catalog of Published Genome-Wide Association Studies.

[4] Hayden E. Technology: The $1,000 genome [Internet]. Nature 2014. Available from: <http://www.nature.com/news/technology-the-1-000-genome-1.14901>.

[5] Ginsburg GS, Willard HF. Genomic and personalized medicine: foundations and applications. Transl Res [Internet] 2009;154(6):277–87. Available from: <http://linkinghub.elsevier.com/retrieve/pii/S1931524409002746>.

[6] Rubinstein WS. Chapter 26. Family history and health risk assessment tools [Internet] Genomic and personalized medicine, Second Edi : Elsevier Inc.; 2013.306.23. Available from: http://dx.doi.org/10.1016/B978-0-12-382227-7.00026-4.

[7] Harrison TA, Hindorff LA, Kim H, Wines RCM, Bowen DJ, McGrath BB, et al. Family history of diabetes as a potential public health tool. Am J Prev Med 2003;24(2):152–9.

[8] Jerrard-Dunne P, Cloud G, Hassan A, Markus HS. Evaluating the genetic component of ischemic stroke subtypes: a family history study. Stroke 2003;36(6):1364–9.

[9] Kaur J. Family history: a vital predictor of cardiovascular health. Int J Heal [Internet] 2014;2(1):17–21. Available from: <http://www.sciencepubco.com/index.php/IJH/article/view/2402>.

[10] Colditz GA, Kaphingst KA, Hankinson SE, Rosner B. Family history and risk of breast cancer: nurses' health study. Breast Cancer Res Treat 2012;133(3):1097–104.

[11] Patel SG, Ahnen DJ. Familial colon cancer syndromes: an update of a rapidly evolving field. Curr Gastroenterol Rep 2012;14(5):428–38.

[12] Meunier Y. Diseases and medical conditions [Internet] Medicine of the future. : Springer International Publishing; 2014.1.132p. Available from: <http://link.springer.com.laneproxy. stanford.edu/chapter/10.1007/978-3-319-07299-9_1>.

[13] U.S. Preventive Services Task Force. Final Recommendation Statement: BRCA-related Cancer: Risk Assessment, Genetic Counseling and Genetic Testing [Internet]. 2013 [cited 2016 Feb 1]. Available from: <http://www.uspreventiveservicestaskforce.org/Page/ Document/RecommendationStatementFinal/brca-related-cancer-risk-assessment-genetic-counseling-and-genetic-testing>.

[14] Centers for Disease Control and Prevention. Lynch Syndrome EGAPP recommendation [Internet]. 2011 [cited 2016 Feb 1]. Available from: <http://www.cdc.gov/genomics/gtesting/ EGAPP/recommend/lynch.htm>.

[15] Agency for Healthcare Research and Quality:Patient Safety Network. National Patient Safety Goals [Internet]. The Joint Commission. 2016 [cited 2015 Feb 8]. Available from: <https:// psnet.ahrq.gov/resources/resource/2230>.

[16] PharmGKB. Dosing Guidelines-CPIC [Internet]. 2016 [cited 2015 Feb 7]. Available from: <https://www.pharmgkb.org/view/dosing-guidelines.do?source=CPIC#>.

[17] Orlando L a, Hauser ER, Christianson C, Powell KP, Buchanan AH, Chesnut B, et al. Protocol for implementation of family health history collection and decision support into primary care using a computerized family health history system. BMC Health Serv Res [Internet] 2011;11(1):264. Available from: <http://www.biomedcentral.com/1472-6963/11/264>.

[18] Obeng AO, Scott SA. Implementation and utilization of genetic testing in personalized medicine. Pharmgenomics Pers Med 2014:227–40.

[19] Antoniou AC, Casadei S, Heikkinen T, Barrowdale D, Pylkäs K, Roberts J, et al. Breast-cancer risk in families with mutations in PALB2. N Engl J Med [Internet] 2014;371(6):497–506. Available from: <http://www.pubmedcentral.nih.gov/articlerender.fcgi?artid=4157599 &tool=pmcentrez&rendertype=abstract>.

[20] Rich EC, Burke W, Heaton CJ, Haga S, Pinsky L, Short MP, et al. Reconsidering the family history in primary care. J Gen Intern Med 2004;19:273–80.

[21] Acheson LS, Wiesner GL, Zyzanski SJ, Goodwin MA, Stange KC. Family history-taking in community family practice: implications for genetic screening. Genet Med [Internet] 2000;2(3):180–5. Available from: <http://www.nature.com/gim/journal/v2/n3/abs/gim2000 243a.html#top>; <http://www.nature.com/gim/journal/v2/n3/pdf/gim2000243a.pdf>.

[22] Feero GW, Bigley MB, Brinner KM. New standards and enhanced utility for family health history information in the electronic health record: an update from the American Health Information Community's Family Health History Multi-Stakeholder Workgroup. J Am Med Inf Assoc 2008;15(6):723–8.

[23] de Hoog CLMM, Portegijs PJM, Stoffers HEJH. Family history tools for primary care are not ready yet to be implemented. A systematic review. Eur J Gen Pract [Internet] 2014; 20(2):125–33. Available from: <http://informahealthcare.com/doi/abs/10.3109/13814788.20 13.840825>.

[24] Facio FM, Feero WG, Linn A, Oden N, Manickam K, Biesecker LG. Validation of My Family Health Portrait for six common heritable conditions. Genet Med [Internet] 2010;12(6):370–5. Available from: <http://www.nature.com/gim/journal/v12/n6/full/gim201059a.html>; <http://www.nature.com/gim/journal/v12/n6/pdf/gim201059a.pdf>.

[25] Yoon PW, Scheuner MT, Jorgensen C, Khoury MJ. Developing Family Healthware, a family history screening tool to prevent common chronic diseases. Prev Chronic Dis [Internet] 2009;6(1):A33. Available from: <http://www.pubmedcentral.nih.gov/articlerender. fcgi?artid=2644613&tool=pmcentrez&rendertype=abstract>.

[26] Powell KP, Christianson CA, Hahn SE, Dave G, Evans LR, Blanton SH, et al. Collection of Family Health History for Assessment of Chronic Disease Risk in Primary. NC Med J 2013;74(4):279–86.

[27] Wu RR, Himmel TL, Buchanan AH, Powell KP, Hauser ER, Ginsburg GS, et al. Quality of family history collection with use of a patient facing family history assessment tool. BMC Fam Pract [Internet] 2014;15(1):31. Available from: <http://www.pubmedcentral.nih.gov/ articlerender.fcgi?artid=3937044&tool=pmcentrez&rendertype=abstract>.

[28] Scheuner MT, de Vries H, Kim B, Meili RC, Olmstead SH, Teleki S. Are electronic health records ready for genomic medicine? Genet Med [Internet] 2009;11(7):510–7. Available from: <http://www.nature.com/doifinder/10.1097/GIM.0b013e3181a53331>.

[29] Baldwin LM, Trivers KF, Andrilla CHA, Matthews B, Miller JW, Lishner DM, et al. Accuracy of ovarian and colon cancer risk assessments by U.S. physicians. J Gen Intern Med 2014;29(5):741–9.

[30] EGAPP. Evaluation of Genomic Applications in Practice and Prevention (EGAPP) [Internet]. 2013. Available from: <http://www.egappreviews.org>.

[31] Ashton-Prolla P, Giacomazzi J, Schmidt AV, Roth FL, Palmero EI, Kalakun L, et al. Development and validation of a simple questionnaire for the identification of hereditary breast cancer in primary care. BMC Cancer 2009;9:283.

[32] Bellcross CA, Lemke AA, Pape LS, Tess AL. Evaluation of a breast/ovarian cancer genetics referral screening tool in a mammography population. Genet Med 2009;11(11):783–9.

[33] Hoskins KF, Zwaagstra A, Ranz M. Validation of a tool for identifying women at high risk for hereditary breast cancer in population-based screening. Cancer 2006;107(8):1769–76.

[34] Goff DC, Lloyd-Jones DM, Bennett G, Coady S, D'Agostino RB, Gibbons R, et al. 2013 ACC/AHA guideline on the assessment of cardiovascular risk: a report of the American college of cardiology/American heart association task force on practice guidelines. Circulation 2014;129(25 SUPPL. 1):2935–9. Available from: <https://www.clinicalkey.com/#!/content/ playContent/1-s2.0-S0735109713060312?returnurl=null&referrer=null>.

[35] James PA, Oparil S, Carter BL, Cushman WC, Dennison-Himmelfarb C, Handler J, et al. Evidence-Based Guideline for the Management of High Blood Pressure in Adults. JAMA [Internet] 2013;1097(5):1–14. Available from: <http://jama.jamanetwork.com/article.aspx? articleid=1791497>; <http://jama.jamanetwork.com/article.aspx?doi=10.1001/jama.2013. 284427>.

[36] Kern LM, Barrón Y, Dhopeshwarkar RV, Edwards A, Kaushal R. Electronic health records and ambulatory quality of care. J Gen Intern Med 2013;28(4):496–503.

[37] Sharaf RN, Myer P, Stave CD, Diamond LC, Ladabaum U. Uptake of genetic testing by relatives of lynch syndrome probands: a systematic review. Clin Gastroenterol Hepatol [Internet]. 2013;11(9):1093–100. Available from: http://dx.doi.org/10.1016/j.cgh.2013.04.044.

[38] Singh H, Schiesser R, Anand G, Richardson PA, El-Serag HB. Underdiagnosis of lynch syndrome involves more than family history criteria. Clin Gastroenterol Hepatol [Internet]. 2010;8(6):523–9. Available from: http://dx.doi.org/10.1016/j.cgh.2010.03.010.

[39] Østbye T, Yarnall KKSH, Krause KM, Pollak KI, Gradison M, Michener JL. Is there time for management of patients with chronic diseases in primary care? Ann Fam Med [Internet] 2005;3(3):209–14. Available from: <http://www.pubmedcentral.nih.gov/articlerender.fcgi? artid=1466884&tool=pmcentrez&rendertype=abstract>.

[40] Wu RR, Orlando LA, Himmel TL, Buchanan AH, Powell KP, Hauser ER, et al. Patient and primary care provider experience using a family health history collection, risk stratification, and clinical decision support tool: a type 2 hybrid controlled implementation-effectiveness trial. BMC Fam Pract 2013;14:111.

[41] Burke W, Emery J. Genetics education for primary-care providers. Nat Rev Genet [Internet] 2002;3(7):561–6. Available from: <papers://a160a322-7748-499f-b1e5-c793de7b7813/Paper/p261>.

[42] Harvey EK, Fogel CE, Peyrot M, Christensen KD, Terry SF, McInerney JD. Providers' knowledge of genetics: a survey of 5915 individuals and families with genetic conditions. Genet Med [Internet] 2007;9(5):259–67. Available from: <http://www.ncbi.nlm.nih.gov/pubmed/17505202>.

[43] NIH National Cancer Institute. BRCA1 and BRCA2: Cancer Risk and Genetic Testing [Internet]. 2015. Available from: <http://www.cancer.gov/about-cancer/causes-prevention/genetics/brca-fact-sheet>.

[44] Bloss CS, Schork NJ, Topol EJ. Direct-to-consumer pharmacogenomic testing assessed in a US-based study. J Med Genet 2014;51:83–9.

[45] Bloss CS, Schork NJ, Topol EJ. Effect of direct-to-consumer genomewide profiling to assess disease risk. New English J Med 2011;364(6):524–34.

[46] Rolnick SJ, Rahm AK, Jackson JM, Nekhlyudov L, Goddard KAB, Field T, et al. Barriers in identification and referral to genetic counseling for familial cancer risk: the perspective of genetic service providers. J Genet Couns 2011;20(3):314–22.

[47] U.S. Food and Drug Administration. Table of Pharmacogenomic Biomarkers in Drug Labeling [Internet]. 2015 [cited 2016 Feb 7]. Available from: <http://www.fda.gov/Drugs/ScienceResearch/ResearchAreas/Pharmacogenetics/ucm083378.htm>.

[48] Charles D, Gabriel M, Searcy T, Carolina N, Carolina S. Adoption of Electronic Health Record Systems among U. S. Non-Federal Acute Care Hospitals: 2008–2014 The Health Information Technology for Economic and Clinical Health (HITECH) Act of 2009 directed the Office of the National Coordinator for Health 2015;4(23):2008–14.

[49] Johnson SG, Terry SF, Green LA. Making personalized health care even more personalized: insights from activities of the IOM genomics roundtable. Ann Fam Med 2015:373–80.

Chapter 6

Pharmacogenetics and Pharmacogenomics

J. Kevin Hicks and Howard L. McLeod

DeBartolo Family Personalized Medicine Institute, Moffitt Cancer Center, Tampa, FL, United States

Chapter Outline

INTRODUCTION

One of the many reasons that medicine remains as much art as science is the extraordinary variation in response to medications. Therapeutic success, adverse drug reactions (ADR), and refractory disease are often apparent among a group of patients with seemingly identical diagnoses (Fig. 6.1). Such differences are often greater across a population than within the same person (or between monozygotic twins) [1]. It is estimated from studies of both large multigeneration families and twins that genetics can account for 20–95% of variability in drug disposition and effects [2,3]. There are now numerous examples where interindividual differences in drug response have been attributed to polymorphisms in genes encoding drug-metabolizing enzymes, drug transporters, or drug targets [2,4]. It is clear that many nongenetic factors such as age, organ function, concomitant pharmacotherapy, and severity of the disease can influence the risk of ADR and therapeutic effects of medications. However, inherited determinants of drug response remain stable for an individual's lifetime, and

Genomic and Precision Medicine. DOI: http://dx.doi.org/10.1016/B978-0-12-800685-6.00004-7

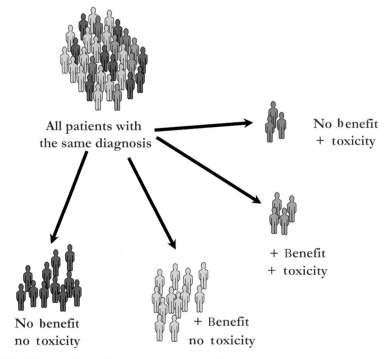

All patients with
the same diagnosis

No benefit
+ toxicity

+ Benefit
+ toxicity

No benefit
no toxicity

+ Benefit
no toxicity

FIGURE 6.1 Population diversity of drug response phenotypes.

the effects can be profound, making them potentially very useful for rational drug prescription strategies, especially given the current environment in which there are usually many available medicines for a given condition but no single best therapeutic strategy. A good example is the treatment of hypertension, where there are nearly 100 different medications currently approved as safe and effective by the USA Food & Drug Administration (FDA), each with a small but real incidence of ADR. In addition, for most patients, there are multiple reasonable treatment strategies, even when the practitioner rigorously adheres to current guidelines and best trial data [5]. There is an opportunity for genetic information to aid both the selection of effective therapy and the avoidance of treatments with an unacceptable risk of ADR.

Inherited differences in drug effects were first documented in terms of drug metabolism in the 1950s [6,7], giving rise to "pharmacogenetics." The field has extended to all aspects of drug disposition (absorption, distribution, and excretion) [8] as well as drug targets and downstream effect mediators. It has also been rediscovered by a broader spectrum of academia and industry, giving birth to "pharmacogenomics." The latter term would apply when broader, genome-wide approaches are used to identify genetic polymorphisms that govern response to specific medications, although in practice the terms are often used interchangeably [4]. The genetic sequence variants of interest to drug therapy also come in many forms, such as variable repeats (where short DNA sequences are repeated a number of times that differs between individuals, e.g., *UGT1A1*28* is defined by an extra TA in the *UGT1A1* promoter region); insertion/deletions (Indels), where some number of bases are either present or absent from the sequence, e.g., the angiotensin-converting enzyme (ACE) indel polymorphism; and single nucleotide polymorphisms (SNPs) [9]. SNPs are the most common, with more than 40 million identified in the initial sequencing of the human genome [10]. The International HapMap project (www.hapmap.org) has identified nearly 6 million SNPs (or one every 600 bp) across multiple world populations [11]. Copy number variation is also important for pharmacogenomics, with important examples of gene amplification already in clinical use [12]. The recent advancements in genomic technology, informatics, and knowledge application have opened endless opportunities to both expand and refine our understanding of pharmacogenetics and consequently its practical applications.

MOLECULAR DIAGNOSTICS FOR OPTIMIZING DRUG THERAPY

The human genes involved in many pharmacogenetic traits have now been identified, their molecular mechanisms defined, their clinical importance more clearly elucidated, and polymorphisms within these genes are now in various stages of making their way into clinical medicine as molecular diagnostics (Table 6.1) [4,13,14]. Because most drug effects are determined by the interplay of multiple gene products that may influence the pharmacokinetics or pharmacodynamics of medications, pharmacogenomics is increasingly focused on polygenic determinants of drug effects involving an entire pathway of genes (Fig. 6.2), including drug targets (e.g., receptors) and those involved in drug disposition (e.g., metabolizing enzymes and transporters). There is also growing awareness that genetic variants in the human leukocyte antigen (HLA) recognition system are a key component of severe hypersensitivity reactions to certain medications, including Stevens–Johnson syndrome [15,16].

Translation of this science into clinical practice is well underway, but will continue to evolve for decades to come. This includes application of pharmacogenomics for genetic discovery, explanation of aberrant phenotype, explanation of drug safety signals, patient inclusion and exclusion, and integration into clinical practice (Fig. 6.3). There are currently several examples where genotype is already being used for the selection of medication doses (Table 6.1). At present, these clinical applications are limited largely to medications with narrow therapeutic indices (e.g., anticancer agents, some antidepressants, and warfarin) or severe ADR (e.g., carbamazepine-induced Stevens–Johnson syndrome), but as additional pharmacogenomic relationships are elucidated, increasing utility for a broader range of medications can be anticipated. Unlike biochemical tests (serum creatinine, bilirubin, etc.), a patient's genotype needs

TABLE 6.1 Examples of Germline Pharmacogenomics that have been Included in FDA Prescribing Recommendations

Category	Genes	Drugs	CLIA Testing
Pharmacokinetics			
	CYP2C9	Warfarin	Y
	CYP2D6	Codeine	Y
		Codeine derivatives	
		5-HT3 receptor antagonists	
		Antidepressants	
		Antipsychotics	
		Atomoxetine	
		Eliglustat	
	CYP2C19	Clopidogrel	Y
	DPYD	Fluoropyrimidines	Y
	TPMT	Mercaptopurine	Y
		Azathioprine	
	UGT1A1	Irinotecan	Y
		Belinostat	
Pharmacodynamics			
	VKORC1	Warfarin	Y
	HLA-B*5701	Abacavir	Y
	HLA-B*1502	Carbamazepine	Y
	IL28B	Interferon α	Y

to be determined only once for any given variant, because it does not change over one's life. Recent advances in genotyping technology have led to rapid and affordable assays that can determine a patient's genotype. Using the amount of DNA that can be isolated from a few milliliters of blood, it is now possible to determine thousands of genotypes. Methodologies are improving so rapidly that a single assay can test for thousands of SNPs. Indeed, the greatest challenge will not be the technology for determining genotype (or haplotype), but rather the strategy for precisely elucidating the genetic determinants of drug response.

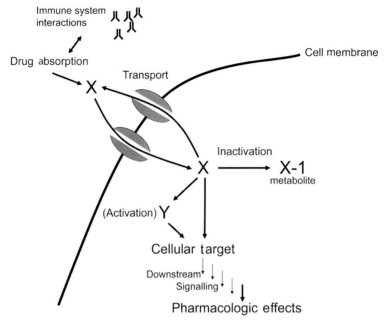

FIGURE 6.2   Schematic of a typical drug pathway, reflecting the pharmacokinetic and pharmacodynamic fate of a medication. The downstream signaling portion is often not well-established.

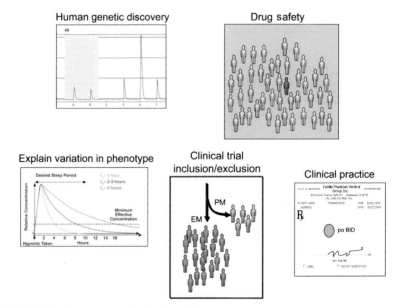

FIGURE 6.3   Distinct areas of application of pharmacogenomics.

Given these technological developments, the process may eventually be to collect a single blood sample from each patient (DNA can be stored for decades), sequence the patient's entire genome, and then filter out the variants known or suspected to be important determinants of drug effects. Patient-specific genotyping results will need to be stored in a secure electronic repository that can be queried as new treatment decisions are made. These new tools will not replace the more conventional biochemical tests that are now routinely used to assess organ function and disease progression. Rather, they will complement those tests, providing additional tools for selecting medications and doses that are optimal for the individual patient. The more we understand the specific pharmacologic characteristics of each patient, the more likely that therapeutic success can be obtained.

DRUG METABOLISM

Many of the initial examples of clinically relevant pharmacogenetics have been altered drug metabolism phenotypes due to genetic polymorphisms. This is not surprising, as every drug-development program in the modern era includes focused in vitro assessment of cytochrome P450 enzymes and extensive pharmacokinetic analysis in the early clinical studies. Nearly all members of more than 30 families of drug-metabolizing enzymes in humans are polymorphic, and many such genetic variants translate into functional changes in the encoded proteins [4,17,18]. In certain instances, multiple SNPs or other genetic alternations such as copy number collectively contribute to a metabolic phenotype [19]. Because drug-metabolizing phenotypes are usually inherited as codominant traits, knowing which genetic variants are interrogated and what allele the alternations are located on is vital for correct interpretation. There are also differences in observed genetic variants among race and ethnicity. As pharmacogenomics is beginning to be applied to clinical practice, efforts are underway to identify the variants that should be interrogated for all patient populations along with standardization of haplotype and phenotype calling [20,21].

One of the best developed examples of pharmacogenetics impacting clinical practice is the genetic polymorphism of thiopurine methyltransferase (TPMT) [22–24]. This enzyme is responsible for degradation of 6-mercaptopurine and azathioprine, which are commonly used for acute leukemia, inflammatory bowel disease, rheumatoid arthritis, and other autoimmune disorders. TPMT molecular genetics have been discussed in detail elsewhere [7]. Patients who inherit complete TPMT deficiency (i.e., two nonfunctional alleles) are at very high risk for severe and potentially fatal hematological toxicity [23], whereas patients who are heterozygotes have an intermediate risk of hematological toxicity [24]. The FDA has now included TPMT genetic information in the package inserts for both 6-mercaptopurine and azathioprine. TPMT genotyping is available from reference laboratories as a Clinical-Laboratory-Improvement-Act-certified molecular diagnostic, and there are now clear dosing guidelines based on TPMT genotype for the treatment of childhood leukemia [24,25], with frequent adaptation to other clinical indications. Most importantly, there is confirmation that dose reductions based on TPMT status will not result in diminution of antileukemia effect [23].

The cytochrome P450 enzyme CYP2D6 is probably the most extensively studied polymorphic drug-metabolizing enzyme in humans and was the first to be characterized at the molecular level. As was common in the pregenomics era, its discovery was in part serendipitous, stemming from an investigator's development of marked hypotension after taking debrisoquine, leading to identification of an inherited deficiency in debrisoquine metabolism [26], a trait discovered independently with sparteine [27]. Population studies demonstrate four distinct groups: poor metabolizers, intermediate metabolizers, normal metabolizers, and ultrarapid metabolizers, reflecting low, medium, average, and high ability to oxidize the drug of interest [21]. Many common medications have been identified as substrates for this enzyme, including analgesics, antidepressants, antipsychotics, and antiemetics. *CYP2D6* genetic polymorphisms have been demonstrated to cause either exaggerated or diminished drug effects, depending on whether it inactivates (e.g., nortriptyline and 5-HT_3 inhibitors) or activates (e.g., codeine family and tamoxifen) the medication of interest [28]. In codeine use, for example, *CYP2D6* genotype delineates individuals with resistance to its effects, as well as those at increased risk of toxicity or overexposure [28–30]. Approximately 10% of patients will receive no pain relief from codeine, oxycodone, or other similar agents due to a poor ability to metabolize as a result of a *CYP2D6* genetic polymorphism [31]. More concerning is the ultrarapid metabolizer situation, where approximately 3% of European white subjects (and as many as 20% of East African subjects) have extra copies of *CYP2D6* (up to 12 extra copies have been reported). Extra copies of *CYP2D6* have been found in case reports of fatalities in children receiving codeine after oral surgery or infants being breastfed while their mother is receiving codeine for postpartum pain [32,33]. This has led to the suggestion that genotype should influence rational codeine prescribing, with poor or ultrarapid metabolizers not receiving the agent [34]. There is now an FDA-approved, commercially available test (Roche Amplichip) for determining *CYP2D6* genotype and copy number status. This has allowed further investigation of the clinical impact of CYP2D6, which is involved in the metabolism of nearly 25% of all FDA-approved drugs. This has included the discovery that a CYP2D6-mediated metabolite of the breast-cancer drug tamoxifen is a potent antiestrogen [35]. Retrospective analysis of breast-cancer patient cohorts has found *CYP2D6* genetic variation to be a predictor of disease-free survival, especially in patients receiving tamoxifen as their only anticancer therapy [35]. Very low active metabolites of tamoxifen have been observed in patients with a poor metabolizer phenotype, whereas normal metabolizer patients had higher blood levels than those with an intermediate phenotype [35]. Prospective intervention studies based on *CYP2D6* genotype have demonstrated that adjusting the dose of tamoxifen in intermediate metabolizer patients (from 20 to 40 mg/day) leads to "normalization" of active metabolite levels [36]. This opens a path for genotype-directed drug dosing to optimize a patient's chance for therapeutic benefit.

The majority of antidepressants, such as the selective serotonin reuptake inhibitors (SSRIs) and tricyclic antidepressants (TCAs), are metabolized by CYP2D6 or the cytochrome P450 enzyme CYP2C19. Certain TCAs, the tertiary amines (e.g., amitriptyline), are metabolized by both CYP2D6 and CYP2C19. Approximately 50% of patients are estimated to fail initial depression treatment, with a significant number of patients experiencing side effects [37,38]. For the SSRIs and TCAs, there is a large body of evidence demonstrating an association between CYP2D6 or CYP2C19 genetic variants and treatment failure or ADR [39,40]. Because of high failure rates and side effects coupled with no "gold standard" for major depressive disorder treatment, pharmacogenomic testing could become an additional routine clinical assay to assist in the selection of the most efficacious treatment [41,42]. Initial studies have demonstrated that those patients undergoing pharmacogenomic analysis have better antidepressant response rates compared to those who are not genotyped, and that genotyping may be cost-effective [43–45]. Dosing guidelines are available for the SSRIs and TCAs based on CYP2D6 or CYP2C19 genotype [39,40]. Generally, for those who are CYP2D6 ultrarapid metabolizers, it is recommended to avoid SSRIs or TCAs metabolized by this enzyme as drug plasma concentrations may be low leading to therapy failure. A 50% dose reduction of TCAs or SSRIs that are metabolized by CYP2D6, or avoidance of CYP2D6 substrates, is recommended for CYP2D6 poor metabolizers due to the elevated risk of high drug plasma concentrations that may cause ADR. Similar recommendations are given for CYP2C19 ultrarapid or poor metabolizers.

There are also examples of warfarin dosing based on CYP2C9 metabolism [46], neutropenia risk assessment for the anticancer drug irinotecan via *UGT1A1* genotype [47], and activation of clopidogrel by CYP2C19 [48]. Use of the CYP2C19 genotype is especially relevant for application of clopidogrel to patients receiving a cardiac stent, with resulting use of an alternate therapy (e.g., prasugrel and ticagrelor) or an alternate clopidogrel dose for CYP2C19 poor metabolizers. Overall, these examples highlight the impact of genetic polymorphism on drug dosing and medication selection.

GENETIC POLYMORPHISMS OF DRUG TARGETS

Evidence is rapidly emerging that genetic variation in drug targets (e.g., receptors) or components of pharmacodynamic pathways can have a profound effect on drug efficacy [4]. Early examples included the β2-adrenoreceptor and response to β2 agonists [49], 5'-lipoxygenase (*ALOX5*) and response to ALOX5 inhibitors, ACE and blood pressure effects of ACE inhibitors [50], the β1-adrenoreceptor and response to beta-blockers [51], serotonin receptors (e.g., HTR2A and HTR2C) and transporters (e.g., SLC6A4) and response to

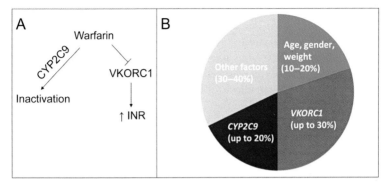

FIGURE 6.4   (A) Flow diagram of warfarin activity. (B) Genetic, pharmacologic, demographic factors that influence dose of warfarin.

antidepressants [52], and vitamin K epoxide-reductase (VKORC1) and warfarin anticoagulation [46]. The case of warfarin pharmacogenetics illustrates both the importance of drug targets and the need to look at the entire drug pathway. It is widely known that many clinical and demographic factors such as age, gender, concomitant drug therapy, and diet impact warfarin dosing [53]. There is also strong evidence that genetic variation contributes to interindividual variability in warfarin dosing. CYP2C9 is the major metabolizing enzyme responsible for the inactivation of warfarin, and VKORC1 is its primary target (Fig. 6.4A).

CYP2C9 has been linked to both toxicity and dosage requirements for optimal anticoagulation [54]. Despite the fact that warfarin dosing can be titrated to a clear coagulation endpoint (i.e., international normalized ratio; INR), data demonstrate that patients with a variant CYP2C9 genotype take a median of 95 days longer to achieve stable dosing compared with the wild-type subjects [55]. Patients with CYP2C9 variants also have a higher risk of acute bleeding complications [55], and patients with the two most common variants (CYP2C9*2 or CYP2C9*3) require 15–30% lower maintenance doses of warfarin to achieve the target INR [54–57]. When added to clinical factors known to impact warfarin dosing, CYP2C9 genotype has been shown to incrementally improve prediction of warfarin dose [54,58].

Another layer of complexity was recently added to this scenario with the identification of the gene for the target of warfarin, VKORC1 [59]. Subsequent to this discovery, polymorphisms within this gene have been found to be associated with warfarin dose requirements [46,60]. The importance of VKORC1 genetic variants in warfarin effect and dosing was observed at both the gene expression level and the optimal warfarin dose [46]. This effect was in addition to the effect of the CYP2C9 variants described above (Fig. 6.4B).

Putting all of these factors together should allow for the construction of a powerful and clinically important tool to improve warfarin therapy. Clinical and demographic variables account for roughly 20% of the interindividual variability in warfarin dose, whereas CYP2C9 genotype makes up 15–20% of individual variability [56]. VKORC1 genotype or haplotype can account for an additional 14% of variability [46,61]. As previous clinical/genetic dosing models (including CYP2C9 but not VKORC1) were able to explain roughly 40% of the total variability in warfarin dose [58], we would expect the addition of the VKORC1 genotype to allow for 50–60% of the total variation in warfarin dose to be accounted for before the drug is ever taken by the patient. This is certainly within the range of clinical significance and provides an important example of how pharmacogenetics has the potential to improve drug therapy— In this case, to reduce the time a patient is receiving an inappropriate warfarin therapy. Clinical studies have been mixed regarding the clinical utility of pharmacogenomic testing to improve warfarin dosing strategies [62–64]. A complicating factor regarding the interpretation of outcome studies is the discovery of CYP2C9 variants that are more common among those of African ancestry but rarely tested for in warfarin pharmacogenomic clinical trials [65,66].

These examples were the result of focused assessment of genetic variants in known targets for these medications and have resulted in unclear utility for clinical practice. Genome-wide assessment technology is now being applied to pharmacogenomics studies. This has the advantage of allowing any region of the genome to indicate a relationship with the interrogated drug effect, rather than candidate genes based on known pharmacology. In some cases, the genome-wide analysis has found the same genes as were initially used in candidate gene pharmacogenetic studies (e.g., warfarin genome-wide association study (GWAS) found CYP2C9 and VKORC1) [67]. However, novel genes and pharmacologic pathways are also being discovered for both drug efficacy and severe ADR. An important example of novel discovery in drug efficacy is with the use of interferon-α for the treatment of hepatitis C. It has been known for decades that 30–45% of patients will not receive benefit from this toxic therapy, with African–American patients being more refractory to interferon treatment. A GWAS in 1137 patients with hepatitis C found a SNP in IL28B that associated with sustained virological response for interferon-α [68]. This association has now been replicated in many clinical studies, and IL28B genotyping has been integrated into clinical management strategies for both interferon-α and the protease inhibitor therapies for hepatitis C treatment, though newer more effective antivirals (e.g., sofosbuvir) are lessening the clinical use of IL28B genotyping for hepatitis C treatment [69]. A similar story of surprise genetic relationships has appeared for severe ADR. Rare hypersensitivity reactions have been known to occur with many medications, without a clear mechanism of action or a method for preemptive prediction of risk for an event. Genome-wide studies have identified distinct alleles in the HLA region of the major histocompatibility complex on chromosome 6 (Table 6.1). In some cases, there was suspicion that an immunoregulatory mechanism was at play, such as for hypersensitivity reactions to the HIV drug abacavir or the occurrence of the severe or fatal skin reaction Stevens–Johnson Syndrome for carbamazepine [15,16,70]. In the case of HLA association with drug-induced liver injury, the unanticipated findings have led to new mechanistic thinking for this rare but devastating iatrogenic event [71].

GLOBAL HEALTH APPLICATIONS

While pharmacogenomics is beginning to take shape in the developed world, the idea of the human genome remains a western toy for the developing world. It is hard to imagine the application of genomic information in countries that often do not have 24 hour electricity or clean water. However, there are approaches being developed to use pharmacogenomics in a public health context. Most diseases in the developing countries have a number of drug treatment options, but limited budget for medications. Selection of a drug formulary is a high-stakes endeavor in the developing world, with a desperate need for ways to optimize the benefit to society of each health care expenditure [72]. Most of the developing countries use the WHO Essential Medicines List (EML) from which to choose the drug formulary for their country. This excellent data source integrates quality of therapeutic evidence, medication cost, and general availability into the final menu of options for a given disease. However, the application in the developing world is limited by the reliance on drugs that were typically approved from studies in Caucasian populations based on a "one dose fits all" FDA/European Medicines Agency (EMEA) safety and dosing guidance paradigm. This practice does not allow local health authorities to use population-specific data to prioritize the drugs with the most favorable safety/effectiveness profile and tailor pharmacovigilance efforts in a manner that best meets the public health needs of their citizens. The Pharmacogenetics for Every Nation Initiative (PGENI) is one strategy that has been using genetics to assist in health decision policy in the developing world. The approach is to analyze populations within a participating country for the frequency of genetic variants which are known to directly cause risk of toxicity or altered efficacy for specific drugs [73]. The risk profile is then compared with that observed in the western populations that are responsible for the majority of participants in the key studies from which safety and efficacy were defined. Any excess genetic risk is used to augment the standard national formulary decision processes, using local data derived to include genetic risk of toxicity or altered efficacy in local populations in developing countries. While the PGENI approach is bringing the genome closer to practical application in the developing world, there are still many opportunities to find innovative uses for the genome to improve global health.

APPLICATION IN DRUG DEVELOPMENT

The area of new drug development presents an important opportunity for pharmacogenomics. An early use of genetics in clinical trials was in the context of interpretation of the population risk around adverse events observed within a given trial. Genetic analysis of cases of hypersensitivity reactions in the early abacavir trials led to identification of approximately 5% of HIV+ patients who should never receive the drug and 95% of patients for whom the chance of experiencing this adverse reaction was zero [70]. This was critical information, as it provided confidence in the prescribing of this effective antiretroviral agent and allowed clinicians to rule out abacavir and focus on other causes in cases where patients had rash and other constitutional symptoms. Ironically, fears that the more complicated pharmacogenomics testing would scare off potential prescribers were unfounded, as the use of abacavir increased dramatically after introduction of testing for toxicity prediction [70].

Pharmacogenomic assessment has also been applied to identify subsets of patients for prospective, genotype-driven trials. Indeed, initial activity of the anticancer drug crizotinib appeared to be confined to nonsmall cell lung cancers that also contained an echinoderm microtubule-associated protein-like 4/anaplastic lymphoma receptor tyrosine kinase (*EML/ALK*) gene fusion. This led to a prospective study of crizotinib that was restricted to patients with this gene fusion, with higher than expected antitumor control and rapid FDA approval [74]. However, this strategy introduces new clinical trial challenges, as screening tumor samples from approximately 1500 patients with nonsmall cell lung cancer for the presence of ALK rearrangements was required to identify the 82 patients with advanced ALK-positive disease who were eligible for the clinical trial.

Pharmacogenetic studies are being included early in the development process for patient selection. In cases where the metabolism of a drug is relatively clear from preclinical studies, pharmacogenetic screening can be conducted to assure a clear signal for whether expensive efficacy trials should be conducted. For example, screening for *CYP2D6* could be conducted to allow enrollment of only patients with a normal metabolizer genotype profile. This would eliminate the extra variation around both the poor metabolizer and ultrarapid metabolizer scenarios and focus on the activity of the drug in this falsely uniform setting. Evidence of drug activity would then trigger a broader trial of all patients (with or without stratification by genotype), while no evidence of activity would allow a company to cut its losses without having to take on the risk of exposing volunteers with altered CYP2D6 metabolism to drug at a point in development where adverse reactions are uncharacterized. Overall, the greatest use of pharmacogenetics and pharmacogenomics is being conducted in the setting of pharmaceutical company trials, with the data being used for important dosing, safety, and efficacy decisions.

CHALLENGES GOING FORWARD

A number of critical issues must be considered as strategies are developed to elucidate inherited determinants of drug effects. A formidable one is that the inherited component of drug response is often polygenic, as described above. Approaches for elucidating these multiple targets include whole-genome scanning for polymorphisms associated with drug effects or candidate gene strategies based on existing knowledge of a medication's mechanism(s) of action, metabolism, or disposition. Each of these strategies has potential utility and limitations [4]. Briefly, the candidate gene strategy has the advantage of focusing resources on a manageable number of genes and polymorphisms that are likely to be important, and has produced encouraging results thus far. Limitations of this approach relate to the incompleteness of our knowledge about a given medication's pharmacokinetics and mechanism(s) of action. Whole-genome approaches overcome this issue by making no assumptions about the importance of a given gene or polymorphism, but suffer from weaknesses that include higher cost, the need for a large number of subjects in the cohort, and difficult statistical analyses.

One of the most important challenges, which cannot be overcome with technological advances in genotyping, is to establish a well-defined pharmacogenetic trait. It is often difficult to have well-characterized patients who have been uniformly treated and systematically evaluated in order to objectively quantify the drug response phenotype. Indeed, most medications have not yet been assessed regarding a role for genetics in the interpatient variation in therapeutic effect or severe adverse reactions. To enhance this ability in the future, the norm should be to obtain genomic DNA from all patients entered into clinical drug trials, along with appropriate consent to permit pharmacogenetic studies. This is now accomplished in most large trials being conducted by pharmaceutical companies, but has not yet become standard for academic or foundation-supported trials.

The challenge is to balance the desire to apply new information and the need to make sure that there is robust data that acting on a pharmacogenomic marker is in the best interest of the patient. The reliance on prospective, randomized, controlled trials as the only way to justify clinical implementation is not practical and guarantees that new information will have a 5–10-year lag, while studies are constructed, conducted, and interpreted. There is also a disconnection between the funding bodies and the prioritization of this type of studies, in terms of financial commitment, clinical trial infrastructure, and agility. There have been several efforts to develop ways to gain confidence in early adoption of pharmacogenomics data, based on the building of consensus around the application of genetic information to drug therapy. One such effort is the Clinical Pharmacogenetics Implementation Consortium (CPIC), which includes participants from >80 institutions across four continents [75]. The CPIC has support from the NIH Pharmacogenetics Research Network and has a practical emphasis on what actions can be justified now, as well as highlighting key next steps for researchers interested in pushing the field forward. A key element to programs such as CPIC are the realization that there are some actions that have robust data on which to benefit patients now, even as the field waits for the "perfect" studies that definitively guide therapy at a broader level.

CPIC guidelines currently encompass about 35 gene–drug pairs, with FDA package inserts containing pharmacogenomic information for over 100 drugs. As research continues, these numbers are expected to greatly increase. A challenge going forward is practitioners remembering which drugs are associated with pharmacogenomic recommendations, if a patient has been previously tested (pharmacogenomic tests in the distant past may still be valid decades later), and what were the results. Electronic health records (EHR) can be leveraged to curate important results and present gene-based dosing recommendations at the point of care [76]. There are technical challenges though, including how to discretely integrate genomic information into the EHR that can be data mined at a later date to trigger clinical decision support. CPIC along with other NIH funded groups are developing methods for capturing genomic data in the EHR that seamlessly integrates with clinical workflows [77–79]. Efforts to implement pharmacogenomics in EHR will enable greatly needed outcome studies.

The endpoints of pharmacogenomics studies have followed a traditional biomarker scheme, trying to explain untoward events, identify low utility, define dose selection, or preemptively predict severe drug reactions. These are important endpoints and should not be neglected in investigational endeavors. However, there are alternate endpoints that are typically considered too boring for including in NIH grants but are major drivers of early adoption for new health care modalities. These include avoidance of 30-day readmission rates, economics of "bundled care," and weighting of choices for the Pharmacy & Therapeutics committee. It will be an emphasis on the "boring" endpoints that will likely drive the implementation of pharmacogenomics into practice, and it is time to be more practical as we move forward.

Although substantial progress has been made in identifying and characterizing pharmacogenetic phenomena, translation of this data into practical clinical application remains slow. A variety of factors contribute to this problem, including a lack of clarity on the amount of data needed to prove clinical utility, the paucity of interventional pharmacogenetic studies, and unresolved practical considerations such as how to establish clear guidelines and how to implement them in the hospital and emergency department. There are also societal factors at play, including social acceptance of widespread genetic testing as well as implications for insurance coverage and liability. These issues will need to be explored and addressed before the promise of genetically customized medicine can become reality.

WEBSITES OF USE

Clinical Pharmacogenetics Implementation Consortium Guidelines
   https://cpicpgx.org/
Pharmacogenetics Knowledgebase
   www.pharmgkb.org
USA Food & Drug Administration Table of Pharmacogenomic Biomarkers in Drug Labels
   http://www.fda.gov/Drugs/ScienceResearch/ResearchAreas/Pharmacogenetics/ucm083378.htm

REFERENCES

[1] Vesell ES. Pharmacogenetic perspectives gained from twin and family studies. Pharmacol Ther 1989;41:535–52.

[2] Evans WE, McLeod HL. Pharmacogenomics—drug disposition, drug targets, and side effects. N Engl J Med 2003;348:538–49.

[3] Kalow W, Tang BK, Endrenyi L. Hypothesis: comparisons of inter- and intra-individual variations can substitute for twin studies in drug research. Pharmacogenetics 1998;8:283–9.

[4] Wang L, Mcleod HL, Weinshilboum RM. Genomics and drug response. N Engl J Med 2011;364:1144–53.

[5] Chobanian AV, Bakris GL, Black HR, Cushman WC, Green LA, Izzo JL, et al. The seventh report of the Joint National Committee on prevention, detection, evaluation, and treatment of high blood pressure: the JNC 7 report. JAMA 2003;289:2560–72.

[6] Kalow W. Familial incidence of low pseudocholinesterase level. Lancet 1956;2:576.

[7] Weinshilboum R. Inheritance and drug response. N Engl J Med 2003;348:529–37.

[8] Meyer UA. Pharmacogenetics and adverse drug reactions. Lancet 2000;356:1667–71.

[9] Feero WG, Guttmacher AE, Collins FS. Genomic medicine—an updated primer. N Engl J Med 2010;362:2001–11.

[10] 1000 Genomes Project Consortium A map of human genome variation from population-scale sequencing. Nature 2010;467:1061–73.

[11] Altshuler DM, Gibbs RA, Peltonen L, Dermitzakis E, Schaffner SF, Yu F, et al. Integrating common and rare genetic variation in diverse human populations. Nature 2010;467:52–8.

[12] He Y, Hoskins JM, Mcleod HL. Copy number variants in pharmacogenetic genes. Trends Mol Med 2011;17:244–51.

[13] Frueh FW, Amur S, Mummaneni P, Epstein RS, Aubert RE, Deluca TM, et al. Pharmacogenomic biomarker information in drug labels approved by the United States food and drug administration: prevalence of related drug use. Pharmacotherapy 2008;28:992–8.

[14] Surh LC, Pacanowski MA, Haga SB, Hobbs S, Lesko LJ, Gottlieb S, et al. Learning from product labels and label changes: how to build pharmacogenomics into drug-development programs. Pharmacogenomics 2010;11:1637–47.

[15] McCormack M, Alfirevic A, Bourgeois S, Farrell JJ, Kasperaviciute D, Carrington M, et al. HLA-A*3101 and carbamazepine-induced hypersensitivity reactions in Europeans. N Engl J Med 2011;364:1134–43.

[16] Chen P, Lin JJ, Lu CS, Ong CT, Hsieh PF, Yang CC, et al. Carbamazepine-induced toxic effects and HLA-B*1502 screening in Taiwan. N Engl J Med 2011;364:1126–33.

[17] Evans WE, Relling MV. Pharmacogenomics: translating functional genomics into rational therapeutics. Science 1999;286:487–91.

[18] Lanfear DE, McLeod HL. Pharmacogenetics: using DNA to optimize drug therapy. Am Fam Physician 2007;76:1179–82.

[19] Hicks JK, Swen JJ, Gaedigk A. Challenges in CYP2D6 phenotype assignment from genotype data: a critical assessment and call for standardization. Curr Drug Metab 2014;15:218–32.

[20] Kalman LV, Agundez J, Appell ML, Black JL, Bell GC, Boukouvala S, et al. Pharmacogenetic allele nomenclature: international workgroup recommendations for test result reporting. Clin Pharmacol Ther 2016;99:172–85.

[21] Caudle KE, Dunnenberger HM, Freimuth RR, Peterson JF, Burlison JD, Whirl-Carrillo M, et al. Standardizing terms for clinical pharmacogenetic test results: consensus terms from the Clinical Pharmacogenetics Implementation Consortium (CPIC). Genet Med in press.

[22] McLeod HL, Siva C. The thiopurine *S*-methyltransferase gene locus—implications for clinical pharmacogenomics. Pharmacogenomics 2002;3:89–98.

[23] Relling MV, Gardner EE, Sandborn WJ, Schmiegelow K, Pui CH, Yee SW, et al. Clinical Pharmacogenetics Implementation Consortium guidelines for thiopurine methyltransferase genotype and thiopurine dosing. Clin Pharmacol Ther 2011;89:387–91.

[24] Relling MV, Hancock ML, Rivera GK, Sandlund JT, Ribeiro RC, Krynetski EY, et al. Mercaptopurine therapy intolerance and heterozygosity at the thiopurine *S*-methyltransferase gene locus. J Natl Cancer Inst 1999;91:2001–8.

[25] Relling MV, Gardner EE, Sandborn WJ, Schmiegelow K, Pui CH, Yee SW, et al. Clinical Pharmacogenetics Implementation Consortium guidelines for thiopurine methyltransferase genotype and thiopurine dosing: 2013 update. Clin Pharmacol Ther 2013;93:324–5.

[26] Mahgoub A, Idle JR, Dring LG, Lancaster R, Smith RL. Polymorphic hydroxylation of Debrisoquine in man. Lancet 1977;2:584–6.

[27] Eichelbaum M, Spannbrucker N, Steincke B, Dengler HJ. Defective N-oxidation of sparteine in man: a new pharmacogenetic defect. Eur J Clin Pharmacol 1979;16:183–7.

[28] Kroemer HK, Eichelbaum M. "It's the genes, stupid". Molecular bases and clinical consequences of genetic cytochrome P450 2D6 polymorphism. Life Sci 1995;56:2285–98.

[29] Gasche Y, Daali Y, Fathi M, Chiappe A, Cottini S, Dayer P, et al. Codeine intoxication associated with ultrarapid CYP2D6 metabolism. N Engl J Med 2004;351:2827–31.

[30] Lotsch J, Skarke C, Liefhold J, Geisslinger G. Genetic predictors of the clinical response to opioid analgesics: clinical utility and future perspectives. Clin Pharmacokinet 2004;43: 983–1013.

[31] Ingelman-Sundberg M. Genetic polymorphisms of cytochrome P450 2D6 (CYP2D6): clinical consequences, evolutionary aspects and functional diversity. Pharmacogenomics J 2005;5:6–13.

[32] Madadi P, Shirazi F, Walter FG, Koren G. Establishing causality of CNS depression in breast-fed infants following maternal codeine use. Paediatr Drugs 2008;10:399–404.

[33] Ciszkowski C, Madadi P, Phillips MS, Lauwers AE, Koren G. Codeine, ultrarapid-metabolism genotype, and postoperative death. N Engl J Med 2009;361:827–8.

[34] Crews KR, Gaedigk A, Dunnenberger HM, Leeder JS, Klein TE, Caudle KE, et al. Clinical Pharmacogenetics Implementation Consortium guidelines for cytochrome P450 2D6 genotype and codeine therapy: 2014 update. Clin Pharmacol Ther 2014;95:376–82.

[35] Hoskins JM, Carey LA, Mcleod HL. CYP2D6 and tamoxifen: DNA matters in breast cancer. Nat Rev Cancer 2009;9:576–86.

[36] Irvin Jr WJ, Walko CM, Weck KE, Ibrahim JG, Chiu WK, Dees EC, et al. Genotype-guided tamoxifen dosing increases active metabolite exposure in women with reduced CYP2D6 metabolism: a multicenter study. J Clin Oncol 2011;29:3232–9.

[37] Hampton LM, Daubresse M, Chang HY, Alexander GC, Budnitz DS. Emergency department visits by adults for psychiatric medication adverse events. JAMA Psychiatry 2014;71:1006–14.

[38] Kennedy SH, Giacobbe P. Treatment resistant depression—advances in somatic therapies. Ann Clin Psychiatry 2007;19:279–87.

[39] Hicks JK, Bishop JR, Sangkuhl K, Muller DJ, Ji Y, Leckband SG, et al. Clinical Pharmacogenetics Implementation Consortium (CPIC) guideline for CYP2D6 and CYP2C19 genotypes and dosing of selective serotonin reuptake inhibitors. Clin Pharmacol Ther 2015;98: 127–34.

[40] Hicks JK, Swen JJ, Thorn CF, Sangkuhl K, Kharasch ED, Ellingrod VL, et al. Clinical Pharmacogenetics Implementation Consortium guideline for CYP2D6 and CYP2C19 genotypes and dosing of tricyclic antidepressants. Clin Pharmacol Ther 2013;93:402–8.

[41] Thompson C, Steven PH, Catriona H. Psychiatrist attitudes towards pharmacogenetic testing, direct-to-consumer genetic testing, and integrating genetic counseling into psychiatric patient care. Psychiatry Res 2015;226:68–72.

[42] Walden LM, Brandl EJ, Changasi A, Sturgess JE, Soibel A, Notario JF, et al. Physicians' opinions following pharmacogenetic testing for psychotropic medication. Psychiatry Res 2015;229:913–8.

[43] Altar CA, Carhart JM, Allen JD, Hall-Flavin DK, Dechairo BM, Winner JG. Clinical validity: combinatorial pharmacogenomics predicts antidepressant responses and healthcare utilizations better than single gene phenotypes. Pharmacogenomics J 2015;15:443–51.

[44] Hall-Flavin DK, Winner JG, Allen JD, Carhart JM, Proctor B, Snyder KA, et al. Utility of integrated pharmacogenomic testing to support the treatment of major depressive disorder in a psychiatric outpatient setting. Pharmacogenet Genomics 2013;23:535–48.

[45] Winner JG, Carhart JM, Altar CA, Goldfarb S, Allen JD, Lavezzari G, et al. Combinatorial pharmacogenomic guidance for psychiatric medications reduces overall pharmacy costs in a 1 year prospective evaluation. Curr Med Res Opin 2015;31:1633–43.

[46] Rieder MJ, Reiner AP, Gage BF, Nickerson DA, Eby CS, Mcleod HL, et al. Effect of VKORC1 haplotypes on transcriptional regulation and warfarin dose. N Engl J Med 2005;352:2285–93.

[47] Hoskins JM, Goldberg RM, Qu P, Ibrahim JG, Mcleod HL. UGT1A1*28 genotype and irinotecan-induced neutropenia: dose matters. J Natl Cancer Inst 2007;99:1290–5.

[48] Scott SA, Sangkuhl K, Gardner EE, Stein CM, Hulot JS, Johnson JA, et al. Clinical Pharmacogenetics Implementation Consortium guidelines for cytochrome P450-2C19 (CYP2C19) genotype and clopidogrel therapy. Clin Pharmacol Ther 2011;90:328–32.

[49] Liggett SB. Beta(2)-adrenergic receptor pharmacogenetics. Am J Respir Crit Care Med 2000;161:S197–201.

[50] Stavroulakis GA, Makris TK, Krespi PG, Hatzizacharias AN, Gialeraki AE, Anastasiadis G, et al. Predicting response to chronic antihypertensive treatment with fosinopril: the role of angiotensin-converting enzyme gene polymorphism. Cardiovasc Drugs Ther 2000;14:427–32.

[51] Lanfear DE, Jones PG, Marsh S, Cresci S, Mcleod HL, Spertus JA. Beta2-adrenergic receptor genotype and survival among patients receiving beta-blocker therapy after an acute coronary syndrome. JAMA 2005;294:1526–33.

[52] Altar CA, Hornberger J, Shewade A, Cruz V, Garrison J, Mrazek D. Clinical validity of cytochrome P450 metabolism and serotonin gene variants in psychiatric pharmacotherapy. Int Rev Psychiatry 2013;25:509–33.

[53] Jonas DE, McLeod HL. Genetic and clinical factors relating to warfarin dosing. Trends Pharmacol Sci 2009;30:375–86.

[54] Gage BF, Eby C, Milligan PE, Banet GA, Duncan JR, Mcleod HL. Use of pharmacogenetics and clinical factors to predict the maintenance dose of warfarin. Thromb Haemost 2004;91:87–94.

[55] Higashi MK, Veenstra DL, Kondo LM, Wittkowsky AK, Srinouanprachanh SL, Farin FM, et al. Association between CYP2C9 genetic variants and anticoagulation-related outcomes during warfarin therapy. JAMA 2002;287:1690–8.

[56] Hillman MA, Wilke RA, Caldwell MD, Berg RL, Glurich I, Burmester JK. Relative impact of covariates in prescribing warfarin according to CYP2C9 genotype. Pharmacogenetics 2004;14:539–47.

[57] Johnson JA, Gong L, Whirl-Carrillo M, Gage BF, Scott SA, Stein CM, et al. Clinical Pharmacogenetics Implementation Consortium guidelines for CYP2C9 and VKORC1 genotypes and warfarin dosing. Clin Pharmacol Ther 2011;90:625–9.

[58] Voora D, Eby C, Linder MW, Milligan PE, Bukaveckas BL, Mcleod HL, et al. Prospective dosing of warfarin based on cytochrome P-450 2C9 genotype. Thromb Haemost 2005;93:700–5.

[59] Li T, Chang CY, Jin DY, Lin PJ, Khvorova A, Stafford DW. Identification of the gene for vitamin K epoxide reductase. Nature 2004;427:541–4.

[60] Klein TE, Altman RB, Eriksson N, Gage BF, Kimmel SE, Lee MT, et al. Estimation of the warfarin dose with clinical and pharmacogenetic data. N Engl J Med 2009;360:753–64.

[61] D'andrea G, D'ambrosio RL, Di Perna P, Chetta M, Santacroce R, et al. A polymorphism in the VKORC1 gene is associated with an interindividual variability in the dose–anticoagulant effect of warfarin. Blood 2005;105:645–9.

[62] Gage BF, Eby C, Johnson JA, Deych E, Rieder MJ, Ridker PM, et al. Use of pharmacogenetic and clinical factors to predict the therapeutic dose of warfarin. Clin Pharmacol Ther 2008;84:326–31.

[63] Kimmel SE, French B, Kasner SE, Johnson JA, Anderson JL, Gage BF, et al. A pharmacogenetic versus a clinical algorithm for warfarin dosing. N Engl J Med 2013;369:2283–93.

[64] Pirmohamed M, Burnside G, Eriksson N, Jorgensen AL, Toh CH, Nicholson T, et al. A randomized trial of genotype-guided dosing of warfarin. N Engl J Med 2013;369:2294–303.

[65] Johnson JA, Cavallari LH. Warfarin pharmacogenetics. Trends Cardiovasc Med 2015;25:33–41.

[66] Nagai R, Ohara M, Cavallari LH, Drozda K, Patel SR, Nutescu EA, et al. Factors influencing pharmacokinetics of warfarin in African–Americans: implications for pharmacogenetic dosing algorithms. Pharmacogenomics 2015;16:217–25.

[67] Cooper GM, Johnson JA, Langaee TY, Feng H, Stanaway IB, Schwarz UI, et al. A genome-wide scan for common genetic variants with a large influence on warfarin maintenance dose. Blood 2008;112:1022–7.

[68] Ge D, Fellay J, Thompson AJ, Simon JS, Shianna KV, Urban TJ, et al. Genetic variation in IL28B predicts hepatitis C treatment-induced viral clearance. Nature 2009;461:399–401.

[69] Clark PJ, Thompson AJ, Mchutchison JG. Genetic variation in IL28B: impact on drug development for chronic hepatitis C infection. Clin Pharmacol Ther 2010;88:708–11.

[70] Mallal S, Phillips E, Carosi G, Molina JM, Workman C, Tomazic J, et al. HLA-B*5701 screening for hypersensitivity to abacavir. N Engl J Med 2008;358:568–79.

[71] Daly AK, Donaldson PT, Bhatnagar P, Shen Y, Pe'er I, Floratos A, et al. HLA-B*5701 genotype is a major determinant of drug-induced liver injury due to flucloxacillin. Nat Genet 2009;41:816–9.

[72] Roederer MW, Sanchez-Giron F, Kalideen K, Kudzi W, Mcleod HL, Zhang W. Pharmacogenetics and rational drug use around the world. Pharmacogenomics 2011;12:897–905.

[73] Roederer MW, McLeod HL. Applying the genome to national drug formulary policy in the developing world. Pharmacogenomics 2010;11:633–6.

[74] Kwak EL, Bang YJ, Camidge DR, Shaw AT, Solomon B, Maki RG, et al. Anaplastic lymphoma kinase inhibition in non-small-cell lung cancer. N Engl J Med 2010;363:1693–703.

[75] Relling MV, Klein TE. CPIC: Clinical Pharmacogenetics Implementation Consortium of the pharmacogenomics research network. Clin Pharmacol Ther 2011;89:464–7.

[76] Hicks JK, Stowe D, Willner MA, Wai M, Daly T, Gordon SM, et al. Implementation of clinical pharmacogenomics within a large health system: from electronic health record decision support to consultation services. Pharmacotherapy 2016;36:940–8.

[77] Hoffman JM, Dunnenberger HM, Kevin Hicks J, Caudle KE, Whirl Carrillo M, Freimuth RR, et al. Developing knowledge resources to support precision medicine: principles from

the Clinical Pharmacogenetics Implementation Consortium (CPIC). J Am Med Inform Assoc 2016;23:796–801.

[78] Rasmussen-Torvik LJ, Stallings SC, Gordon AS, Almoguera B, Basford MA, Bielinski SJ, et al. Design and anticipated outcomes of the eMERGE-PGx project: a multicenter pilot for preemptive pharmacogenomics in electronic health record systems. Clin Pharmacol Ther 2014;96:482–9.

[79] Weitzel KW, Alexander M, Bernhardt BA, Calman N, Carey DJ, Cavallari LH, et al. The IGNITE network: a model for genomic medicine implementation and research. BMC Med Genomics 2016;9:1.

Chapter 7

Hypertension

Patricia B. Munroe and Helen R. Warren
Queen Mary University of London, London, United Kingdom

Chapter Outline

ABBREVIATIONS

AGEN-BP Asian Genetic Epidemiology Network Blood Pressure
BP blood pressure
CAD coronary artery disease
CHARGE Cohorts for Heart and Aging Research in Genome Epidemiology
CHD coronary heart disease
CVD cardiovascular disease
DBP diastolic blood pressure
DNA deoxyribonucleic acid
ENCODE Encyclopedia of DNA elements consortium
GRS genetic-risk score
GWAS genome-wide association studies
ICBP International Consortium for Blood Pressure
KARE Korean Association Resource
LD linkage disequilibrium
MAF minor allele frequency

Genomic and Precision Medicine. DOI: http://dx.doi.org/10.1016/B978-0-12-800685-6.00007-2

MAP	mean arterial pressure
nsSNPs	nonsynonymous single-nucleotide polymorphisms
PP	pulse pressure
SBP	systolic blood pressure
SNPs	single-nucleotide polymorphisms

INTRODUCTION

Elevated blood pressure (BP) or hypertension [≥140 mmHg systolic blood pressure (SBP) and/or ≥ 90 mmHg diastolic blood pressure (DBP)] is estimated to have caused 9.4 million deaths, and 7% of disease burden in 2010 (as measured in disability-adjusted life years) [1]. Hypertension is a major risk factor for cardiovascular disease (CVD), and if left uncontrolled, it causes myocardial infarction, stroke, cardiac failure, and renal failure. The global prevalence was around 20% in 2014 in adults aged 18 or over [2]. A recent study using electronic health records from 1.25 million individuals (30 years or older, with no CVD and follow up over a 5 year period) observed CVD risk differs with age and across different cardiovascular conditions, extending our understanding of BP as a CVD risk factor [3]. Associations between morbidity and systolic BP were found to be strongest for intracerebral and subarachnoid hemorrhage and stable angina, while the weakest associations were for abdominal aortic aneurysms. The study also indicated that a 30-year old with hypertension had a lifetime risk of 63.3% (95% confidence interval, 62.9–63.8) for a cardiovascular event compared to 46.1% (45.5–46.8) for a normotensive individual. Thus, the lifetime burden of hypertension is substantial. The existing antihypertensive treatments are effective at the population level, reducing the risk of developing coronary heart disease (CHD) events (fatal and nonfatal) by 25% and the risk of stroke by 33%, per 10 mmHg reduction in SBP and 5 mmHg reduction in DBP achieved, independent of pretreatment BP level [4]. However, at the individual level, BP is often poorly controlled, and many patients do not achieve <140 mmHg SBP and <90 mmHg DBP targets. A better understanding of BP associations with cardiovascular endpoints is required, and the ultimate goal is to identify new strategies for the management and treatment of the hypertensive patient.

The main causes of hypertension are well known, and lifestyle and genetic effects are both influential. The most important lifestyle risk factors are excess dietary sodium intake, body weight, increased alcohol consumption, psychological stress, and lack of exercise [5]. Evidence for a genetic component comes from studies of families and twins, and a recent study in twins suggests that the heritability (the fraction of BP variance contributed by genetic factors) for both SBP and DBP is circa 50% [6]. However, heritability studies do not identify which genetic differences are important or by what mechanisms they exert their effects on BP. Recent advances in human genetics offer the opportunity to discover hitherto-unknown mechanisms and pathways affecting BP, which could in principle be targeted by novel therapeutic approaches and thus improve treatment of hypertension and prevention of CVD.

BLOOD PRESSURE GENE DISCOVERY

From 2005, with the advent and rapid technological advances of genome-wide association studies (GWAS) [7], there has been remarkable progress in the discovery of BP loci. Initially, single-study GWAS were performed with either hypertension as the outcome or SBP and DBP as continuous traits. These studies provided limited new findings, and few were replicated, primarily because they were underpowered due to small sample sizes. The first genome-wide significant association ($p \leq 5 \times 10^{-8}$) for SBP and DBP was for a single-nucleotide polymorphism (SNP) located near the *ATP2B1* gene. This was reported in 2009 by investigators of the Korean Association Resource project [8]. In 2011, a GWAS in only 4000 individuals using an "extreme case/control" study for hypertension identified and validated one novel SNP at *UMOD* [9]. With only modest findings of individual GWAS for BP traits, the natural next step was to meta-analyse results from multiple GWAS. This has since been the main strategy, with also further successes from using advanced analytical strategies and meta-analysis of association studies [10], and with genotypes from bespoke chips in increasingly larger sample sizes. All of these analyses have led to the identification of 125 distinct loci, each harboring one or more genetic variants with robust and validated association with BP traits. A timeline of BP gene discoveries and approaches from 2009 to 2016 is shown in Fig. 7.1.

Large-scale European GWAS Meta-analyses

The first large-scale meta-analysis of GWAS results for SBP, DBP, and case/control hypertension were published in 2009 [11,12]. Two consortia analyzed approximately 2.5 M SNPs in large numbers of individuals (~30,000) of European ancestry, and followed up their top ten independent signals from each scan by performing a simultaneous reciprocal exchange of association results. Overall, a total of 13 novel loci were discovered with genome-wide significant associations. Subsequently, an analysis of a single large cohort of 23,019 individuals from the Women's Genome Health Study led to the discovery of an association near the *BLK-GATA4* genes [13]. The majority of the genome-wide significant SNPs were associated with both SBP and DBP and risk of hypertension, with the same direction of effect. The associated SNPs were mostly common, with a minor allele frequency (MAF) $\geq 5\%$, and modest effect sizes of ≤ 1.0 mmHg for SBP and ≤ 0.5 mmHg for DBP.

The International Consortium for Blood Pressure (ICBP) reported two of the largest meta-analyses of GWAS in 2011 using 2.5 M SNPs, one for associations with SBP and DBP, the other for mean arterial pressure (MAP) and pulse pressure (PP) phenotypes. The meta-analysis of GWAS for SBP and DBP was performed in 69,395 individuals of European ancestry, and was followed by a three-stage validation study using 133,661 additional individuals of European ancestry [14]. Similar sample sizes were deployed in the MAP and PP study.

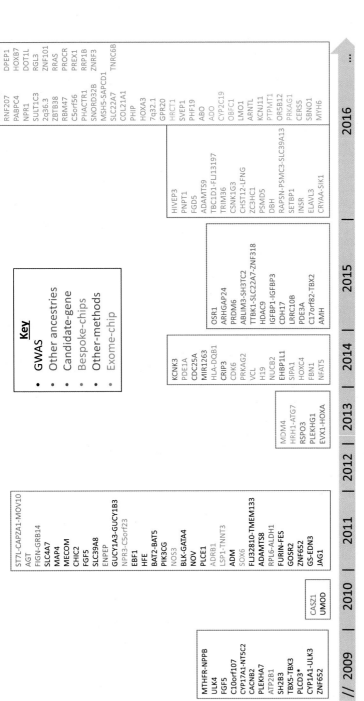

FIGURE 7.1 **Timeline of discovery for BP associated loci.** All published blood-pressure-associated loci are listed along the timeline in boxes according to the year they were first published. Within each year's box, loci are listed in chr:pos order. The color coding of loci, defined in the key, annotates the loci to the type of analysis strategy that they were discovered from "GWAS": Genome-Wide Association Studies, predominantly with European subjects; "Other ancestries": association analyses performed in cohorts of individuals of non-European ancestry; "Candidate-Gene": candidate gene association analysis where only a limited number of variants at genetic loci were tested; "Bespoke-Chips": association analyses performed using bespoke custom-designed gene-centric chip arrays, e.g., HumanCVD BeadChip (aka "CardioChip") with ~50k SNPs, or Cardio-MetaboChip with ~200k SNPs; "New Methods": analyses using advanced methodological strategies, e.g., gene*age interaction analyses, age-stratified analyses, Long-Term-Averaging analyses of BP, joint analysis of multiple correlated traits; "Exome-chip": association analysis using genotyped Exome-chip array variants with the aim of investigating rare, coding variants. NB: Each locus is only listed once, i.e., the identification of any secondary SNPs at previously known loci is not mapped on the timeline.

These GWAS meta-analyses identified 16 novel genome-wide significant associations with SBP and BP, and 8 for MAP and PP [15].

GWAS of Non-European Ancestries

From 2009 to 2011, many of the BP-SNP associations discovered in individuals of European ancestry were replicated in samples of different ancestries (East Asian, South Asian, and African) [16]. The ICBP consortium also showed that some of the 29 SNPs associated with SBP and DBP had significant associations in populations of East- or South-Asian ancestry (or both) after correction for multiple testing [14].

Large GWAS have also been performed in cohorts of non-European ancestry. A GWAS meta-analysis discovered a SNP located near *CASZ1* in samples of East Asian (Japanese) ancestry in 2010 [17]. In 2011, the Asian Genetic Epidemiology Network Blood Pressure consortium identified six significant associations with BP from GWAS meta-analyses of 19,608 individuals of East Asian ancestry, some of which overlapped the ICBP GWAS associations [18].

The largest BP discovery meta-analysis in African Americans was published in 2013 by Franceschini et al. [19], with 29,378 individuals from the Continental Origins and Genetic Epidemiology Network, follow-up in the ICBP dataset, and further replication in additional African Americans and East Asian samples (Total ~29,000 individuals). Overall, three new BP loci (*RSPO3*, *PLEKHG1*, and *EVX1-HOXA*) were identified from the combined meta-analyses.

Kato et al. in 2015 reported a large transethnic meta-analyses of GWAS for SBP, DBP, MAP and hypertension with 99,994 individuals (31,516 East Asians, 35,352 Europeans, and 33,126 South Asians) from the International Genomics of BP Consortium [20], which conducted a meta-analysis with the ICBP dataset and further replication in 133,052 additional samples (48,268 East Asian, 68,456 European, and 16,328 South Asian). Overall, 19 novel SNPs were identified with little heterogeneity observed between the different ethnic groups, although two variants identified in ethnic specific analyses did not validate. In 2012, it was observed that a quarter of the BP loci (8/34) were common across ethnic groups (excluding African Americans), with the remaining 26 loci showing BP-trait associations in only two ethnic groups or a single ethnic group [16]. With larger meta-analyses of GWAS now being performed across different ethnic groups, our knowledge of shared BP loci will become clearer.

Candidate Gene Studies

Loci have also been discovered via candidate gene studies, which only test a selection of genetic variants. For example, an analysis of SNPs at 30 genes that were known targets for antihypertensive drugs provided genome-wide significant associations at two loci, *AGT* and *ADRB1* [21].

Longitudinal Data, Gene–Lifestyle Interactions and Multi-trait Analyses

GWAS utilizing different methodological approaches have also been used for discovery of BP loci. These include using long-term averages of repeated BP measures within longitudinal cohorts [22] and performing an analysis of gene–age interactions or stratified analyses within different age subgroups [23]. Furthermore, a study has performed a multi-trait analysis by analyzing all BP traits simultaneously [24]. These analyses have identified seven new loci (*KCNK3*, *CRIP3* [22]; *MIR1263*, *CDC25A*, *EHBP1L1* [23]; and *IGFBP1-IGFBP3* and *CDH17* [24]), validated previously reported loci, and identified additional variants within known regions.

Bespoke Genotyping Arrays

Alongside genome-wide arrays and imputation methods for the discovery of trait-associated loci, the scientific community has developed genotyping arrays with bespoke content. The general purpose of these arrays is to provide cost-effective genotyping of prespecified SNPs. Two arrays have been used for analyses of BP and other cardiovascular traits; these are the gene-centric HumanCVD BeadChip and the Cardio-MetaboChip array.

The HumanCVD BeadChip contains approximately 50,000 SNPs which provide dense coverage of ~2,000 genes considered to be more likely to have functional effects on cardiovascular traits [25]. An analysis of 25,118 individuals of European ancestry for all BP traits with follow up of the most associated signals in a further 59,349 individuals led to the discovery of eight BP loci in 2011 [26]. Four of the associated loci were simultaneously reported from the ICBP consortium [14]. This study analyzed fewer SNPs in smaller sample numbers ($N\sim25,000$) than the larger meta-analyses of GWAS and indicated that discovery analyses in independent samples will discover new loci because the BP genetic architecture involves possibly hundreds of variants, and each study will detect more or less random subsets of BP variants. Other BP loci have also been identified from the HumanCVD BeadChip: two further loci (*MDM4* and *HRH1ATG7*) were reported in 2013 [27], and in 2014, eleven novel loci were reported in a discovery sample of 87,736 Europeans and independent replication in 68,368 Europeans [28]. An analysis of HumanCVD BeadChip genotypes in up to 8,600 African American individuals (a relatively small discovery sample) did not yield any new BP loci [29], although three previously known loci were validated.

The Cardio-MetaboChip array comprises 196,725 variants, including ~5,000 SNPs with nominal ($p < 0.016$) evidence of BP association from previous GWAS meta-analyses [14]. It also includes several genomic regions for fine mapping of selected loci [14,30]. A meta-analyses of 201,529 individuals of European ancestry was performed, including 109,096 individuals genotyped on Cardio-MetaboChip and 92,433 individuals with imputed genotype data at SNPs overlapping the variants on Cardio-MetaboChip [31]. Sixty-seven loci attained genome-wide significance, of which 18 were novel. To validate the novel SNPs, a further meta-analyses combining the association summary statistics with an additional 140,886 individuals was performed, which validated 17 of the 18 loci. As the Cardio-MetaboChip had high-density coverage across 21 published BP loci and 3 newly identified loci, a fine-mapping analysis was performed to refine the localization of likely functional variants. Using a Bayesian approach, a credible set of variants was defined at each locus, with 99% probability for the set containing or tagging the causal variant. The 99% credible sets included only the index variants at three BP loci (*SLC39A8*, *ZC3HC1*, and *PLCE1*).

Rare Genetic Variants and BP

The contribution of low frequency and rare SNPs to BP traits has been largely unexplored by GWAS. However, these SNPs may explain some of the missing phenotypic variance, by having a larger phenotypic effect than common SNPs [32]. The Exome chip is a bespoke genotyping array which consists of ~250,000 mostly rare (MAF ≤ 0.01) and low frequency (0.01 < MAF < 0.05) nonsynonymous coding variants. These were selected after sequencing the genomes or exomes of ~18,000 genes in 12,000 individuals, primarily of European ancestry. The array also includes other content (e.g., tagging variants from GWAS and ancestry informative markers).

Two large-scale Exome chip meta-analyses for BP traits have recently been undertaken. One was a European-led consortium which genotyped up to 193,000 individuals (165,276 Europeans and 27,487 South Asians) and assessed association of SNPs with SBP, DBP, PP and hypertension [33]. The consortium selected lead variants from 80 novel loci for follow up in an independent set of samples from The Cohorts for Heart and Aging Research in Genomic Epidemiology+ (CHARGE+) Exome Chip BP Working Group (*N*~147,000), and validated 30 novel loci associations. The associations included rare nonsynonymous SNPs (nsSNPs) in three novel genes: *COL21A1*, *RBM47*, and *RRAS*. The other Exome chip meta-analyses were undertaken by the CHARGE+ Exome Chip BP Working Group for SBP, DBP, MAP, PP, and hypertension [34]. The discovery stage included up to 147,000 individuals comprising European, African, and Hispanic individuals; SNPs were followed-up in meta-analyses of 180,726 samples from the European-led consortium. The analysis identified 31 novel loci and included three low-frequency nsSNPs in the novel genes *SVEP1* and *PTPMT1*, and the previously implicated BP gene *NPR1* [35]. In total, 52 distinct novel BP loci were identified using the Exome chip across both consortia.

Each consortium also performed gene-based tests using the Burden test and the sequence kernel association test (SKAT). Burden tests detect associations when all variants contribute to effects in a concordant direction [36], and SKAT detects effects of alleles that collectively contribute to higher or lower BP effects [37]. These tests have increased statistical power to detect associations in genes harboring rare variants. Analysis was restricted to coding variants with MAF <5% or MAF <1%. One gene was reported by the European consortium to be significantly associated with hypertension (*A2ML1*), containing multiple rare variant associations. The CHARGE + Consortium identified three significant genes; two of these (*DBH* and *NPR1*) overlap loci identified using single SNP testing, *PTPMT1* was a novel locus.

A review of the effect sizes across BP SNPs with comparative MAF (≥5%) indicates the majority to have small-effect sizes (mmHg, mean, and s.d.) on BP individually [SBP: 0.51 (0.22); DBP: 0.35 (0.23), and PP: 0.31 (0.09)]. There are four SNPs that have relatively larger effect sizes. These include the three rare nsSNPs discovered by Exome chip analyses: variant rs61760904 at the *RRAS* locus, associated with systolic BP, with an effect size of 1.5 mmHg per allele; rs200999181 at the *COL21A1* locus associated with PP, with an effect size of 3.14 mmHg per allele and r235529250 at the *RBM47* locus associated with systolic BP, with an effect size of 1.61 mmHg per allele, and rs16833934 at the *MIR1263* locus [23]. The fourth SNP, rs16833934 at the *MIR1263* locus is associated with DBP. It is a common variant (MAF = 26%) and the effect size is 1.63 mmHg per allele. This association was discovered using a gene × age interaction analysis for BP. The SNP was significantly associated with BP amongst 20–29 year olds, within one of the specific age–bin-stratified subgroup analyses. Hence, the effect is more extreme within a particular age group, rather than at an overall population level [23].

In 2011, the percentage of BP variance explained for 29 BP SNPS was <1% for SBP and DBP [14]. In order to show the added contribution of the recent genetic discoveries to the percentage of variance explained, we have firstly recalculated these results from the 29 SNP genetic-risk score (GRS) and then extended the GRS model to include all currently known BP SNPs, using data from the 1958 Birth Cohort (*N* = 5,639). Of the 29 SNPs from the ICBP GRS, 26 were available in our data. For the full GRS, we included 143 linkage disequilibrium (LD)-filtered (r^2 < 0.2) variants that were available from the 163 known BP SNPs. The GRS was constructed according to the trait-increasing alleles and weighted by effect estimates from summary data. The 26 SNP GRS explained 0.67, 0.72, and 0.22% of the variance for SBP, DBP, and PP, respectively. This increases to 1.48, 1.66, and 0.47%, respectively, showing at least a twofold increase in the percentage variance explained due to the variants identified since 2011. We note that the percentage variance explained results in the 1958 Birth Cohort and those reported in the literature are variable, as results are population specific and differ according to the sample size tested. In general, we conclude, however, that the percentage variance explained overall is still very low, consistent with large numbers of common variants with weak effects.

NEW BP GENES AND MOLECULAR MECHANISMS

Over a hundred genomic regions have now been identified and one of the post-GWAS challenges is to move from a statistical SNP association to the identification of causal gene(s) and biological mechanism(s). This can be difficult due to LD, where SNPs are correlated with each other, so it is not clear which SNP is directly causing the phenotype. A first step is to annotate the associated SNPs and their proxies (SNPs in LD, $r^2 > 0.8$) using bioinformatics tools. The discovery of nsSNPs that are predicted to be damaging by more than one bioinformatics algorithm provides strong support for it being a causal candidate gene. The meta-analyses of Exome chip genotypes discovered three rare predicted damaging nsSNPs in *RRAS*, *RBM47*, and *COL21A1*. *RRAS* encodes a small GTPase, and it has been implicated in angiogenesis and actin cytoskeleton remodeling [38]. *RBM47* encodes the RNA-binding motif protein 47 and is responsible for posttranscriptional regulation of RNA [39]. *COL21A1* encodes the collagen alpha1 chain precursor of type XXI collagen. Type XX1 collagen is an extracellular matrix component of blood vessel walls [40]. Each gene presents as an excellent candidate for functional analysis.

Gene-based testing using Exome chip genotypes have revealed four further candidate causal genes. DBH encodes dopamine beta hydroxylase. This catalyzes the transformation of dopamine to norepinephrine, which plays an important role in the regulation of BP. *NPR1* encodes the receptor for atrial and B-type natriuretic peptides, which are known regulators of blood volume and BP. *PTPMT1* encodes a mitochondrial protein tyrosine phosphatase 1, which mediates the dephosphorylation of mitochondria proteins, and therefore, it has an important role in ATP production. *A2ML1* encodes alpha-2-macroglobulin like 1, which is thought to act as an inhibitor of several proteases, and mutations in this gene are associated with some cases of Noonan syndrome, a condition with cardiac abnormalities. Several other BP loci from published papers have also been found to contain SNPs that are predicted to affect protein coding.

The majority of BP-associated SNPs however are intronic or intergenic suggesting they are likely to exert their phenotypic effects by regulating the expression of nearby gene(s). Analysis of DNase I Hypersensitivity sites (DHSs), cis-acting expression SNPs (eSNPs), and methylation are some of the methods being deployed to prioritize candidate genes. A DHS is a mark of chromatin accessibility and is correlated with the presence of regulatory elements such as promoters and enhancers [41]. The Cardio-MetaboChip study investigators performed an analysis testing for enrichment of 66 BP SNPs (new and established loci) in a cell specific manner using the Epigenomics Roadmap and Encyclopedia of deoxyribonucleic acid (DNA) elements Consortium datasets [42]. The analysis indicated 7 out of the 10 cell types with the greatest relative enrichment of BP SNPs mapped to DHSs from blood vessels (vascular or microvascular endothelial cell-lines or cells), suggesting that functional variation in these cells may affect endothelial permeability or vascular smooth muscle cell contractility via multiple pathways.

The identification of eSNPs associated with transcript levels for nearby genes also provides insight into potential mechanisms of disease. These analyses do not distinguish the functional SNP but implicate the gene or genes that may be associated with the measured phenotype. The candidate SNPs identified in promoters or enhancers will require experimental work to prove causality. The majority of BP genetic studies have provided results from eSNP analyses; however, until very recently, there was a paucity of analyses that included cardiovascular tissues. With the Genotype-Tissue Expression (GTEx) project, we now have access to a broader range of tissues. We have performed an analysis including the lead SNPs at 125 BP loci using cis-eSNP data from GTEx. This analysis indicated that the lead SNP at 94/125 loci was an eSNP across a range of different tissues ($N = 44$).There were forty-four eSNPs in arterial tissues (aorta, coronary and tibia), out of which, these 14 were associated with expression levels of only one gene at the locus (*AMH, ARHGAP24, ARHGAP42, ARL3, CLCN6, FES, HRCT1, IRAK1BP1, IRF5, PDXK, PHACTR1, PSMD5, SLC5A11*, and *ULK4*), indicating possible candidate genes at these loci.

DNA methylation silences gene expression if located in promoter regions; therefore, association of testing of genetic variants with DNA methylation can also provide insights into mechanisms. The trans-ethnic GWAS by Kato *et al.* in 2015 tested for a relationship between BP SNPs and local DNA methylation (using 1 Mb window at the SNP) in peripheral blood from 1,904 South Asian individuals. They observed a two-fold enrichment for association between BP SNPs and methylation, with 28/35 BP SNPs having associations with one or more methylation markers. These data provide some support for a role of DNA methylation in BP regulation.

Alongside annotation of BP loci, several detailed functional studies have been done on new BP genes using cellular systems and animal models. These include analyses of: uromodulin (*UMOD*) [43], rho GTPase activating protein 42 (*ARHGAP42*) [44], and solute carrier family 39 (Zinc Transporter), member 8 (*SLC39A8*) [45]. These studies are providing new insights into BP pathways and mechanisms; uromodulin in sodium homeostasis, ARHGAP42 in cytoskeletal remodeling, by inhibiting RhoA in vascular smooth muscle cells, and SLC39A8 via an effect on intracellular cadmium accumulation and cell toxicity. We expect many more functional studies over the next few years.

BP VARIANTS AND ASSOCIATION WITH OTHER TRAITS AND OUTCOMES

A Heatmap plot illustrating SNP associations across the different quantitative BP traits (SBP, DBP and PP) is shown in Fig. 7.2. These results are from a pairwise-independent set of 68 of the 163 previously reported BP SNPs available

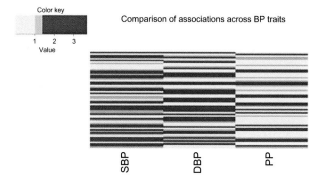

Comparison of associations across BP traits

Color key

1 2 3
Value

SBP   DBP   PP

FIGURE 7.2   Comparison of associations across different blood pressure traits. A Heatmap plot comparing the −log10(*p*-value) association results for systolic blood pressure (SBP), diastolic BP (DBP) and pulse pressure (PP). Association analysis was performed within 1958BC for $N = 5,639$ European individuals for 157 available SNPs from 1000 genomes imputed data. The Heatmap plot is restricted to 68 SNPs with $p < 0.1$ for at least one BP trait from the 1958BC analyses and furthermore restricted to 61 SNPs which are pairwise-LD-independent according to LD (Linkage Disequilibrium) of $r^2 < 0.2$. The SNPs on the Heatmap are ordered according to genomic position. The legend indicates the Heatmap color coding on the −log10 scale, corresponding to: Red = most significant ($0 < p < 0.05$), Orange = weak significance ($0.05 < p < 0.1$), Yellow = not significant ($p > 0.1$). Note: These *p*-value significance thresholds take into account the statistical power, with results only taken from one single study, in order to show the relative associations across traits.

in a single study with $p < 0.1$ ($N = 5,800$), and filtered by LD ($r^2 < 0.2$). We observe that some of the SNPs are associated with multiple BP traits, whereas the associations at other loci are trait specific. The three BP traits are highly correlated, but they measure partly distinct physiological features including cardiac output, vascular resistance, and arterial stiffness [46]. The analysis of related quantitative BP traits has led to an increased discovery of BP loci, which may in part be due to our limited knowledge about inter-individual variation of each trait. Furthermore, it is possible that the genetic associations that are observed with specific BP traits may implicate slightly different underlying biology and mechanisms.

Large-scale GWAS across many traits permits the cross checking of SNPs with a range of different phenotypes, so called "Phenome Scanning." This type of analysis may also provide some insights into disease pathways and biology [47]. We have performed a look up of the 163 reported BP SNPs at 125 BP loci using Phenoscanner (the published SNP only). We observed BP SNPs at 52 loci to be associated with at least one other trait at a genome-wide significance level ($p < 5 \times 10^{-8}$). The BP-associated variant at *SH2B3* (rs3184504) was the most pleiotropic, with 17 trait associations (CHD, myocardial infarction, type 1 diabetes, lipid levels, eosinophils, hypothyroidism, beta 2 microglubulin plasma levels, platelet counts, red blood cell traits, celiac disease, cystatin C, hematocrit levels, hemoglobin, juvenile idiopathic arthritis, generalized vitiligo, primary biliary cirrhosis, and selective immunoglobulin levels). To get an overview of BP-SNP associations across traits, we created five broad disease categories (Fig. 7.3). We observe several SNPs associated with traits in the CVD category as expected: 10 SNP associations with CHD or MI (*HDAC9, ZC3HC1, NOS3, ATP2B1, PHACTR1, CYP17A1, SH2B3, RPL6, FURIN-FES,* and *PROCR*), supporting the known causal relationship with BP, and one SNP association with stroke (*HDAC9*). There were five SNP associations with renal traits (*BAT2-BAT5, PRAKG2, SH2B3, UMOD,* and *TBX2*). Several of the BP SNPs are also associated with other CV risk factors, such as lipid levels and body mass index. Interestingly, there were also associations with non-cardiovascular traits such as autoimmune diseases, cancer, and schizophrenia. The multiple trait-SNP associations mirror the work by Andreassen et al. who also observed polygenic overlap between SBP and several related disorders [48]. These analyses suggest, there may be shared etiological relationship between phenotypes, although with overlapping samples across many of the GWAS it is also possible that lifestyle or non genetic factors are accounting for some of these relationships.

GRSs can also be used to assess the relationship between BP and other cardiovascular outcomes. Whilst each common variant individually only has a small-effect size, the overall predictive information can be increased by combining the effects of multiple genetic variants together into a risk score.

For example, even though single SNPs usually show less significant associations with hypertension compared to continuous BP traits, due to lower statistical power from case-control analyses, a GRS constructed from BP-associated variants is highly significant for association with hypertension as the binary clinical outcome. This was illustrated by ICBP in 2011 with the GRS of 29 SNPs associated with hypertension in an independent cohort ($p = 3.1 \times 10^{-33}$) [14]. We have created a GRS plot (Fig. 7.4) from 1958 Birth Cohort data incorporating all currently known BP variants (143 pairwise-independent available SNPs) using the average of the SBP-GRS and DBP-GRS (as used for percentage variance explained above). This plot demonstrates a positive trend between the increase in genetic risk and the proportion of hypertensive individuals within the population. As hypertension is defined according to thresholds of SBP and DBP levels, the association between a BP-GRS and hypertension may not seem surprising, and certainly as BP can be easily measured, genetic variants would never be needed as a diagnostic biomarker for hypertension. Similarly, as all known SNPs collectively still only explain a small proportion of the trait variance, there is limited predictive power to use a GRS to predict hypertension at the population level. However, we do see a large-fold increase in risk of hypertension between the extreme tails of the GRS distribution comparing the lowest risk group vs the highest risk group (Fig. 7.4). Indeed ICBP showed that the 29 SNP GRS is associated with an OR of 1.8 when comparing the top vs bottom quintiles of the GRS distribution, which increases to 2.09 for comparison of the deciles [14]. Therefore there is greater potential for clinical utility for

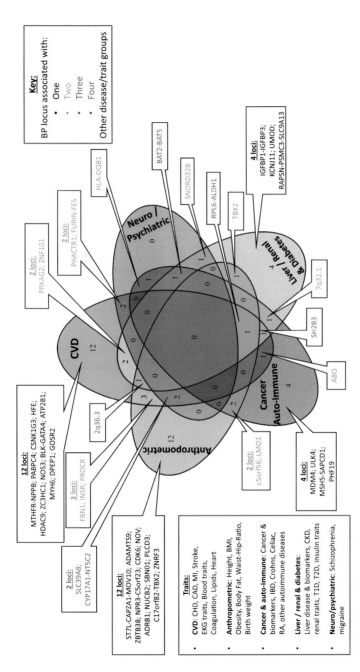

FIGURE 7.3   A Venn Diagram showing the associations of blood pressure loci with other disease traits. The Venn Diagram is divided into five different disease trait categories, which are defined in the "Traits" legend box. All 163 blood pressure (BP) associated variants were looked up directly in PhenoScanner (exact SNP only, no proxies) and 52 loci are found to be associated with other disease traits genome-wide significance level ($p < 5 \times 10^{-8}$). The PhenoScanner traits were restricted to disease-related traits which were associated with more than one BP locus. The key defines the color coding, which shows how many different disease trait groups each set of loci is associated with.

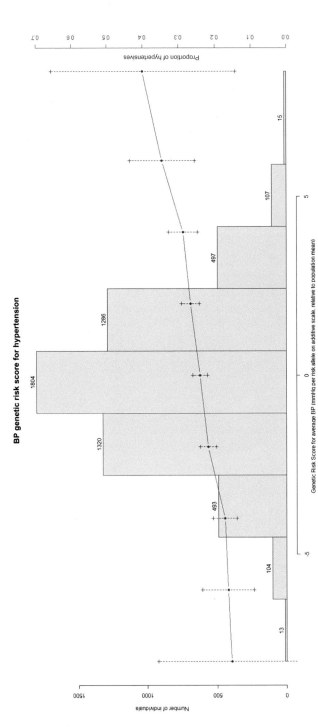

FIGURE 7.4 GRS and hypertension. The histogram shows the frequency distribution (left-hand y-axis) of the blood pressure (BP) genetic-risk sore (GRS) within N = 5,639 individuals from the 1958 Birth Cohort. Of the 163 currently known BP-associated variants, 158 were available from 1958 Birth Cohort 1000 Genomes imputed data, and these were further filtered by LD ($r^2 < 0.2$) to a total of 143 variants. Two GRS, for SBP and DBP, were constructed according to trait-increasing alleles and weighed by effect size estimates, and the mean of these scores was used as the average BP-GRS, which was centered at mean zero (x-axis). Within each GRS group the proportion of hypertensive individuals (right-hand y-axis) is indicated together with standard errors, in order to show the increase in hypertension risk as genetic risk increases.

either screening subgroups of high-risk individuals, or for use in GRS models for stratified risk prediction to identify patients at high risk of hypertension earlier in life, which may enable lifestyle intervention strategies. Nevertheless, recent research from a longitudinal study has shown that a BP-GRS is less associated with increases in BP levels over time and hence less predictive of incident hypertension in an older population [49]. Genetic-risk scores for BP have also been investigated for associations with other cardiovascular traits. ICBP showed that the 29-SNP GRS was associated with risk of stroke and coronary artery disease (CAD) [14], which under Mendelian Randomization assumptions provides evidence of a causal relationship between BP and stroke/CAD, and these associations have been further confirmed from the Cardio-MetaboChip analyses with an extended 66-SNP GRS [31]. However an analysis within a Swedish cohort showed that the same 29-SNP GRS adds very little improvement to the prediction of ischemic stroke over and above the traditional non genetic risk factors such as hypertension, diabetes, and smoking [50].

In contrast, neither the ICBP or Cardio-MetaboChip analyses have observed a significant association between a BP-GRS and kidney disease or renal function [14,31], implying a much weaker causal effect of BP on renal traits than on stroke or CAD. This could suggest that the associations previously reported from observational epidemiological studies may have been either biased by confounding factors, or actually been due to reverse causation, i.e., kidney function instead potentially causing changes in BP.

CONCLUSION

Modern GWAS and meta-analyses that combine data from tens of thousands of individuals have robustly identified and validated 163 genetic variants associated with BP. The majority of SNPs are common (MAF $\geq$ 5%), with relatively small-effect sizes (mostly ≤ 1 mmHg for SBP and ≤ 0.5 mmHg for DBP). The analyses thus far analyzing rare SNPs with BP traits have been disappointing, although we have only interrogated a small proportion of rare coding SNPs. The costs of whole genome sequencing are continually reducing, with costs in late 2015 being $1,500 per genome (National Health Genome Research Institute), and these are expected to decrease further with the significant influence on price from projects such as 100,000 Genomes Project [51], hence the analysis of whole genomes in larger populations will soon be a reality. There are also National Biobanks being established (for example in Estonia [52] and the United Kingdom [53] amongst others), which will permit large well-powered studies for the analysis of rare variants, and interactions. Gene–lifestyle interactions, relevant to underlying pathobiology, also offer an exciting opportunity to discover new BP loci. There are analyses underway in the CHARGE Gene–Lifestyle Interactions Project (>800,000 individuals) for BP vs smoking, alcohol intake and physical activity. The rare inherited disease program in the 100,000 Genomes Project will help to uncover further candidate BP genes, and from this, there is the expectation of new diagnostic tests in the shorter term, and the ultimate future goal of potential new drug development based on the function and pathways of genes which had no previously known links to BP or CVD.

ACKNOWLEDGMENTS

We wish to acknowledge the support of the NIHR Cardiovascular Biomedical Research Unit at Barts and The London, Queen Mary University of London, United Kingdom.

REFERENCES

[1] WHO, *Global status report on noncommunicable diseases.* 2014. p. 67–77.

[2] Lim SS, et al. A comparative risk assessment of burden of disease and injury attributable to 67 risk factors and risk factor clusters in 21 regions, 1990–2010: a systematic analysis for the Global Burden of Disease Study 2010. Lancet 2012;380(9859):2224–60.

[3] Rapsomaniki E, et al. Blood pressure and incidence of twelve cardiovascular diseases: lifetime risks, healthy life-years lost, and age-specific associations in 1.25 million people. Lancet 2014;383(9932):1899–911.

[4] Law MR, Morris JK, Wald NJ. Use of blood pressure lowering drugs in the prevention of cardiovascular disease: meta-analysis of 147 randomised trials in the context of expectations from prospective epidemiological studies. BMJ 2009;338:b1665.

[5] Chiong JR. Controlling hypertension from a public health perspective. Int J Cardiol 2008;127(2):151–6.

[6] Menni C, et al. Heritability analyses show visit-to-visit blood pressure variability reflects different pathological phenotypes in younger and older adults: evidence from UK twins. J Hypertens 2013;31(12):2356–61.

[7] Hirschhorn JN, Daly MJ. Genome-wide association studies for common diseases and complex traits. Nat Rev Genet 2005;6(2):95–108.

[8] Cho YS, et al. A large-scale genome-wide association study of Asian populations uncovers genetic factors influencing eight quantitative traits. Nat Genet 2009;41(5):527–34.

[9] Padmanabhan S, et al. Genome-wide association study of blood pressure extremes identifies variant near UMOD associated with hypertension. PLoS Genet 2010;6(10):e1001177.

[10] Cabrera CP, et al. Exploring hypertension genome-wide association studies findings and impact on pathophysiology, pathways, and pharmacogenetics. Wiley Interdiscip Rev Syst Biol Med 2015;7(2):73–90.

[11] Newton-Cheh C, et al. Genome-wide association study identifies eight loci associated with blood pressure. Nat Genet 2009;41(6):666–76.

[12] Levy D, et al. Genome-wide association study of blood pressure and hypertension. Nat Genet 2009;41(6):677–87.

[13] Ho JE, et al. Discovery and replication of novel blood pressure genetic loci in the Women's Genome Health Study. J Hypertens 2011;29(1):62–9.

[14] Ehret GB, et al. Genetic variants in novel pathways influence blood pressure and cardiovascular disease risk. Nature 2011;478(7367):103–9.

[15] Wain LV, et al. Genome-wide association study identifies six new loci influencing pulse pressure and mean arterial pressure. Nat Genet 2011;43(10):1005–11.

[16] Kato N. Ethnic differences in genetic predisposition to hypertension. Hypertens Res 2012;35(6):574–81.

[17] Takeuchi F, et al. Blood pressure and hypertension are associated with 7 loci in the Japanese population. Circulation 2010;121(21):2302–9.

[18] Kato N, et al. Meta-analysis of genome-wide association studies identifies common variants associated with blood pressure variation in east Asians. Nat Genet 2011;43(6):531–8.

[19] Franceschini N, et al. Genome-wide association analysis of blood-pressure traits in African-ancestry individuals reveals common associated genes in African and non-African populations. Am J Hum Genet 2013;93(3):545–54.

[20] Kato N, et al. Trans-ancestry genome-wide association study identifies 12 genetic loci influencing blood pressure and implicates a role for DNA methylation. Nat Genet 2015;47(11):1282–93.

[21] Johnson AD, et al. Association of hypertension drug target genes with blood pressure and hypertension in 86,588 individuals. Hypertension 2011;57(5):903–10.

[22] Ganesh SK, et al. Effects of long-term averaging of quantitative blood pressure traits on the detection of genetic associations. Am J Hum Genet 2014;95(1):49–65.

[23] Simino J, et al. Gene-age interactions in blood pressure regulation: a large-scale investigation with the CHARGE, Global BPgen, and ICBP Consortia. Am J Hum Genet 2014;95(1):24–38.

[24] Zhu X, et al. Meta-analysis of correlated traits via summary statistics from GWASs with an application in hypertension. Am J Hum Genet 2015;96(1):21–36.

[25] Keating BJ, et al. Concept, design and implementation of a cardiovascular gene-centric 50 k SNP array for large-scale genomic association studies. PLoS One 2008;3(10):e3583.

[26] Johnson T, et al. Blood pressure loci identified with a gene-centric array. Am J Hum Genet 2011;89(6):688–700.

[27] Ganesh SK, et al. Loci influencing blood pressure identified using a cardiovascular gene-centric array. Hum Mol Genet 2013;22(8):1663–78.

[28] Tragante V, et al. Gene-centric meta-analysis in 87,736 individuals of European ancestry identifies multiple blood-pressure-related loci. Am J Hum Genet 2014;94(3):349–60.

[29] Fox ER, et al. Association of genetic variation with systolic and diastolic blood pressure among African Americans: the Candidate Gene Association Resource study. Hum Mol Genet 2011;20(11):2273–84.

[30] Voight BF, et al. The metabochip, a custom genotyping array for genetic studies of metabolic, cardiovascular, and anthropometric traits. PLoS Genet 2012;8(8):e1002793.

[31] Ehret G, et al. The genomics of blood pressure regulation and its target organs from association studies in 342,415 individuals. Nat Genet 2016;48:1171–84.

[32] Schork NJ, et al. Common vs. rare allele hypotheses for complex diseases. Curr Opin Genet Dev 2009;19(3):212–9.

[33] Suvendran P, et al. Large scale trans-ethnic meta-analyses identify novel rare and common variants associated with blood pressure and hypertension. Nat Genet 2016;48:1151–61.

[34] Liu C, et al. Meta-analysis identifies common and rare variants influencing blood pressure and overlapping with metabolic trait loci. Nat Genet 2016;48(10):1162–70.

[35] Oliver PM, et al. Natriuretic peptide receptor 1 expression influences blood pressures of mice in a dose-dependent manner. Proc Natl Acad Sci U S A 1998;95(5):2547–51.

[36] Li B, Leal SM. Methods for detecting associations with rare variants for common diseases: application to analysis of sequence data. Am J Hum Genet 2008;83(3):311–21.

[37] Wu MC, et al. Rare-variant association testing for sequencing data with the sequence kernel association test. Am J Hum Genet 2011;89(1):82–93.

[38] Wozniak MA, et al. R-Ras controls membrane protrusion and cell migration through the spatial regulation of Rac and Rho. Mol Biol Cell 2005;16(1):84–96.

[39] Guan R, et al. rbm47, a novel RNA binding protein, regulates zebrafish head development. Dev Dyn 2013;242(12):1395–404.

[40] Tuckwell D. Identification and analysis of collagen alpha 1(XXI), a novel member of the FACIT collagen family. Matrix Biol 2002;21(1):63–6.

[41] Cockerill PN. Structure and function of active chromatin and DNase I hypersensitive sites. FEBS J 2011;278(13):2182–210.

[42] Tak YG, Farnham PJ. Making sense of GWAS: using epigenomics and genome engineering to understand the functional relevance of SNPs in non-coding regions of the human genome. Epigenetics Chromatin 2015;8:57.

[43] Padmanabhan S, et al. Uromodulin, an emerging novel pathway for blood pressure regulation and hypertension. Hypertension 2014;64(5):918–23.

[44] Bai X, et al. The smooth muscle-selective RhoGAP GRAF3 is a critical regulator of vascular tone and hypertension. Nat Commun 2013;4:2910.

[45] Zhang R, et al. A blood pressure-associated variant of the SLC39A8 gene influences cellular cadmium accumulation and toxicity. Hum Mol Genet 2016.

[46] Franklin SS, et al. Single versus combined blood pressure components and risk for cardiovascular disease: the Framingham Heart Study. Circulation 2009;119(2):243–50.

[47] Staley JR, et al. PhenoScanner: a database of human genotype-phenotype associations. Bioinformatics 2016.

[48] Andreassen OA, et al. Identifying common genetic variants in blood pressure due to polygenic pleiotropy with associated phenotypes. Hypertension 2014;63(4):819–26.

[49] Niiranen TJ, et al. Prediction of blood pressure and blood pressure change with a genetic risk score. J Clin Hypertens (Greenwich) 2016;18(3):181–6.

[50] Fava C, et al. A genetic risk score for hypertension associates with the risk of ischemic stroke in a Swedish case-control study. Eur J Hum Genet 2015;23(7):969–74.

[51] Peplow M. The 100,000 genomes project. BMJ 2016;353:i1757.

[52] Keis A. Biobanking in Estonia. J Law Med Ethics 2016;44(1):20–3.

[53] Sudlow C, et al. UK biobank: an open access resource for identifying the causes of a wide range of complex diseases of middle and old age. PLoS Med 2015;12(3):e1001779.

Chapter 8

Coronary Artery Disease and Myocardial Infarction

Themistocles L. Assimes

Stanford University School of Medicine, Stanford, CA, United States

Chapter Outline

INTRODUCTION

Coronary artery disease (CAD) is a complex trait whose incidence has fallen substantially over the last four decades due to major advances in the prevention, diagnosis, and treatment of this condition. These advances have largely resulted from the identification of the major "traditional" risk factors of CAD, the implementation of coronary care units, and the development of effective

Genomic and Precision Medicine. DOI: http://dx.doi.org/10.1016/B978-0-12-800685-6.00008-4
127

pharmacological and mechanical interventions [1]. Despite these advances, the overall prevalence of CAD is expected to continue to increase over the next couple of decades in both developed and developing countries due to the aging population in the former and the completion of the epidemiological transition in the latter [1]. Thus, many people will continue to suffer from the consequences of CAD and the condition is expected to remain the leading cause of morbidity and mortality worldwide [1]. Furthermore, the case-fatality rate of CAD remains high with approximately one in three subjects dying from their first myocardial infarction (MI) even in the developed nations [2].

Recent progress in genomic medicine steered by astonishing breakthroughs in laboratory technology and computing power provide us with an opportunity to build on the advances made to date through a better understanding of the genetic underpinnings of this condition. Such an understanding is expected to ultimately translate not only to an improved ability to identify individuals at high risk but also to develop novel therapies capable of further curbing the incidence of this lethal condition. In this chapter, we review the remarkable progress made in the identification of the genetic determinants of CAD and its complications in the last decade, providing a historical context where appropriate.

PATHOPHYSIOLOGY OF CAD AND ITS MAJOR COMPLICATIONS

CAD results from excessive cholesterol buildup within the wall of the major arteries that supply blood to the heart [1]. The process of plaque formation begins in adolescence in many different parts of the coronary tree concurrently [1]. However, the rate of plaque buildup differs substantially among individuals and is driven by several well-established risk factors including older age, male sex, smoking, poor dietary patterns, obesity, adverse blood lipid profile, elevated blood pressure, diabetes, and physical inactivity [1,3]. Individuals who do not have any of the modifiable subset of risk factors usually develop very few plaques during their lifetimes and their plaques remain relatively small. On the other hand, individuals with multiple risk factors develop more plaques that tend to also grow at a faster rate putting them at risk of complications later in life.

A dreaded complication of CAD is an MI, which occurs when the fibrous cap of a plaque that has reached a critical size ruptures and spills some or all of its contents into the blood stream [1]. These contents do not only include cholesterol crystals but also a variety of proteins that may vigorously activate the coagulation system leading to the formation of an occlusive thrombosis [1]. If blood flow is not restored either spontaneously or through a medical intervention within about 4 h of an acute atherothrombotic occlusion, death of cardiac muscle tissue will ensue [1]. Fatal MIs occur when plaques coincidentally rupture in a critical region of the arterial tree (e.g., near the top of one of the three arteries) leading to either overwhelming pump failure or a lethal arrhythmia within ischemic myocardial tissue [1]. It is currently not possible to predict when a specific plaque will rupture and its consequences, but—in general—the more plaques you have, and the greater the volume of these plaques, the higher the probability that at least one will rupture and lead to an MI [1]. The pathophysiology of coronary atherosclerosis can thus be broadly divided into three phases: (1) deposition of plaque within the arterial wall that occurs over several decades, (2) the episodic rupture of the wall overlying a plaque, and (3) the hemostatic response to that rupture. For decades, scientists have been interested in identifying genetic susceptibility to one or more of these phases of coronary atherosclerosis with the hope that such knowledge will eventually translate to more effective ways of preventing the clinical complications of CAD [1]. The proof of principle in this respect was established early for CAD and MI through the identification of the key role of cholesterol metabolism in the pathophysiology of CAD in the 1950s and 1960s, the realization that genetic variants within this pathway modify levels of cholesterol as well as the risk of CAD in the 1960s and 1970s, and the demonstration that cholesterol altering agents such as the statins were capable of substantially lowering not only cholesterol levels but also the risk of CAD in the 1970s and 1980s [4].

THE GENETIC BASIS OF CAD

Evidence of Heritability

Families demonstrating a strong penetrance of complications related to CAD with a Mendelian or near-Mendelian mode of inheritance exist but are not common [5]. For many of these families, the manifestation of CAD is linked to one or more damaging mutations within genes that are key regulators of circulating levels of low-density lipoprotein (LDL) (i.e., familial hypercholesterolemia related genes) [5]. These mutations lead to profoundly elevated LDL levels from birth followed by premature clinical CAD. However, a vast majority of the burden of CAD occurs outside of this setting within families where the penetrance for any given causal genetic variant is substantially lower than that observed for Mendelian traits [5].

A range of methods has been used to estimate the fraction of the variation in rates of CAD that can be attributed to inherited susceptibility (i.e., genetic) factors versus the environment. Early studies of migrants in the 1960s such as the Ni–Hon–San Study documented how disease rates among migrants of similar genetic background quickly approached those of the host population [6]. Unfortunately, these observations were misinterpreted for many years as evidence that CAD had little to no genetic basis.

Investigators subsequently estimated the contribution of genetic factors by comparing rates of CAD among subjects with a family history of CAD in a first degree relative to subjects without such a history through "familial aggregation" studies [7,8]. Several of these studies documented a threefold to fourfold increase in the risk of CAD among individuals with a family history, which was mitigated to about 1.5 to 2.5 fold after adjusting for traditional risk factors suggesting the presence of a genetic basis to CAD that is completely independent of that related to heritable risk factors. These studies also provided compelling evidence that a stronger family history or an earlier age of onset of disease in the family member further increased risk for the individual.

A chief criticism of familial aggregation studies is their potential to overestimate the contribution of genetic factors due to confounding from a shared environment (e.g., smoking, diet, pollution, and exercise) [9]. In response to this concern, several twin studies were conducted [10–12]. The major advantage of twin studies is their ability to largely neutralize the effect of the common familial environment by comparing the concordance rates of disease among monozygotic twins, which share 100% of their genetic makeup, to that among dizygotic twins, which share only 50% of their genetic makeup [9]. Differences in concordance rates can then be transformed to unbiased estimates of heritability assuming the degree to which environmental determinants are shared between each set of twins is approximately equal [9]. In general, twin studies of northern Europeans have estimated the heritability of CAD at approximately 50% in both males and females with a higher heritability for clinical disease that occurs at younger age. Furthermore, this heritability is only partially a consequence of the heritability of traditional risk factors.

Population Attributable Risk versus Heritability

The population attributable risk (PAR) indicates the number (or fraction) of cases that would be prevented in a population if a risk factor (or a set of risk factors) was eliminated [13]. Some studies have reported a very high PAR for all "modifiable" traditional risk factors of CAD suggesting that CAD can be eradicated simply by modifying all of these risk factors [14,15]. These studies inadvertently imply that the search for additional risk factors for CAD is unnecessary as a majority of risk factors have already been uncovered [15]. Several caveats to this interpretation need to be considered [16]. First, the PAR can be strongly influenced by how the reference group is defined for a "modifiable" trait. Thus, the PAR for cholesterol can be arbitrarily greatly increased simply by defining the target level of subjects for cholesterol as the group with extremely low cholesterol. For example, the PAR when the bottom decile of LDL is selected as the reference "healthy" group can quickly approach 100% simply because this reference group rarely suffers from CAD. Second, several of the risk factors are highly heritable themselves, and it is unrealistic to assume we can modify some of these risk factors through lifestyle changes alone in a way that would result in the entire population moving into the reference group. More likely, lifelong pharmacotherapy of most of the population would be required to accomplish this feat. Third, a high PAR for a single, or a collection, of risk factors is still compatible with the presence of many yet undiscovered genetic and environmental risk factors. Traditional risk factors appear necessary but often are not sufficient to result in clinical disease given most events occurs in subjects with one only one or two of these risk factors and many subjects with at least one risk factor do not suffer from a clinical event [17]. As a result, the sum of the PAR fraction of individual risk factors, or independent sets of risk factors, can equal well over 100 and the high PAR for traditional risk factors should not be misinterpreted as a lack of presence of other risk factors including genetic risk factors. A substantial fraction of the variability of risk of CAD remains unexplained even after considering all established risk factors.

Linkage and Candidate Gene Association Studies

Studies aimed at the identification of genetic determinants of complex traits can be classified into two broad categories, linkage and association [18]. Linkage studies examine the pattern of segregation of a section of the genome with disease among family members while association studies focus on the correlation between the frequency of a variant and disease in cases and controls (or noncases) and typically involve unrelated individuals [18]. Traditional linkage studies are not well suited to the study of traits lacking a very strong genetic effect at one or more loci (i.e., risk ratio >4) [18]. The moderate heritability of CAD is a result of allelic variation at multiple loci each with small to moderate effects [5]. The principal initial observation that supported this theory was the inability of linkage studies performed in the 1980s to detect novel susceptibility loci for CAD with strong effects [5]. This initial observation was subsequently confirmed through findings of modest effects observed in large-scale association studies described in the next section. Provided certain assumptions are met, association studies comparing unrelated cases to controls represent the most efficient way to detect the modest genotypic effects expected in complex traits [18].

Despite the theoretical advantages of association over linkage in detecting small to modest genetic effects of complex traits, most candidate gene case–control association studies of complex traits conducted in the late 1990s and early 2000s just prior to the Genome-Wide Association Study (GWAS) era were disappointing as many initially positive reports could not be replicated in other cohorts [18]. The reasons for the overall failure of candidate gene studies are many, but stemmed primarily from the very low pretest probability that a given variant or gene examined contributed to the susceptibility of a complex trait combined with sample sizes that were underpowered to detect modest genetic effects and inadequate adjustment of p value thresholds for multiple testing [18].

The Genome-Wide Association Study (GWAS) Era

The GWAS design calls for the use of high-throughput genotyping platforms to genotype as many single nucleotide polymorphisms (SNPs) in the genome as possible in a set of cases and controls irrespective of the location of these SNPs relative to genes [18]. In this respect, GWAS is similar to linkage because no prior information on gene function is necessary to select a region to be studied.

However, GWAS is unlike linkage because it maintains the ability to detect modest genotypic effects. Furthermore, it localizes initial genetic signals to a much smaller genomic region compared to linkage. Starting about 2005, the GWAS design became analytically and economically feasible primarily as a result of whole genome resequencing efforts by the International HapMap project in a large number of racially diverse subjects [19]. These efforts produced genotypes, frequencies, and assay information on close to 6 million SNPs out of the estimated 11–15 million SNPs in the human genome with a minor allele frequency of >1%. In turn, this information was mobilized by two companies, Affymetrix and Ilumina, to develop high-throughput and highly parallel genotyping technology that allowed for up to 1 million SNPs to be genotyped in a single individual at a fraction of the cost per genotype compared to other reliable platforms used for candidate gene studies [20]. More recent resequencing efforts in thousands of individuals have led to the cataloging of many more millions of SNPs that are less common in frequency [21]. A subset of these SNPs has subsequently been incorporated into second and third generation chips that include up to 5 million SNPs.

Many reports of successful GWA studies localizing common genetic variants influencing various complex traits have been published over the last decade [22]. Although studies have differed in size and genotyping platforms used, the major findings have been convincingly reproduced. The main reasons for this success stem from the early adoption by the scientific community of genetic epidemiologists, statistical geneticists, and journal editors of strict criteria for establishing positive replication of associations uncovered through GWAS in the context of testing hundred of thousands of SNPs concurrently [23]. These criteria included the need for very large sample sizes and stringent p value cutoffs that adjusted for at least 1 million independent tests ($p < 5 \times 10^{-8}$) [23]. Thus, they collectively represent an important proof of principle that GWAS can identify common variants that contribute to common disease.

The first two GWAS for CAD were published in 2007 [24,25]. Both groups of investigators identified common polymorphisms at a locus on 9p21 near the genes *CDKN2A* and *CDKN2B* to be associated with clinical complications of CAD as well as subclinical coronary atherosclerosis. The variants were found to increase the risk of clinical CAD by approximately 20–30% for each high-risk allele with higher risks observed in subjects with onset of disease at a younger age. Furthermore, the same variants were not found to be associated with any traditional risk factors. These associations were subsequently widely replicated in European as well as in East Asian populations and still represent the strongest associations identified to date between common polymorphisms and CAD through GWAS [26–29].

Several consortia of investigators including the Welcome Trust Case-Control Consortium, the German MI Family Study, and the Myocardial Infarction Genetics Consortium subsequently identified an additional 11 loci harboring at least one common genetic variant reaching genome-wide significance through meta-analysis in European populations conducted between 2007 and 2009 [30–34]. These studies were followed in 2011 by the identification of another 17 loci by

The Coronary Artery Disease (C4D) Genetics and the Coronary ARtery DIsease Genome-wide Replication and Meta-analysis (CARDIoGRAM) consortia [35,36]. Both of these consortia then joined to form the CARDIoGRAMplusC4D consortium, which conducted further large-scale meta-analysis that included many additional samples genotyped with the Metabochip, a low cost custom chip that incorporated the most promising signals from the CARDIoGRAM study [37]. These efforts resulted in the identification of an additional 15 loci for clinical CAD in 2013 involving up to 63,746 cases and 130,681 controls [38]. Last, the CARDIoGRAMplusC4D consortium recently identified an additional 8 loci using the additive model and 2 loci using the recessive model through a meta-analysis of 48 studies involving 60,801 cases and 123,504 controls after further improving the imputation of all common SNPs using the 1000 genomes reference panel [39]. Of note, effect sizes of new polymorphisms reaching genome-wide significance have on average decreased as the sample sizes for discovery have increased with recent discoveries largely clustered around odds ratios of 1.05 [36,39].

The 58 susceptibility loci for CAD identified to date by these consortia have been identified in populations of European and, to a lesser degree, South Asian ancestry. Arguably, the next best-studied race/ethnic group is the East Asians where at least five GWAS have been reported involving Chinese, Japanese, and Korean populations [27–29,40,41]. All studies used modest sized discovery cohorts with sample sizes ranging from ~200 to 2100 cases and ~1300 to 5000 controls followed by targeted replication of both known and novel SNPs in several thousand additional cases and controls. The first East-Asian locus was reported in 2011 at *C6orf105* on chromosome 6p24 among a Chinese Han population. The effect size reported for the lead SNP at this locus remains the highest observed to date among CAD loci with an estimated odds ratio of 1.51 [42]. However, this finding could not be replicated in an independent GWAS that was very well powered to detect this effect and in general has not been widely replicated in any race/ethnic group [29]. Out of 12 additional loci reaching genome-wide significance to date, at least six overlap substantially with loci identified in European populations including *CDKN2A-CDKN2B, BRAP-ALDH2, GUCY1A3, TCF21, PHACTR1, and ATP2B1*. Lee et al. uncovered one additional novel locus near *MYL2*, although the independence of this locus is questionable given its proximity to *BRAP* [27]. Thus, both ethnic specific as well as cross-ethnic loci have been uncovered to date despite very modest discovery sample sizes among the East Asian studies. The yield of CAD loci among admixed race/ethnic groups such as African-Americans and Hispanics has been substantially less with only one locus reported to date through GWAS in African Americans near *CDK14* [43]. To date, this locus has not been reported in other race/ethnic groups. While evidence firmly implicating most CAD loci uncovered in European populations is largely lacking in non-European race/ethnic groups at the present time, larger scale studies of non-European populations are expected to eventually identify susceptibility polymorphisms within a large fraction of all GWAS loci for CAD uncovered to date [44]. A summary of all CAD loci uncovered to date as well as the effect sizes can be found in Table 8.1.

TABLE 8.1 Susceptibility Loci for Coronary Artery Disease and/or Myocardial Infarction Identified to Date Through Genome Wide Association Studies

CHR	Closest Gene(s)[a]	Possible Mechanisms of Association[b]	Lead SNP[c]	High Risk Allele Frequency	Odds Ratio	Race/Ethnic Group
Cholesterol/Lipoprotein Metabolism						
1	*PCSK9*	Regulation of LDL receptor recycling, lipid GWAS	rs11206510	0.85	1.08	European/South Asian
1	*SORT1*	Regulate apoB secretion and LDL catabolism, lipid GWAS	rs599839	0.78	1.11	European/South Asian
2	*APOB*	Major apolipoprotein of LDL, lipid GWAS	rs515135	0.78	1.07	European/South Asian
2	*ABCG5/G8*	Cholesterol absorption and secretion, lipid GWAS	rs6544713	0.29	1.05	European/South Asian
6	*SLC22A3-LPAL2-LPA*	Lipoprotein(a), lipid GWAS	rs2048327	0.35	1.06	European/South Asian
8	*LPL*	Lipolysis of TG-rich lipoproteins	rs264	0.85	1.06	European/South Asian
8	*TRIB1*	TG, MAPK signaling, SMC proliferation, lipid GWAS	rs2954069	0.55	1.04	European/South Asian
9	*ABO*	DVT, Factor VIII, vWF, ICAM-1, IL-6, E-selectin, LDL-C levels, lipid GWAS	rs579459	0.21	1.08	European/South Asian
6	*ANKS1A*	May inhibit PDGF-induced mitogenesis, lipid GWAS	rs17609940	0.82	1.03	European/South Asian
11	*ZNF259-APOA5-APOC3*	TG-rich lipoprotein metabolism, lipid GWAS	rs964184	0.18	1.05	European/South Asian
12	*SH2B3*	Negative regulator of cytokine signaling, lipids GWAS	rs3184504	0.42	1.07	European/South Asian

	Gene	Function	SNP			Population
12	BRAP-ALDH2	Immune system, signal transduction by GPCR, lipid GWAS	rs671	0.23	1.43	East Asian-Japanese
19	LDLR	LDL clearance, lipid GWAS	rs1122608	0.77	1.08	European/South Asian
19	APOE	LDL and VLDL clearance	rs4420638	0.17	1.1	European/South Asian
Blood Pressure						
4	GUCY1A3	Nitric oxide signaling, BP GWAS	rs7692387	0.8	1.07	European/South Asian
4	GUCY1A3	Nitric oxide signaling, BP GWAS	rs1842896	0.76	1.14	East Asian-Chinese Han
7	ZC3HC1	Encodes NIPA, regulator of cell proliferation	rs11556924	0.69	1.08	European/South Asian
7	NOS3	Production of nitric oxide, BP GWAS	rs3918226	0.06	1.14	European/South Asian
10	CYP17A1-CNNM2-NT5C2	CYP17A1: steroidogenic pathway, BP GWAS	rs12413409	0.89	1.08	European/South Asian
12	ATP2B1	Intracellular calcium homeostasis, BP GWAS	rs7136259	0.43	1.04	European/South Asian
12	ATP2B1	Intracellular calcium homeostasis, BP GWAS	rs7136259	0.39	1.11	East Asian-Chinese Han
12	SH2B3	Negative regulator of cytokine signaling, BP GWAS	rs3184504	0.42	1.07	European/South Asian
12	BRAP-ALDH2	Immune system, signal transduction by GPCR, aldehyde dehydrogenase (NAD) activity, BP GWAS	rs671	0.23	1.43	East Asian-Japanese
15	FURIN	Endoprotease—TGF-β1 precursor and type I MMP, BP GWAS	rs17514846	0.44	1.05	European/South Asian

(Continued)

TABLE 8.1 Susceptibility Loci for Coronary Artery Disease and/or Myocardial Infarction Identified to Date Through Genome Wide Association Studies (Continued)

CHR	Closest Gene(s)[a]	Possible Mechanisms of Association[b]	Lead SNP[c]	High Risk Allele Frequency	Odds Ratio	Race/Ethnic Group
Coagulation						
6	PLG	Fibrinolysis	rs4252120	0.74	1.03	European/South Asian
9	ABO	DVT, Factor VIII, vWF, ICAM-1, IL-6, E-selectin, LDL levels	rs579459	0.21	1.08	European/South Asian
Obesity						
12	KSR2	Suppressor of Ras2–cell proliferation; hyperphagia, obesity, insulin resistance	rs11830157	0.36	1.12[d]	European/South Asian
18	PMAIP1-MC4R	PMAIP1: HIF1A-induced proapoptotic gene; MC4R: leptin signaling, BMI GWAS	rs663129	0.26	1.06	European/South Asian
Arterial Vessel Wall—Smooth Muscle Cell						
4	REST-NOA1	maintains VSMCs in quiescent state, mitochodrial respiration and apoptosis	rs17087335	0.21	1.06	European/South Asian
4	EDNRA	Receptor for endothelin—vasoconstriction	rs1878406	0.16	1.06	European/South Asian
6	TCF21	Transcriptional regulator	rs12190287	0.64	1.06	European/South Asian

6	TCF21	Transcriptional regulator	rs12524865	0.61	1.11	East Asian-Chinese Han
7	HDAC9	Represses MEF2 activity/beige adipogenesis	rs2023938	0.1	1.08	European/South Asian
9	CDKN2BAS	Cellular proliferation, platelet function, abdominal aortic and intracranial aneurysms GWAS	rs10757274	0.48	1.21	European/South Asian
9	CDKN2BAS	Cellular proliferation, platelet function	rs10757274	0.46	1.37	East Asian-Chinese Han
10	CYP17A1-CNNM2-NT5C2	CYP17A1: steroidogenic pathway, intracranial aneurysm GWAS	rs12413409	0.89	1.08	European/South Asian
11	SWAP70	Leukocyte and VSMC migration and adhesion	rs10840293	0.55	1.06	European/South Asian
15	ADAMTS7	Proliferative response to vascular injury	rs7173743	0.56	1.08	European/South Asian
Arterial Vessel Wall—Endothelial Cell						
6	PHACTR1	Regulates protein phosphatase 1 activity, migraine GWAS	rs12526453	0.71	1.1	European/South Asian
6	PHACTR1	Regulates protein phosphatase 1 activity	Rs9349379	0.74	1.15	East Asian-Chinese Han
13	FLT1	VEGFR family; angiogenesis	Rs9319428	0.31	1.04	European/South Asian
15	MFGE8-ABHD2	VEGF dependent neovascularization secreted by macrophages	rs8042271	0.9	1.1	European/South Asian
17	BCAS3	Rudhira—endothelial cell polarity and angiogenesis	rs7212798	0.15	1.08	European/South Asian

(Continued)

TABLE 8.1 Susceptibility Loci for Coronary Artery Disease and/or Myocardial Infarction Identified to Date Through Genome Wide Association Studies (Continued)

CHR	Closest Gene(s)[a]	Possible Mechanisms of Association[b]	Lead SNP[c]	High Risk Allele Frequency	Odds Ratio	Race/Ethnic Group
Cell Growth/Differentiation/Apoptosis						
1	IL6R	IL-6 receptor, immune response	rs4845625	0.46	1.05	European/South Asian
2	TTC32-WDR35	Cell cycle progression, signal transduction, apoptosis and gene regulation	rs2123536	0.39	1.12	East Asian–Chinese Han
2	ZEB2-AC074093.1	ZEB2-transcriptional repressor	rs2252641	0.44	1.03	European/South Asian
3	MRAS	Cell growth and differentiation	rs9818870	0.15	1.07	European/South Asian
4	REST-NOA1	Maintains VSMCs in quiescent state, mitochodrial respiration and apoptosis	rs17087335	0.21	1.06	European/South Asian
4	GUCY1A3	Nitric oxide signaling	rs7692387	0.8	1.07	European/South Asian
4	GUCY1A3	Nitric oxide signaling	rs1842896	0.76	1.14	East Asian–Chinese Han
7	CDK14	Serine/threonine-protein kinase involved in the control of the eukaryotic cell cycle	rs1859023	0.69	1.39	African American
9	CDKN2BAS	Cellular proliferation, platelet function	rs10757274	0.48	1.21	European/South Asian
9	CDKN2BAS	Cellular proliferation, platelet function	rs10757274	0.46	1.37	East Asian–Chinese Han

Chr	Gene	Function	SNP			Population
11	PDGFD	Role in SMC proliferation	rs974819	0.33	1.07	European/South Asian
15	SMAD3	Downstream mediator of TGF-β signaling	rs56062135	0.79	1.07	European/South Asian
17	UBE2Z	Protein ubiquination; apoptosis	rs46522	0.51	1.04	European/South Asian
Intercellular Matrix						
1	MIA3	Collagen secretion	rs17465637	0.73	1.08	European/South Asian
13	COL4A1/A2	Type IV collagen chain of basement membrane	rs4773144	0.43	1.05	European/South Asian
			rs9515203	0.76	1.07	European/South Asian
Inflammation/Immune System/Cell Migration–Adhesion						
1	PPAP2B	Regulation of cell–cell interactions	rs17114036	0.92	1.13	European/South Asian
1	IL6R	IL-6 receptor, immune response	rs4845625	0.46	1.05	European/South Asian
6	C6orf10-BTNL2	T-cell costimulatory molecule, immune related diseases including Kawasaki's vasculitis	rs9268402	0.59	1.16	East Asian-Chinese Han
6	HLA, DRB-DQB	Major Histocompatibility complex region	rs11752643	0.06	1.26	East Asian-Japanese
10	CXCL12	Endothelial regeneration; neutrophil migration	rs501120	0.81	1.08	European/South Asian
10	KIAA1462	Component of endothelial cell–cell junctions	rs2505083	0.4	1.07	European/South Asian
10	LIPA	Monocyte Intracellular hydrolysis of cholesteryl esters	rs1412444	0.37	1.07	European/South Asian
11	SWAP70	Leukocyte and VSMC migration and adhesion	rs10840293	0.55	1.06	European/South Asian

(Continued)

TABLE 8.1 Susceptibility Loci for Coronary Artery Disease and/or Myocardial Infarction Identified to Date Through Genome Wide Association Studies (Continued)

CHR	Closest Gene(s)[a]	Possible Mechanisms of Association[b]	Lead SNP[c]	High Risk Allele Frequency	Odds Ratio	Race/Ethnic Group
Other						
2	LINC00954	LncRNA of unknown function	rs16986953	0.07	1.09	European/South Asian
2	VAMP5/8-GGCX	Intracellular vesicle trafficking	rs1561198	0.47	1.06	European/South Asian
2	WDR12	Component of nucleolar protein complex	rs6725887	0.14	1.14	European/South Asian
3	PLCL2	Calcium ion binding and phosphoric diester hydrolase activity	rs4618210	0.44	1.1	East Asian-Japanese
5	LOC105374626-LOC105374627	Unknown	rs11748327	0.76	1.25	East Asian-Japanese
5	SLC22A4/A5	Organic cation transporter	rs273909	0.14	1.06	European/South Asian
6	KCNK5	Potassium channel protein	rs10947789	0.78	1.05	European/South Asian
14	HHIPL1	Unknown	rs2895811	0.41	1.04	European/South Asian

17	SMG6	Role in nonsense mediated RNA decay	rs216172	0.35	1.05	European/South Asian
17	RAI1-PEMT-RASD1	PEMT encoded protein converts phosphatidylethanolamine to phosphatidylcholine	rs12936587	0.61	1.03	European/South Asian
19	AP3D1-DOT1L	Binding and protein transporter activity, transcription factor binding, histone methyltransferase activity	rs3803915	0.19	1.12	East Asian-Japanese
19	ZNF507-LOC400684	Unknown	rs12976411	0.91	1.49[d]	European/South Asian
21	KCNE2	Maintains cardiac electric stability	rs9982601	0.13	1.12	European/South Asian
22	POM121L9P-ADORA2A	ncRNA POM121, Adenosine A2a receptor: infarct-sparing effects	rs180803	0.97	1.2	European/South Asian

Adapted from McPherson R, Tybjaerg-Hansen A. Circ Res 2016; 118(4): 564–578, with permission.

[a]This indicates one to three of the closest genes in the region around the lead SNP. In some regions and for some race/ethnic groups, many more genes can be mapped to one or more lead SNPs in the region that are associated with CAD. The actual causal gene for many loci is not definitively known. For some loci, there may be more than one causal gene in a region affected by the genetic variation. Some loci are listed more than once either because they have been identified in more than one race/ethnic group or because they may have more than one potential mechanism of action.

[b]These mechanisms reflect was is known about the function of the mapped genes as well as results for the region and/or lead SNP in GWAS of traits related to CAD including established risk factors. Some of the loci can be classified into multiple potential mechanisms of action. For some of these loci, the causal gene may end up being further away and not at all reflect the listed potential mechanism.

[c]The lead SNP in a region frequently changes when additional studies are added or new SNPs are tested through genotyping or imputation. For the signals in European/South Asians, the lead SNPs largely reflect the lead SNPs for these regions reported by CARDIOGRAMplusC4D report published in 2015. Although the lead SNP may change they are generally highly correlated to each other. Non-European/South Asian loci are highlighted in dark red font.

[d]These ORs were significant only in the recessive model.

Heritability Revisited

The location of the "missing" heritability of complex traits has been hotly debated during the last few years [45,46]. For most traits including CAD, the proportion of variation of risk explained by validated GWAS loci is small in relation to the trait's heritability estimated from family studies. The fraction of the heritability explained by the lead SNPs at the CAD loci that have reached genome-wide significance to date is only approximately 10–15% [39]. Some researchers have questioned once again whether prior family-based estimates of heritability including twin studies have substantially overestimated the heritability of complex traits, but recent estimates using unrelated subjects appear similar to those derived from family-based studies [47].

Two opposing theories of the location of this missing-heritability have been proposed [48]. The first implicates the presence of many rare causal variants with modest effect sizes that are poorly tagged by common variants on GWAS arrays. The second purports that many more common variants exist with very weak effects that have not yet reached genome-wide significance. Recent empirical evidence from multiple sources including estimates of GWAS "chip heritability" as well as targeted resequencing studies of GWAS loci support a reality that lies somewhere in between these two extremes with a large tail of relatively common SNPs that have very modest effects likely contributing substantially to the overall genetic variance [49–52]. Analysis of the most recent large-scale GWAS of CAD also supports these conclusions [39].

MECHANISTIC INSIGHTS OF NEWLY IDENTIFIED LOCI

Relationship of GWAS Susceptibility Loci of CAD with Traditional Risk Factors

The identification of pleiotropic effects of some of the CAD susceptibility has provided key insights into the pathophysiology of CAD even in the absence of extensive experimental evidence shedding light on the mechanistic basis of these associations. For example, about two-thirds of the CAD susceptibility loci appear to be operating independent of established risk factors based on their strength of association in GWAS for traditional risk factors [36,38,39]. Signals for lipoprotein metabolism and blood pressure have been particularly strong with several of them having reached genome-wide significance for both CAD and the risk factor [36,38,39]. One of the strongest GWAS signals shared between CAD and risk factors for CAD to date has been a region of the genome within SORT1. This region was subsequently found to influence the expression of this gene in hepatocytes leading to changes in plasma LDL and very LDL (VLDL) particle levels through the regulation of both apolipoprotein B secretion and LDL catabolism [53].

The Arterial Wall as a Risk Factor for CAD

Several of the susceptibility loci appear to be influencing the development and/or progression of CAD through mechanisms that are specific to the arterial vessel wall. Perhaps, one of the most compelling findings supporting this hypothesis was the discovery that the same CAD susceptibility SNPs at the 9p21 locus involving *CDKN2BAS* are among the strongest association signals not only for abdominal aortic aneurysms but also for intracranial saccular (berry) aneurysms [54,55]. The latter discovery was crucial in pinpointing the effects of the causal gene at this locus to the arterial wall given saccular aneurysms form largely in the absence of atherosclerosis. A second example of this type of pleiotropy was uncovered in 2011 when another polymorphism in near *CNNM2* on chromosome 10 already reported as a susceptibility variant for intracranial aneurysm was found to also influence the risk of CAD by the CARDIoGRAM consortium [36,56]. Thus, some pathways active in cells of the arterial vessel wall appear to not only be responsible for the vessel wall's structural integrity but also its predisposition to the deposition of atherosclerotic plaque. Additional persuasive examples of susceptibility loci that may be exclusively active in the arterial wall and operating independent of traditional risk factors include two loci on chromosome 6 involving *PHARTR1* and *TCF21*. For the former, the same SNPs have been found not only to predispose to subclinical coronary atherosclerosis but also to migraines without aura [57,58], a condition whose exact etiology remains poorly understood but leading hypotheses include neuronal and/or vascular dysfunction. Interestingly, a recent cross-phenotype analysis of two large-scale GWAS of migraine and CAD revealed a significant overlap of the top genetic risk loci between the two phenotypes driven by the subtype of migraines without aura [59]. For the locus overlapping *TCF21*, we note that this gene was already known to function during the embryonic development of the coronary vasculature and in organs containing epithelial-lined vessels, such as kidney and lungs. *TCF21* functions in the origin of coronary artery smooth muscle cells (SMC) and cardiac fibroblasts, but its expression persists in the mature heart [60,61]. Interestingly, *TCF21* is one of the few CAD loci operating completely independent of traditional risk factor that has not been found to be associated with complications of atherosclerosis in noncoronary vascular beds [62].

Susceptibility to Plaque Development versus Plaque Rupture and/or Atherothrombosis

Whether individual CAD loci discovered to date not operating through traditional risk factors predispose primarily to the development and/or deposition of atherosclerosis versus the occurrence of plaque rupture and/or atherothrombosis remains to be determined. The major challenge in making this determination is the inability to reliably and noninvasively quantify both the burden of atherosclerosis as well as the number of plaque ruptures that have occurred in a given individual. A key design principle in case-control studies is the requirement for controls to be at risk of the outcome [63]. Thus, it is difficult to identify loci that predispose to plaque rupture and/or thrombosis if controls are not at risk of these complications because they have minimal or no underlying atherosclerosis. In this respect, it is important to note that approximately 50% of participants of cardiovascular disease cohorts sponsored by the NHLBI with no clinical history of CAD who underwent a coronary artery calcium scan during follow up had a coronary artery calcification score (CAC) of 0, a score that is associated with an extremely low rate of events and a very low prevalence of advanced subclinical disease on angiography [64–66]. This situation is further aggravated when genetic studies limit their samples to younger subjects where up to 80% of noncases may have a CAC of 0 [64].

A more efficient design to identify loci associated with plaque rupture and/or thrombosis is to compare subjects with a critical amount of coronary atherosclerosis and one or more well-documented MIs ("CAD+, MI− cases") to subjects with a similar amount of CAD but no history of MI (CAD+, MI− "controls"). While such a design represents a step in the right direction, a substantial loss of power due to the misclassification of "controls" may still occur because many ruptures and thrombotic events occur in the absence of clinical symptoms [67–69]. Such a design has been used to identify the ABO locus as a locus that potentially may specifically predispose to MI [70]. However, the same locus has also been found to associate with LDL as well as CAD in more standard case–control studies of CAD [36,71]. These observations make it difficult to conclude with certainty whether the ABO locus predisposes to MI independent of its ability to promote atherosclerosis.

Another approach to distinguishing between effects on plaque development versus plaque rupture and/or thrombosis is to examine associations between established loci for clinical CAD with early lesions of atherosclerosis that are not yet prone to rupture and/or thrombosis. If CAD loci identified to date are associated with the presence of early and uncomplicated lesions to a comparable degree with more advanced and complicated lesion, then it can be inferred that their predominant mechanism of action involves the formation of atherosclerosis. Salfati et al. adopted this approach and found that a genetic risk score (GRS) of 49 susceptibility variants for clinical CAD tracked to a comparable degree with CAC in subjects >35 years of age as well as early and uncomplicated raised lesions on postmortem examination of the right coronary artery in subjects aged 15 to 35 years at the time of death not related CAD [64]. Similar results were observed when the GRS was restricted to 32 non–risk factor SNPs suggesting that the associations observed were not driven by the effects of SNPs on traditional risk factors (Fig. 8.1) [64]. Overall, these findings support the hypothesis that a majority of the newly identified GWAS loci of CAD predispose to the development of atherosclerosis. However, this study was poorly powered to draw conclusions on a SNP by SNP basis. Thus, additional longitudinal and

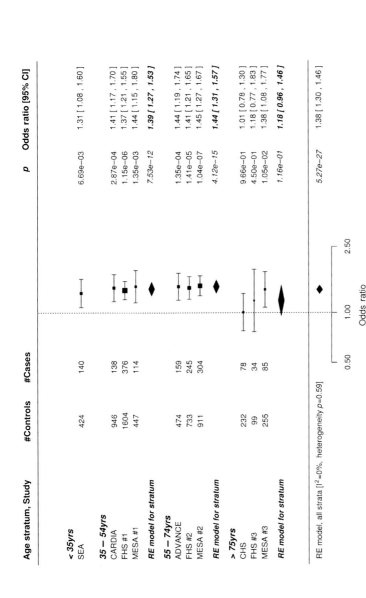

Age stratum, Study	#Controls	#Cases		p	Odds ratio [95% CI]
< 35yrs					
SEA	424	140		6.69e−03	1.31 [1.08 , 1.60]
35 − 54yrs					
CARDIA	946	138		2.87e−04	1.41 [1.17 , 1.70]
FHS #1	1604	376		1.15e−06	1.37 [1.21 , 1.55]
MESA #1	447	114		1.35e−03	1.44 [1.15 , 1.80]
RE model for stratum				*7.53e−12*	*1.39 [1.27 , 1.53]*
55 − 74yrs					
ADVANCE	474	159		1.35e−04	1.44 [1.19 , 1.74]
FHS #2	733	245		1.41e−05	1.41 [1.21 , 1.65]
MESA #2	911	304		1.04e−07	1.45 [1.27 , 1.67]
RE model for stratum				*4.12e−15*	*1.44 [1.31 , 1.57]*
> 75yrs					
CHS	232	78		9.66e−01	1.01 [0.78 , 1.30]
FHS #3	99	34		4.50e−01	1.18 [0.77 , 1.83]
MESA #3	255	85		1.05e−02	1.38 [1.08 , 1.77]
RE model for stratum				*1.16e−01*	*1.18 [0.96 , 1.46]*
RE model, all strata [I²=0%, heterogeneity p=0.59]				5.27e−27	1.38 [1.30 , 1.46]

Odds ratio: 0.50 1.00 2.50

FIGURE 8.1 Forest plot of odds ratio for being in the top quartile of degree of subclinical atherosclerosis per SD increase of the weighted genetic risk score constructed with 32 single nucleotide polymorphisms associated with clinical coronary artery disease in the CARDIoGRAMplusC4D study but not associated with traditional risk factors. Odds ratios are adjusted for age and sex. The magnitude of association with early and uncomplicated lesions in SEA was found to be similar to more advance lesions later in life. ADVANCE indicates Atherosclerotic Disease, Vascular Function, and Genetic Epidemiology; CARDIA, Coronary Artery Risk Development in Young Adults Study; CHS, Cardiovascular Health Study; CI, confidence interval; FHS, Framingham Heart Study; MESA, MultiEthnic Study of Atherosclerosis; and SEA, Single Nucleotide Polymorphisms and the Extent of Atherosclerosis. *Reproduced from Salfati E, Nandkeolyar S, Fortmann SP, Sidney S, Hlatky MA, Quertermous T, et al. Susceptibility loci for clinical coronary artery disease and subclinical coronary atherosclerosis throughout the life-course. Circ Cardiovasc Genet 2015; 8(6): 803–811, with permission from Wolters Kluwer Health, Inc.*

mechanistic studies are warranted before definitive conclusions can be drawn on a locus-by-locus basis.

Pathway Analyses

The conclusion that many more common susceptibility variants for CAD remain to be discovered suggests that many real genetic signals lurk within the set of polymorphisms that have shown suggestive association with CAD to date in GWAS but have not quite reached genome-wide significance ($p < 5 \times 10^{-8}$) because of their small effect size, allele frequency, and overall sample size. However, even weakly associated variants may provide crucial insights on the pathophysiology of a complex trait when they are clustered within a common pathway or a module of gene products working together. One approach to pathway-based analysis of genomic data that has provided some insight on pathways involved in CAD is gene set enrichment analysis (GSEA) [72,73]. GSEA aims to identify a subset of variants within or nearby set of genes collectively demonstrate strong association with a trait of interest even if the SNPs individually exhibit relatively modest or nonsignificant association. Importantly, pathway analysis can also place the set of validated SNPs for a trait of interest into a broader and clearer biological context. Pathways or modules of genes that can be tested in this manner include those that are well established and understood, as cataloged by the Reactome, Biocarta, and KEGG databases, as well as those that are formed through experiments involving human or animal tissues demonstrating coexpression of genes either at baseline or after an intervention.

Three such analyses have been published using various GSEA algorithms applied to the CARDIoGRAM and CARDIoGRAMplusC4D consortium GWAS results, which represent one of the largest and most powerful meta-analyses to date [38,74,75]. These studies have not only been able to confirm the importance of known CAD associated gene networks, such as those related to lipid metabolism, but also clarified the role of other pathways long suspected to be involved in the pathogenesis of CAD but whose causal role has been debated. For example, multiple pathways related to the immune response and inflammation have been ranked high through GSEA analyses suggesting that a critical mass of causal variants may be inherited within a subset of these genes (Fig. 8.2). Additional strong signals have been observed for processes regulating cellular growth, migration, and proliferation including the *NOTCH*, *PDGF*, and TGF-β receptor complex signaling pathways, genes related to the degradation of the extracellular matrix, and axon guidance pathways (Fig. 8.2). These findings are highly plausible given the known roles of these pathways in atherosclerosis. For example, PDGF is expressed in every cell type of the atherosclerotic arterial wall, as well as in infiltrating inflammatory cells [76]. This protein also plays a key role in the migration of vascular smooth muscle cells from the media into the intima and their subsequent proliferation [76]. The TGF-β receptor complex signaling pathway is also known to be involved in major aspects

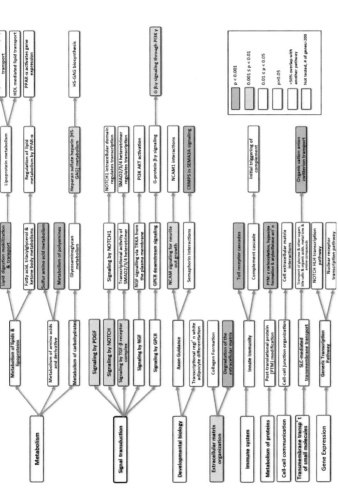

FIGURE 8.2 Replicated Reactome pathways for CAD using gene set enrichment analysis with i-GSEA4GWAS. Replicated pathways are represented in a hierarchical Reactome pathway diagram. Top-level pathways, representing core biological processes, are listed to the left, and sublevels corresponding to each top level are illustrated progressively to the right. The 9 top-level pathways that contain at least 1 replicated pathway (top-level or sublevels) are shown. No sublevel pathways are shown to the right of the last replicated pathway. Pathways are color coded according to their gene-set enrichment P value from the replication stage as indicated in the legend. A $P < 0.05$ corresponds to a false discovery rates <12.5%. Pathways containing <10 or greater than 200 genes were not tested. Replicated pathways with >50% overlap of genes with other replicated pathways are also identified as indicated in the legend. HDL indicates high-density lipoprotein; NCAM, neural cell adhesion molecule; and TGF, transforming growth factor. *Reproduced from Ghosh S, Vivar J, Nelson CP, Willenborg C, Segre AV, Mäkinen VP, et al. Systems genetics analysis of genome-wide association study reveals novel associations between key biological processes and coronary artery disease. Arterioscler Thromb Vasc Biol 2015; 35(7): 1712–1722, with permission from Wolters Kluwer Health, Inc.*

of the atherosclerotic process including cell proliferation, cell migration, matrix synthesis, wound contraction, calcification, and the immune response [77]. Lastly, the axon guidance pathways are of particular interest as they modulate a diverse biological phenomena, including cellular adhesion, migration, proliferation, differentiation, survival, and synaptic plasticity through the participation of highly conserved families of guidance molecules, including netrins, slits, semaphorins, and ephrins, and their cognate receptors [78]. The GSEA findings for this pathway strongly support an existing body of experimental evidence demonstrating important roles of neural guidance cues such as netrin-1 and semaphorins outside of the nervous system including the chemokine-directed migration of human monocytes through the arterial wall and the inhibition of macrophage foam cell emigration out of atherosclerotic lesions [79,80].

Gene-set enrichment analyses provide a solid basis for generating testable hypothesis, but they are limited by the knowledge of the pathways being tested and have been criticized for their lack of specificity given some of the higher order pathways for CAD have been highlighted in multiple other traits and contain a large number of genes. Because of this variability, extensive mechanistic and functional validation of pathway-derived networks at multiple levels will be essential to validate findings from GSEA [38,74,75].

MENDELIAN RANDOMIZATION STUDIES IN CAD

A fundamental problem in observational epidemiology is determining the nature of the relationship between two variables, most often an exposure (or risk factor) and an outcome [81]. Understanding whether there is a cause and effect relationship between two variables is vital as it allows us to more reliably assess the potential impact of a health intervention on the exposure [81].

Mendelian Randomization (MR) studies estimate the causal effect of an exposure on a trait or a disease in nonexperimental settings by leveraging the well-established statistical method of instrumental variable (IV) analysis [81]. IV analysis has been used for several decades in the field of econometrics to help deal with issues of confounding, reverse causality, and regression dilution bias (more often referred to collectively as "endogeneity" in econometrics) [81]. An MR study is simply an IV analysis that uses one or more genetic variants as the instrument [81].

The concept of MR appears to have been first introduced in medicine by MB Katan who in 1986 argued that variants within the *APOE* could be used to decipher the direction of the association between blood levels of LDL and cancer [82]. Shortly after the completion of the Human Genome project, the concept was further developed by several scientists anticipating the delivery of a plethora of new genetic instruments through large-scale GWAS [83]. A key prerequisite for MR studies is that genetic variants assort independently at meiosis and a subset exists that influence lifetime levels of the risk factor without being associated with any other risk factors for the outcome of interest [84]. Once

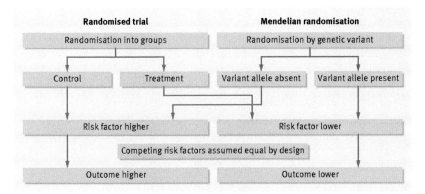

FIGURE 8.3 Comparison of Mendelian Randomization and randomized trial. *Reproduced from Burgess S, Butterworth A, Malarstig A, Thompson SG. Use of Mendelian randomisation to assess potential benefit of clinical intervention. BMJ 2012; 345: e7325, with permission from BMJ Publishing Group Ltd.*

identified, one or more of these genetic variants can then be used to completely deconfound the relationship between two variables through an MR study in much the same way we use randomization in clinical trials to safeguard against similar biases (Fig. 8.3) [84,85]. The validity of the conclusions reached by MR studies is dependent on the power of the study, which itself is dependent on the number of events as well as the strength of the instrument [86]. The validity is also dependent on the assumption that the genetic instrument is not associated with any confounder of the exposure-outcome association and can only affect the outcome through its association with the exposure [84].

MR studies have been widely applied to study atherosclerotic related cardiovascular disease and have provided us with some surprising insights into the causal relationship between cardio metabolic risk factors and CAD outcomes. For example, MR studies suggest a noncausal relationship between CAD and blood levels of HDL, hsCRP, fibrinogen, Lp-PLA2, and homocysteine [87]. These findings imply that pharmacological interventions developed specifically to modify the levels of these biomarkers are not likely to translate to corresponding changes in the risk of CAD, a fact that has been born out by the failure of cholesterol ester transfer protein (CETP), Lp-PLA2 inhibitors, and folic acid supplementation trials to reduce the risk of CAD despite remarkable changes in blood levels of HDL, LpPLA2, and homocysteine by these agents [88–90]. On the other hand, MR studies have implied that associations between CAD and LDL, Lp(a), triglycerides, body mass index, blood pressure, and diabetes are causal in nature [87]. These conclusions are supported by the beneficial effects on the risk of CAD of LDL lowering drugs including statins and ezetemide, blood pressure lowering medications, and interventions resulting in substantial and sustained weight loss [87,91]. Not all findings between MR studies and corresponding clinical trials have been concordant indicating that MR studies cannot be trusted entirely to predict the effectiveness of related pharmacologic interventions. For example, drugs with substantial effects on blood levels of triglycerides including niacin, fibrates, and omega-3 fatty acids have no clear beneficial effects on the risk of CAD when statin drugs are already in use [88,92]. Furthermore, the most recent CETP inhibitor trial involving evacetrapib marks the first time a drug has demonstrated an inability to lower the risk of CAD despite substantially lowering LDL [93]. The reasons for these discrepancies remain unclear but are believe to be a consequence of an incomplete knowledge of all biological pathways influencing blood levels of a biomarker of interest. In the presence of more than one pathway influencing a biomarker level of interest, it is possible that some are causal and some are not causal in relation to predisposition to CAD.

MR studies have also been leveraged to better understand the relationship between dietary exposures and CAD. A large MR study involving a meta-analysis of over 50 prospective studies involving ~262,000 individuals and ~20,250 CHD events examined the relationship between moderate alcohol use and CAD outcomes using a polymorphism in the alcohol dehydrogenase 1B gene (*ADH1B*) [94]. This study has called into question the cardiovascular health benefits of moderate alcohol given carriers of the minor allele of SNP in *ADH1B* were found to drink less alcohol than noncarriers and to have an ~10% reduction in the risk of CAD.

Methodological aspects of MR studies continue to mature and, consequently, a better understanding of their strengths and limitations has evolved [95,96]. New methods have been recently developed to identify and remove problematic genetic instruments that may result in a violation of one or more of the major assumptions through unrecognized pleiotropy or other less well-understood mechanisms [97]. Furthermore, mega observational studies involving various large biobanks around the world are expected to dramatically improve the power of MR studies and minimize the chances of weak instrument bias, a bias that can result in an MR study falsely concluding that the observational association is causal due to residual confounding [84]. Thus, MR studies are expected to continue to play an important role in assessing the nature of the associations between CAD-related traits and to help determine the potential value of developing a therapeutic intervention intended to specifically modify a risk factor of interest.

GENETIC RISK PREDICTION IN CAD

GRSs provide a means to aggregate the health-related risk of a collection of genetic alleles into a single number. They were originally developed with the intention of improving risk prediction in clinical practice [98–100]. A major strength of the GRS as a risk biomarker is its reproducibility and validity. A substantial body of evidence has recently emerged to suggest that contemporary GRSs incorporating established loci for CAD predicts coronary heart disease

(CHD) outcomes in independent prospective studies even after adjusting for all major risk factors [101–106]. Translating these observations to clinical utility has proven more challenging given the impression that improvements in risk prediction with GRSs are not clinically meaningful. These impressions are largely a consequence of a lack of a substantial increase in traditional discrimination statistics such as the C-statistic by the addition of a GRS. This is particularly true for CAD where the baseline C-statistic of traditional risk factors is typically in the range of 0.7 to 0.8. However, recent work by reputable biostatisticians suggests that even well-established model performance metrics like the C-statistic may be unacceptably conservative [107].

The minimal degree of improvement needed to make any biomarker including a GRS clinical useful depends on the characteristics of the population being tested, which model performance metric is being embraced, and the risk-benefit trade off of the therapeutic intervention being considered. For atherosclerotic related outcomes, the degree of improvement in prediction may not have to be substantial given the overall prevalence of disease and the availability of relatively safe and effective interventions such as the use of statins and aspirin [108,109]. Ultimately, large clinical trials may be needed to prove clinical utility of a GRS for CAD. In the setting of a mega-cohort accruing a large number of events annually, the incremental benefit of incorporating a GRS for CAD into a standard clinical risk score may become clear within a couple of years of follow up. An example of how a GRS can be used to help update one's 10-year risk of CAD as well as potentially modify therapeutic options is illustrated in Fig. 8.4 [101].

GENE–GENE AND GENE–ENVIRONMENT INTERACTIONS

The GWAS approach, while successful, has relied on the analysis of each SNP individually, but as a substantial amount of the genetic variance of most complex traits remains unexplained, focus has turned to higher order analyses including interactions between SNPs or their loci [110,111]. Such interactions are often referred to as "epistasis".

Researchers have provided empirical evidence and made strong arguments in support of the presence as well as in the absence of higher order interactions contributing to the heritability of complex traits [112–115]. What is almost certainly true is that variance from higher order genetic effects likely accounts for a minority of the variance of complex traits including CAD, it will be challenging to identify them in the absence of large sample sizes and the adoption of methods that intelligently reduce the number of statistical tests.

Investigators are also now exploring in considerable depth the interplay between environmental factors and SNPs or GRSs [116]. For example, several GRS-by-environment interaction studies have emerged in the study of obesity, diabetes, and lipids [117–120]. For CAD-related outcomes, a recent report not only replicated the main effect of a GRS on the risk of CAD, but also found that individuals at highest genetic risk for CAD benefited the most from statins in both primary and secondary prevention settings [102]. The investigators hypothesized that the increased benefit may simply reflect a higher burden of disease in those with a higher GRS [102]. Robust and replicable interactions between individual CAD susceptibility SNPs and environmental factors have not yet been reported.

Your risk score

Based on the traditional Framingham risk score, your risk of coronary heart disease over the next 10 years is approximately 5.5%.

We tested for a total of 90 possible risk variants or alleles. Out of these 90, you carry 49 variants that are associated with higher risk. Your genetic profile puts you in the 89 percentile for risk. This means 89% of the general population have a genetic risk score more favorable than you and 11% have a genetic risk score less favorable than you.

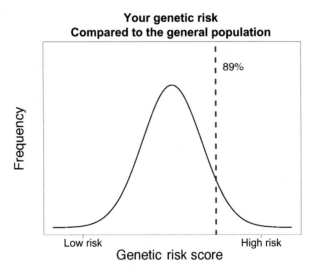

Based on the traditional Framingham risk score plus the genetic risk score, your risk of coronary heart disease over the next 10 years is approximately 7.6%.

Your 10 year risk of coronary heart disease risk is ≥7.5% when considering your genetic risk. This information may be discussed with your physician in terms of what would be recommended as most appropriate management given your estimated risk.

FIGURE 8.4 Example of how a genetic risk score can be used to help update one's 10-year risk of CAD as well as potentially modify therapeutic options. *Reproduced from Frontiers in genetics 2015 Jul 7; 7: 231 (open access).*

WHOLE EXOME SEQUENCING ASSOCIATION STUDIES OF CAD

Genotyping arrays used in GWAS generally do not include probes for less common and rare variants [20]. The few arrays that include such variants only include the subset that has been previously observed in other individuals. The only way to reliably identify all such variants in a given sample set is through genome sequencing. Next generation sequencing technologies developed over the last decade have led to profound reductions in the cost of sequencing and have facilitated large-scale genome sequencing projects [121].

Genome sequencing of all exons in a genome, referred to as whole exome sequencing (WES), became economically feasible toward the end of the last decade. In 2010, the NHLBI funded several groups through the Exome Sequencing Project (ESP) to perform WES for the identification of novel coding variants associated with heart, lung, and blood traits including MI (http://www.ncbi.nlm. nih.gov/bioproject/165957). ESP investigators collaborated with other investigators in the United States and abroad to perform large-scale WES association studies of MI and CAD with replication of promising signals in additional samples using custom genotyping chips referred to as "exome chips" [122]. Overall, the discoveries to date have been somewhat limited because effect sizes of rare variants for most complex traits including CAD were found to be not much higher than the effect sizes observed for more common variants identified through GWAS. To improve power, rare variants within genes have been examined collectively and large-scale meta-analyses have been necessary. The major discoveries to date using WES have been limited to genes previously known to be involved in lipoprotein metabolism including *LDLR*, *APOA5*, *APOC3*, *NPC1L1*, *LPL*, and *ANGPTL4* where loss- or gain-of-function mutations in these genes have been associated with a change in either LDL and/or triglyceride levels as well as a corresponding change in the risk of CAD [123–126]. Despite the limited novelty in terms of mechanism of disease, the findings have provided important clues on novel therapeutic targets in metabolic pathways that lead to changes in the levels of circulating triglycerides. In the case of *NPC1L1*, WES studies predicted a benefit in CAD outcomes with the use of ezetimide, a drug that inactivates the protein encoded by *NPC1L1*, which was subsequently confirmed in a large clinical trial [91].

CONCLUSION AND FUTURE DIRECTIONS

We have witnessed remarkable progress in the field of the genetic epidemiology of CAD over the last decade building off important observations made prior to the GWAS era. As a result, several key inferences on the genetics of CAD can be now be confidently made. First, recent studies of densely genotyped unrelated individuals confirm the high heritability of CAD identified by prior twin studies. Second, the genetic architecture of CAD is dominated by a large number of relatively common susceptibility variants with very modest effects. Third, large-scale GWAS of CAD not only confirm the central role of established risk factors in the pathogenesis of CAD but also point to other causal biological mechanisms involving a variety of cells active in the vessel wall of the coronary including fibroblasts, smooth muscle cells, and immune related cells. Fourth, the majority of CAD loci uncovered to date likely predispose to the formation of coronary atherosclerosis. Reliably identifying a subset of susceptibility loci that specifically predispose to plaque rupture and/or thrombosis remains challenging. Fifth, GWAS of risk factors of CAD have facilitated the conduct of MR studies, which in turn have provided important clues on the causality of CAD risk factors. Lastly, the clinical utility of incorporating a GRS of high-risk alleles at established loci of CAD in risk prediction algorithms remains uncertain. The GRS clearly predicts CAD events independent of other risk factors but this prediction translates to a very modest change in traditional model performance metrics whose clinical significance remains unclear. However, the predictive power of the GRS will continue to improve as more susceptibility loci are identified and the decreasing cost of genotyping should facilitate the conduct of randomized trials that will clarify the role of GRSs in the management of subjects at risk of CAD.

Major challenges that remain over the next several years include the identification of the remaining susceptibility loci of CAD, proving the predictive value of this genomic information through its efficient integration into the electronic health record (EHR) for clinical decision support, and gaining a solid understanding of the mechanisms responsible for association among established loci. The first task is expected to require exceptionally large sample sizes, which will be fulfilled by mega-cohort studies involving bio banks [127,128]. Sequencing, including whole genome sequencing, will undoubtedly play an increasing role in cataloging the complete spectrum of variation contributing to the susceptibility of CAD within coding and regulatory regions of the genome [129]. The same infrastructure used for large-scale biobank studies could also be leveraged to conduct pragmatic trials to prove the predictive value of this genomic information in clinical practice. Naturally, such proof will need to be complemented with the continued development of structured EHR systems capable of efficiently storing and integrating genomic information that can be ported through rules-based decision-support engines while at the same time providing enough information to allow clinicians to comfortably interpret genomic data [130]. The final task will require careful integration of genetic association signals identified at the population level with the functional annotation of the human genome being mapped out by several NIH initiatives such as ENCODE, Roadmap Epigenomics, and GTEx projects [131]. These and other genomic projects around the world are slowly but surely unraveling the regulatory landscape of noncoding regions of many different human cell types allowing researchers to more reliably identify which SNPs and genes within an established susceptibility locus are causal for CAD [132,133]. Further mechanistic insights will be gleaned through targeted genome engineering of cells, tissues, and/or animal models involving the susceptibility locus. These latter experimental techniques have recently received a major boost of efficiency through the development of the CRISPR/Cas9 gene editing protocol [134]. Collectively, this body of experimental evidence is expected to pave the way for the development of novel therapeutic agents. Such progress will not occur overnight but recent examples within the field of genetics of CAD such as the development of *PCSK9* inhibitors provide some reassurance that the pace of translation will be substantially faster than it has been in the past [135].

GLOSSARY TERMS, ACRONYMS, ABBREVIATIONS

CAD	coronary artery disease
LDL	low-density lipoprotein
MI	myocardial infarction
PAR	population attributable risk
GWAS	genome wide association study(ies)
GWA	genome wide association
SNP(s)	single nucleotide polymorphism(s)
WTCCC	Welcome Trust Case-Control Consortium
MIGEN	Myocardial Infarction Genetics Consortium
CARDIoGRAM	Coronary ARtery DIsease Genome wide Replication and Meta-analysis Consortium
C4D	Coronary Artery Disease Genetics Consortium
CARDIoGRAM plusC4D	Coronary ARtery DIsease Genome wide Replication and Meta-analysis Consortium plus the Coronary Artery Disease Genetics Consortium
VLDL	very low-density lipoprotein GRS(s): genetic risk score(s)
CAC	coronary artery calcification
GSEA	gene set enrichment analysis
KEGG	Kyoto Encyclopedia of Genes and Genomes
MR	Mendelian Randomization
HDL	high-density lipoprotein
hsCRP	high sensitivity C-reactive protein
Lp-PLA2	lipoprotein-associated phospholipase A2
BMI	body mass index
CETP	cholesteryl ester transfer protein
CHD	coronary heart disease
WES	whole exome sequencing
NHLBI	National Heart Lung Blood Institute
ESP	Exome Sequencing Project
HER	electronic health record
ENCODE	Encyclopedia of DNA Elements project
GTex	Genotype-Tissue Expression project
CRISPR/Cas9	Clustered Regularly Interspaced Short Palindromic Repeats/CRISPR-associated protein-9 nuclease
CHR	chromosome
DVT	deep vein thrombosis
TG	triglyceride

BP blood pressure
SMC smooth muscle cell
VSMC vascular smooth muscle cell
VEGF vascular endothelial growth factor
VEGFR VEGF receptor
TGF-β transforming growth factor beta
RNA ribonucleic acid
lncRNA long noncoding RNA
ncRNA noncoding RNA

REFERENCES

[1] Mann DL, Zipes DP, Libby P, Bonow RO, Braunwald E. Braunwald's heart disease: a text-book of cardiovascular medicine, Tenth edition. ed. Philadelphia, PA: Elsevier/Saunders; 2015.

[2] Smolina K, Wright FL, Rayner M, Goldacre MJ. Incidence and 30-day case fatality for acute myocardial infarction in England in 2010: national-linked database study. Eur J Public Health 2012;22(6):848–53.

[3] McMahan CA, Gidding SS, Malcom GT, Tracy RE, Strong JP, McGill Jr. HC. Pathobiological Determinants of Atherosclerosis in Youth Research G. Pathobiological determinants of atherosclerosis in youth risk scores are associated with early and advanced atherosclerosis. Pediatrics 2006;118(4):1447–55.

[4] Endo A. A historical perspective on the discovery of statins. Proc Jpn Acad Ser B Phys Biol Sci 2010;86(5):484–93.

[5] Musunuru K, Kathiresan S. Genetics of coronary artery disease. Annu Rev Genomics Hum Genet 2010;11:91–108.

[6] Robertson TL, Kato H, Rhoads GG, Kagan A, Marmot M, Syme SL, et al. Epidemiologic studies of coronary heart disease and stroke in Japanese men living in Japan, Hawaii and California. Incidence of myocardial infarction and death from coronary heart disease. Am J Cardiol 1977;39(2):239–43.

[7] Friedlander Y. Familial clustering of coronary heart disease: a review of its significance and role as a risk factor for the disease Goldbourt U, De Faire U, Berg K, editors. Genetic factors in coronary heart disease. Dordrecht; Boston: Kluwer Academic; 1994. p. 37–53.

[8] Lloyd-Jones DM, Nam BH, D'Agostino Sr. RB, Levy D, Murabito JM, Wang TJ, et al. Parental cardiovascular disease as a risk factor for cardiovascular disease in middle-aged adults: a prospective study of parents and offspring. JAMA 2004;291(18):2204–11.

[9] de Faire U, Pedersen N. Studies of twins and adoptees in coronary heart disease Goldbourt U, De Faire U, Berg K, editors. Genetic factors in coronary heart disease. Dordrecht; Boston: Kluwer Academic; 1994. p. 55–68.

[10] Zdravkovic S, Wienke A, Pedersen NL, Marenberg ME, Yashin AI, De Faire U. Heritability of death from coronary heart disease: a 36-year follow-up of 20966 Swedish twins. J Intern Med 2002;252(3):247–54.

[11] Wienke A, Holm NV, Skytthe A, Yashin AI. The heritability of mortality due to heart diseases: a correlated frailty model applied to Danish twins. Twin Res 2001;4(4):266–74.

[12] Marenberg ME, Risch N, Berkman LF, Floderus B, de Faire U. Genetic susceptibility to death from coronary heart disease in a study of twins. N Engl J Med 1994;330(15):1041–6.

[13] Rothman KJ, Greenland S, Lash TL. Modern epidemiology, 3rd ed. Philadelphia: Wolters Kluwer Health/Lippincott Williams & Wilkins; 2008.

[14] Stamler J, Stamler R, Neaton JD, Wentworth D, Daviglus ML, Garside D, et al. Low risk-factor profile and long-term cardiovascular and noncardiovascular mortality and life expectancy: findings for 5 large cohorts of young adult and middle-aged men and women. JAMA 1999;282(21):2012–8.

[15] Yusuf S, Hawken S, Ounpuu S, Dans T, Avezum A, Lanas F, et al. Effect of potentially modifiable risk factors associated with myocardial infarction in 52 countries (the INTERHEART study): case–control study. Lancet 2004;364(9438):937–52.

[16] Greenland S, Robins JM. Conceptual problems in the definition and interpretation of attributable fractions. Am J Epidemiol 1988;128(6):1185–97.

[17] Greenland P, Knoll MD, Stamler J, Neaton JD, Dyer AR, Garside DB, et al. Major risk factors as antecedents of fatal and nonfatal coronary heart disease events. JAMA 2003;290(7):891–7.

[18] Palmer L, Burton PR, Smith GD. An introduction to genetic epidemiology. Bristol: Policy Press; 2011.

[19] International HapMap Consortium A haplotype map of the human genome. Nature 2005;437(7063):1299–320.

[20] Ragoussis J. Genotyping technologies for genetic research. Annu Rev Genomics Hum Genet 2009;10:117–33.

[21] Genomes Project C, Auton A, Brooks LD, Durbin RM, Garrison EP, Kang HM, et al. A global reference for human genetic variation. Nature 2015;526(7571):68–74.

[22] Welter D, MacArthur J, Morales J, Burdett T, Hall P, Junkins H, et al. The NHGRI GWAS catalog, a curated resource of SNP-trait associations. Nucleic Acids Res 2014;42(Database issue):D1001–6.

[23] Chanock SJ, Manolio T, Boehnke M, Boerwinkle E, Hunter DJ, Thomas G, et al. Replicating genotype-phenotype associations. Nature 2007;447(7145):655–60.

[24] Helgadottir A, Thorleifsson G, Manolescu A, Gretarsdottir S, Blondal T, Jonasdottir A, et al. A common variant on chromosome 9p21 affects the risk of myocardial infarction. Science (New York, NY) 2007;316(5830):1491–3.

[25] McPherson R, Pertsemlidis A, Kavaslar N, Stewart A, Roberts R, Cox DR, et al. A common allele on chromosome 9 associated with coronary heart disease. Science (New York, NY) 2007;316(5830):1488–91.

[26] Schunkert H, Gotz A, Braund P, McGinnis R, Tregouet DA, Mangino M, et al. Repeated replication and a prospective meta-analysis of the association between chromosome 9p21.3 and coronary artery disease. Circulation 2008;117(13):1675–84.

[27] Lee JY, Lee BS, Shin DJ, Woo Park K, Shin YA, Joong Kim K, et al. A genome-wide association study of a coronary artery disease risk variant. J Hum Genet 2013;58(3):120–6.

[28] Takeuchi F, Yokota M, Yamamoto K, Nakashima E, Katsuya T, Asano H, et al. Genome-wide association study of coronary artery disease in the Japanese. Eur J Hum Genet 2012;20(3):333–40.

[29] Lu X, Wang L, Chen S, He L, Yang X, Shi Y, et al. Genome-wide association study in Han Chinese identifies four new susceptibility loci for coronary artery disease. Nat Genet 2012;44(8):890–4.

[30] Kathiresan S, Voight BF, Purcell S, Musunuru K, Ardissino D, Mannucci PM, et al. Genome-wide association of early-onset myocardial infarction with single nucleotide polymorphisms and copy number variants. Nat Genet 2009;41(3):334–41.

[31] Samani NJ, Deloukas P, Erdmann J, Hengstenberg C, Kuulasmaa K, McGinnis R, et al. Large scale association analysis of novel genetic loci for coronary artery disease. Arterioscler Thromb Vasc Biol 2009;29(5):774–80.

[32] Erdmann J, Grosshennig A, Braund PS, Konig IR, Hengstenberg C, Hall AS, et al. New susceptibility locus for coronary artery disease on chromosome 3q22.3. Nat Genet 2009.

[33] Wellcome Trust Case Control Consortium Genome-wide association study of 14,000 cases of seven common diseases and 3,000 shared controls. Nature 2007;447(7145):661–78.

[34] Samani NJ, Erdmann J, Hall AS, Hengstenberg C, Mangino M, Mayer B, et al. Genomewide association analysis of coronary artery disease. N Engl J Med 2007;357(5):443–53.

[35] Peden JF, Hopewell JC, Saleheen D, Chambers JC, Hager J, Soranzo N, et al. A genome-wide association study in Europeans and South Asians identifies five new loci for coronary artery disease. Nat Genet 2011.

[36] Schunkert H, Konig IR, Kathiresan S, Reilly MP, Assimes TL, Holm H, et al. Large-scale association analysis identifies 13 new susceptibility loci for coronary artery disease. Nat Genet 2011;43(4):333–8.

[37] Voight BF, Kang HM, Ding J, Palmer CD, Sidore C, Chines PS, et al. The metabochip, a custom genotyping array for genetic studies of metabolic, cardiovascular, and anthropometric traits. PLoS Genet 2012;8(8):e1002793.

[38] CARDIoGRAMplusC4D C, Deloukas P, Kanoni S, Willenborg C, Farrall M, Assimes TL, et al. Large-scale association analysis identifies new risk loci for coronary artery disease. Nat Genet 2013;45(1):25–33.

[39] Nikpay M, Goel A, Won HH, Hall LM, Willenborg C, Kanoni S, et al. A comprehensive 1,000 genomes-based genome-wide association meta-analysis of coronary artery disease. Nat Genet 2015;47(10):1121–30.

[40] Aoki A, Ozaki K, Sato H, Takahashi A, Kubo M, Sakata Y, et al. SNPs on chromosome 5p15.3 associated with myocardial infarction in Japanese population. J Hum Genet 2011;56(1):47–51.

[41] Hirokawa M, Morita H, Tajima T, Takahashi A, Ashikawa K, Miya F, et al. A genome-wide association study identifies PLCL2 and AP3D1-DOT1L-SF3A2 as new susceptibility loci for myocardial infarction in Japanese. Eur J Hum Genet 2015;23(3):374–80.

[42] Wang F, Xu CQ, He Q, Cai JP, Li XC, Wang D, et al. Genome-wide association identifies a susceptibility locus for coronary artery disease in the Chinese Han population. Nat Genet 2011;43(4):345–9.

[43] Barbalic M, Reiner AP, Wu C, Hixson JE, Franceschini N, Eaton CB, et al. Genome-wide association analysis of incident coronary heart disease (CHD) in African Americans: a short report. PLoS Genet 2011;7(8):e1002199.

[44] Marigorta UM, Navarro A. High trans-ethnic replicability of GWAS results implies common causal variants. PLoS Genet 2013;9(6):e1003566.

[45] Manolio TA, Collins FS, Cox NJ, Goldstein DB, Hindorff LA, Hunter DJ, et al. Finding the missing heritability of complex diseases. Nature 2009;461(7265):747–53.

[46] Visscher PM, Brown MA, McCarthy MI, Yang J. Five years of GWAS discovery. Am J Hum Genet 2012;90(1):7–24.

[47] Stahl EA, Wegmann D, Trynka G, Gutierrez-Achury J, Do R, Voight BF, et al. Bayesian inference analyses of the polygenic architecture of rheumatoid arthritis. Nat Genet 2012;44(5):483–9.

[48] Gibson G. Rare and common variants: twenty arguments. Nat Rev Genet 2011;13(2):135–45.

[49] Maher MC, Uricchio LH, Torgerson DG, Hernandez RD. Population genetics of rare variants and complex diseases. Hum Hered 2012;74(3–4):118–28.

[50] Agarwala V, Flannick J, Sunyaev S, Altshuler D. Evaluating empirical bounds on complex disease genetic architecture. Nat Genet 2013;45(12):1418–27.

[51] Park JH, Gail MH, Weinberg CR, Carroll RJ, Chung CC, Wang Z, et al. Distribution of allele frequencies and effect sizes and their interrelationships for common genetic susceptibility variants. Proc Natl Acad Sci USA 2011;108(44):18026–31.

[52] Lee SH, Yang J, Chen GB, Ripke S, Stahl EA, Hultman CM, et al. Estimation of SNP heritability from dense genotype data. Am J Hum Genet 2013;93(6):1151–5.

[53] Strong A, Ding Q, Edmondson AC, Millar JS, Sachs KV, Li X, et al. Hepatic sortilin regulates both apolipoprotein B secretion and LDL catabolism. J Clin Invest 2012.

[54] Helgadottir A, Thorleifsson G, Magnusson KP, Gretarsdottir S, Steinthorsdottir V, Manolescu A, et al. The same sequence variant on 9p21 associates with myocardial infarction, abdominal aortic aneurysm and intracranial aneurysm. Nat Genet 2008;40(2):217–24.

[55] Bilguvar K, Yasuno K, Niemela M, Ruigrok YM, von Und Zu Fraunberg M, van Duijn CM, et al. Susceptibility loci for intracranial aneurysm in European and Japanese populations. Nat Genet 2008;40(12):1472–7.

[56] Yasuno K, Bilguvar K, Bijlenga P, Low SK, Krischek B, Auburger G, et al. Genome-wide association study of intracranial aneurysm identifies three new risk loci. Nat Genet 2010;42(5):420–5.

[57] Anttila V, Winsvold BS, Gormley P, Kurth T, Bettella F, McMahon G, et al. North American Brain Expression C, Consortium UKBE, International Headache Genetics C. Genome-wide meta-analysis identifies new susceptibility loci for migraine. Nat Genet 2013;45(8): 912–7.

[58] Freilinger T, Anttila V, de Vries B, Malik R, Kallela M, Terwindt GM, et al. Genome-wide association analysis identifies susceptibility loci for migraine without aura. Nat Genet 2012;44(7):777–82.

[59] Winsvold BS, Nelson CP, Malik R, Gormley P, Anttila V, Vander Heiden J, et al. Genetic analysis for a shared biological basis between migraine and coronary artery disease. Neurol Genet 2015;1(1):e10.

[60] Acharya A, Baek ST, Huang G, Eskiocak B, Goetsch S, Sung CY, et al. The bHLH transcription factor Tcf21 is required for lineage-specific EMT of cardiac fibroblast progenitors. Development 2012;139(12):2139–49.

[61] Braitsch CM, Combs MD, Quaggin SE, Yutzey KE. Pod1/Tcf21 is regulated by retinoic acid signaling and inhibits differentiation of epicardium-derived cells into smooth muscle in the developing heart. Dev Biol 2012;368(2):345–57.

[62] Dichgans M, Malik R, Konig IR, Rosand J, Clarke R, Gretarsdottir S, et al. International Stroke Genetics C. Shared genetic susceptibility to ischemic stroke and coronary artery disease: a genome-wide analysis of common variants. Stroke 2014;45(1):24–36.

[63] Wacholder S, McLaughlin JK, Silverman DT, Mandel JS. Selection of controls in case-control studies. I. Principles. Am J Epidemiol 1992;135(9):1019–28.

[64] Salfati E, Nandkeolyar S, Fortmann SP, Sidney S, Hlatky MA, Quertermous T, et al. Susceptibility loci for clinical coronary artery disease and subclinical coronary atherosclerosis throughout the life-course. Circ Cardiovasc Genet 2015;8(6):803–11.

[65] Sarwar A, Shaw LJ, Shapiro MD, Blankstein R, Hoffmann U, Cury RC, et al. Diagnostic and prognostic value of absence of coronary artery calcification. JACC Cardiovasc Imaging 2009;2(6):675–88.

[66] de Carvalho MS, de Araujo Goncalves P, Garcia-Garcia HM, de Sousa PJ, Dores H, Ferreira A, et al. Prevalence and predictors of coronary artery disease in patients with a calcium score of zero. Int J Cardiovasc Imaging 2013;29(8):1839–46.

[67] Burke AP, Kolodgie FD, Farb A, Weber DK, Malcom GT, Smialek J, et al. Healed plaque ruptures and sudden coronary death: evidence that subclinical rupture has a role in plaque progression. Circulation 2001;103(7):934–40.

[68] Hong MK, Mintz GS, Lee CW, Kim YH, Lee SW, Song JM, et al. Comparison of coronary plaque rupture between stable angina and acute myocardial infarction: a three-vessel intravascular ultrasound study in 235 patients. Circulation 2004;110(8):928–33.

[69] Hong MK, Mintz GS, Lee CW, Park KM, Lee BK, Kim YH, et al. Plaque ruptures instable angina pectoris compared with acute coronary syndrome. Int J Cardiol 2007;114(1):78–82.

[70] Reilly MP, Li M, He J, Ferguson JF, Stylianou IM, Mehta NN, et al. Identification of ADAMTS7 as a novel locus for coronary atherosclerosis and association of ABO with myocardial infarction in the presence of coronary atherosclerosis: two genome-wide association studies. Lancet 2011;377(9763):383–92.

[71] Teslovich TM, Musunuru K, Smith AV, Edmondson AC, Stylianou IM, Koseki M, et al. Biological, clinical and population relevance of 95 loci for blood lipids. Nature 2010;466(7307):707–13.

[72] Zhang K, Cui S, Chang S, Zhang L, Wang J. i-GSEA4GWAS: a web server for identification of pathways/gene sets associated with traits by applying an improved gene set enrichment analysis to genome-wide association study. Nucleic Acids Res 2010;38(Web Server issue):W90–5.

[73] Holden M, Deng S, Wojnowski L, Kulle B. GSEA-SNP: applying gene set enrichment analysis to SNP data from genome-wide association studies. Bioinformatics 2008;24(23):2784–5.

[74] Ghosh S, Vivar J, Nelson CP, Willenborg C, Segre AV, Makinen VP, et al. Systems genetics analysis of genome-wide association study reveals novel associations between key biological processes and coronary artery disease. Arterioscler Thromb Vasc Biol 2015;35(7):1712–22.

[75] Makinen VP, Civelek M, Meng Q, Zhang B, Zhu J, Levian C, et al. Integrative genomics reveals novel molecular pathways and gene networks for coronary artery disease. PLoS Genet 2014;10(7):e1004502.

[76] Raines EW. PDGF and cardiovascular disease. Cytokine Growth Factor Rev 2004;15(4):237–54.

[77] Toma I, McCaffrey TA. Transforming growth factor-beta and atherosclerosis: interwoven atherogenic and atheroprotective aspects. Cell Tissue Res 2012;347(1):155–75.

[78] Schmidt EF, Strittmatter SM. The CRMP family of proteins and their role in Sema3A signaling. Adv Exp Med Biol 2007;600:1–11.

[79] Wanschel A, Seibert T, Hewing B, Ramkhelawon B, Ray TD, van Gils JM, et al. Neuroimmune guidance cue semaphorin 3E is expressed in atherosclerotic plaques and regulates macrophage retention. Arterioscler Thromb Vasc Biol 2013;33(5):886–93.

[80] van Gils JM, Ramkhelawon B, Fernandes L, Stewart MC, Guo L, Seibert T, et al. Endothelial expression of guidance cues in vessel wall homeostasis dysregulation under proatherosclerotic conditions. Arterioscler Thromb Vasc Biol 2013;33(5):911–9.

[81] Burgess S, Thompson SG. Mendelian randomization: methods for using genetic variants in causal estimation. Boca Raton: Chapman & Hall/CRC; 2014.

[82] Katan MB. Commentary: Mendelian randomization, 18 years on. Int J Epidemiol 2004;33(1):10–11.

[83] Davey Smith G, Ebrahim S. Mendelian randomization': can genetic epidemiology contribute to understanding environmental determinants of disease? Int J Epidemiol 2003;32(1):1–22.

[84] Burgess S, Thompson SG. Mendelian randomization: methods for using genetic variants in causal estimation. Boca Raton, Florida: CRC Press/Taylor & Francis Group; 2015.

[85] Burgess S, Butterworth A, Malarstig A, Thompson SG. Use of Mendelian randomisation to assess potential benefit of clinical intervention. BMJ 2012;345:e7325.

[86] Brion MJ, Shakhbazov K, Visscher PM. Calculating statistical power in Mendelian randomization studies. Int J Epidemiol 2013;42(5):1497–501.

[87] Jansen H, Samani NJ, Schunkert H. Mendelian randomization studies in coronary artery disease. Eur Heart J 2014;35(29):1917–24.

[88] Keene D, Price C, Shun-Shin MJ, Francis DP. Effect on cardiovascular risk of high density lipoprotein targeted drug treatments niacin, fibrates, and CETP inhibitors: meta-analysis of randomised controlled trials including 117,411 patients. BMJ 2014;349:g4379.

[89] Talmud PJ, Holmes MV. Deciphering the causal role of sPLA2s and Lp-PLA2 in coronary heart disease. Arterioscler Thromb Vasc Biol 2015;35(11):2281–9.

[90] Yang HT, Lee M, Hong KS, Ovbiagele B, Saver JL. Efficacy of folic acid supplementation in cardiovascular disease prevention: an updated meta-analysis of randomized controlled trials. Eur J Intern Med 2012;23(8):745–54.

[91] Cannon CP, Blazing MA, Giugliano RP, McCagg A, White JA, Theroux P, et al. Investigators I-I. Ezetimibe added to statin therapy after acute coronary syndromes. N Engl J Med 2015;372(25):2387–97.

[92] Walz CP, Barry AR, Koshman SL. Omega-3 polyunsaturated fatty acid supplementation in the prevention of cardiovascular disease. Can Pharm J (Ott) 2016;149(3):166–73.

[93] Nicholls SJ. Assessment of clinical effects of cholesteryl ester transfer protein inhibition with evacetrapib in patients at a high risk for vascular outcomes—ACCELERATE American College of Cardiology Annual Scientific Session 2016; 04/03/2016; Chicago, IL; 2016.

[94] Holmes MV, Dale CE, Zuccolo L, Silverwood RJ, Guo Y, Ye Z, et al. Association between alcohol and cardiovascular disease: Mendelian randomisation analysis based on individual participant data. BMJ 2014;349:g4164.

[95] Evans DM, Davey Smith G. Mendelian randomization: new applications in the coming age of hypothesis-free causality. Annu Rev Genomics Hum Genet 2015;16:327–50.

[96] Burgess S, Small DS, Thompson SG. A review of instrumental variable estimators for Mendelian randomization. Stat Methods Med Res 2015;0(0):1–26.

[97] Burgess S, Harshfield E. Mendelian randomization to assess causal effects of blood lipids on coronary heart disease: lessons from the past and applications to the future. Curr Opin Endocrinol Diabetes Obes 2016;23(2):124–30.

[98] Morrison AC, Bare LA, Chambless LE, Ellis SG, Malloy M, Kane JP, et al. Prediction of coronary heart disease risk using a genetic risk score: the Atherosclerosis Risk in Communities Study. Am J Epidemiol 2007;166(1):28–35.

[99] Purcell SM, Wray NR, Stone JL, Visscher PM, O'Donovan MC, Sullivan PF, et al. Common polygenic variation contributes to risk of schizophrenia and bipolar disorder. Nature 2009;460(7256):748–52.

[100] Thanassoulis G, Vasan RS. Genetic cardiovascular risk prediction: will we get there? Circulation 2010;122(22):2323–34.

[101] Goldstein BA, Knowles JW, Salfati E, Ioannidis JP, Assimes TL. Simple, standardized incorporation of genetic risk into non-genetic risk prediction tools for complex traits: coronary heart disease as an example. Front Genet 2014;5:254.

[102] Mega JL, Stitziel NO, Smith JG, Chasman DI, Caulfield MJ, Devlin JJ, et al. Genetic risk, coronary heart disease events, and the clinical benefit of statin therapy: an analysis of primary and secondary prevention trials. Lancet 2015;385(9984):2264–71.

[103] Krarup NT, Borglykke A, Allin KH, Sandholt CH, Justesen JM, Andersson EA, et al. A genetic risk score of 45 coronary artery disease risk variants associates with increased risk of myocardial infarction in 6041 Danish individuals. Atherosclerosis 2015;240(2): 305–10.

[104] Tikkanen E, Havulinna AS, Palotie A, Salomaa V, Ripatti S. Genetic risk prediction and a 2-stage risk screening strategy for coronary heart disease. Arterioscler Thromb Vasc Biol 2013 sep; 33(9):2261–6.

[105] Ganna A, Magnusson PK, Pedersen NL, de Faire U, Reilly M, Arnlov J, et al. Multilocus genetic risk scores for coronary heart disease prediction. Arterioscler Thromb Vasc Biol 2013;33(9):2267–72.

[106] Tada H, Melander O, Louie JZ, Catanese JJ, Rowland CM, Devlin JJ, et al. Risk prediction by genetic risk scores for coronary heart disease is independent of self-reported family history. Eur Heart J 2016;37(6):561–7.

[107] Pepe MS, Kerr KF, Longton G, Wang Z. Testing for improvement in prediction model performance. Stat Med 2013;32(9):1467–82.

[108] Dorresteijn JA, Visseren FL, Ridker PM, Paynter NP, Wassink AM, Buring JE, et al. Aspirin for primary prevention of vascular events in women: individualized prediction of treatment effects. Eur Heart J 2011;32(23):2962–9.

[109] Dorresteijn JA, Visseren FL, Ridker PM, Wassink AM, Paynter NP, Steyerberg EW, et al. Estimating treatment effects for individual patients based on the results of randomised clinical trials. BMJ 2011;343:d5888.

[110] Cordell HJ. Detecting gene–gene interactions that underlie human diseases. Nat Rev Genet 2009;10(6):392–404.

[111] Wei WH, Hemani G, Haley CS. Detecting epistasis in human complex traits. Nat Rev Genet 2014;15(11):722–33.

[112] Hemani G, Shakhbazov K, Westra HJ, Esko T, Henders AK, McRae AF, et al. Detection and replication of epistasis influencing transcription in humans. Nature 2014;508(7495):249–53.

[113] Taylor MB, Ehrenreich IM. Higher-order genetic interactions and their contribution to complex traits. Trends Genet 2015;31(1):34–40.

[114] Polderman TJ, Benyamin B, de Leeuw CA, Sullivan PF, van Bochoven A, Visscher PM, et al. Meta-analysis of the heritability of human traits based on fifty years of twin studies. Nat Genet 2015.

[115] Zhu Z, Bakshi A, Vinkhuyzen AA, Hemani G, Lee SH, Nolte IM, et al. Dominance genetic variation contributes little to the missing heritability for human complex traits. Am J Hum Genet 2015;96(3):377–85.

[116] Smith JA, Ware EB, Middha P, Beacher L, Kardia SL. Current applications of genetic risk scores to cardiovascular outcomes and subclinical phenotypes. Curr Epidemiol Rep 2015;2(3):180–90.

[117] Goni L, Cuervo M, Milagro FI, Martinez JA. A genetic risk tool for obesity predisposition assessment and personalized nutrition implementation based on macronutrient intake. Genes Nutr 2015;10(1):445.

[118] Cole CB, Nikpay M, Lau P, Stewart AF, Davies RW, Wells GA, et al. Adiposity significantly modifies genetic risk for dyslipidemia. J Lipid Res 2014;55(11):2416–22.

[119] Qi Q, Chu AY, Kang JH, Jensen MK, Curhan GC, Pasquale LR, et al. Sugar-sweetened beverages and genetic risk of obesity. N Engl J Med 2012;367(15):1387–96.

[120] Langenberg C, Sharp SJ, Franks PW, Scott RA, Deloukas P, Forouhi NG, et al. Gene–lifestyle interaction and type 2 diabetes: the EPIC interact case-cohort study. PLoS Med 2014;11(5):e1001647.

[121] van Dijk EL, Auger H, Jaszczyszyn Y, Thermes C. Ten years of next-generation sequencing technology. Trends Genet 2014;30(9):418–26.

[122] Grove ML, Yu B, Cochran BJ, Haritunians T, Bis JC, Taylor KD, et al. Best practices and joint calling of the HumanExome BeadChip: the CHARGE Consortium. PLoS One 2013;8(7):e68095.

[123] Do R, Stitziel NO, Won HH, Jorgensen AB, Duga S, Angelica Merlini P, et al. Exome sequencing identifies rare LDLR and APOA5 alleles conferring risk for myocardial infarction. Nature 2015;518(7537):102–6.

[124] Tg, Hdl Working Group of the Exome Sequencing Project NHL Blood I, Crosby J, Peloso GM, Auer PL, Crosslin DR, et al. Loss-of-function mutations in APOC3, triglycerides, and coronary disease. N Engl J Med 2014;371(1):22–31.

[125] Myocardial Infarction Genetics Consortium I, Stitziel NO, Won HH, Morrison AC, Peloso GM, Do R, et al. Inactivating mutations in NPC1L1 and protection from coronary heart disease. N Engl J Med 2014;371(22):2072–82.

[126] Myocardial Infarction G, Investigators CAEC. Coding variation in ANGPTL4, LPL, and SVEP1 and the risk of coronary disease. N Engl J Med 2016;374(12):1134–44.

[127] Panagiotou OA, Willer CJ, Hirschhorn JN, Ioannidis JP. The power of meta-analysis in genome-wide association studies. Annu Rev Genomics Hum Genet 2013;14:441–65.

[128] Iacoviello L, De Curtis A, Donati MB, de Gaetano G. Biobanks for cardiovascular epidemiology and prevention. Future Cardiol 2014;10(2):243–54.

[129] Altman RB, Prabhu S, Sidow A, Zook JM, Goldfeder R, Litwack D, et al. A research roadmap for next-generation sequencing informatics. Sci Trans Med 2016;8(335):335ps10.

[130] Kannry JL, Williams MS. Integration of genomics into the electronic health record: mapping terra incognita. Genet Med 2013;15(10):757–60.

[131] Tak YG, Farnham PJ. Making sense of GWAS: using epigenomics and genome engineering to understand the functional relevance of SNPs in non-coding regions of the human genome. Epigenetics Chromatin 2015;8:57.

[132] Nurnberg ST, Zhang H, Hand NJ, Bauer RC, Saleheen D, Reilly MP, et al. From loci to biology: functional genomics of genome-wide association for coronary disease. Circ Res 2016;118(4):586–606.

[133] Braenne I, Civelek M, Vilne B, Di Narzo A, Johnson AD, Zhao Y, et al. Leducq Consortium CADGd. Prediction of causal candidate genes in coronary artery disease loci. Arterioscler Thromb Vasc Biol 2015;35(10):2207–17.

[134] Haeussler M, Concordet JP. Genome editing with CRISPR-Cas9: can it get any better? J Genet Genomics 2016;43(5):239–50.

[135] Rodriguez F, Knowles JW. PCSK9 inhibition: current concepts and lessons from human genetics. Curr Atheroscler Rep 2015;17(3):487.

Chapter 9

Lung Cancer

Yaron B. Gesthalter, Ehab Billatos and Hasmeena Kathuria
Boston University School of Medicine, Boston, MA, United States

Chapter Outline

INTRODUCTION

Lung cancer is the most common cause of cancer-related death worldwide, with over 1.8 million lung cancer deaths yearly. While cigarette smoking is the leading cause of lung cancer, there has been an increase in nonsmoking related lung cancers, suggesting additional genetic and environmental interactions influencing individual susceptibility to develop lung cancer. Unfortunately, survival rates for newly diagnosed lung cancer remain at 15%, largely unchanged for the past 4–5 decades.

Genomic and Precision Medicine. DOI: http://dx.doi.org/10.1016/B978-0-12-800685-6.00009-6
165

This review summarizes recent progress in understanding the molecular pathogenesis of lung cancer, and attempts to describe how gene expression profiling of human lung cancer and airway specimens is leading to new approaches in lung cancer screening, classification, diagnosis, prognosis, and treatment. The review focuses mainly on genomic studies of lung cancer. Studies using single-nucleotide polymorphism (SNP) and protein arrays are beyond the scope of this review, as are studies of in-vitro cell lines and mouse models. Although we have attempted to be as inclusive as possible, our list of studies reviewed is likely to be incomplete.

EARLY DIAGNOSIS/SCREENING OF LUNG CANCER

Smoking, Lung Cancer, and Genetics

Approximately 80–85% of lung cancer patients have a smoking history, though only 10–15% of smokers develop lung cancer suggesting that hereditary factors play a significant role in lung tumorigenesis. It is well established that smokers with a family history of lung cancer have a two to threefold increased risk of developing lung cancer [1].

In smokers, there are racial and ethnic differences in lung cancer incidence. In the United States, Native American, African American, Native Hawaiian, and other Polynesians have a higher incidence of lung cancer than whites, while Latinos and Japanese-Americans have a lower incidence than whites [2]. Studies suggest that some of these differences in susceptibility may be related to nicotine metabolism where racial/ethnic groups with higher rates of lung cancer have slower nicotine metabolism [3].

A never-smoker is defined as an individual who has had a lifetime exposure of fewer than 100 cigarettes. Worldwide, lung cancer in never-smokers is more prevalent among women. In Asia, 60–80% of women with lung cancer are never-smokers whereas 15% of women in the United States are never-smokers [4].

Among never-smokers, risk factors for lung cancer include exposures to secondhand smoke, domestic coal use for heating, high-dose vitamin E supplementation, indoor air pollution from solid fuel use, cannabis smoking, cooking fumes, asbestos, radon, chromium, arsenic, cadmium, silica, and nickel as well as outdoor air pollutants, previous lung disease, and dietary factors [4]. Germline transmission of the epidermal growth factor receptor (EGFR) gene mutation confers a one-in-three chance of developing lung cancer in never-smokers [5].

Genome-wide association studies (GWAS) have also been instrumental in identifying some of the key lung cancer susceptibility loci at 15q25, 6p21, and 5p15.33. A confirmatory study by Truong et al. in more than 26,000 lung cancer case and control subjects validated an association between lung cancer risk and 15q25 and 5p15, but not 6p21. Interestingly, associations between 15q25 and the risk of lung cancer were replicated in white smokers, but not in never-smokers or in Asians [6].

Identifying Smokers at Risk

Smoking cessation carries the highest potential in curbing lung cancer related mortality [7]. However, lung cancer risk persists years after quitting, and when considering the high rates of ever smokers, identifying those at greatest risk remains an important public health issue. Current strategies target the early detection of lung cancer. The largest lung cancer screening trial to date, National Lung Screening Trial (NLST) concluded with a 20% decrease in lung cancer related mortality among participants screened with low-dose computer tomography (LDCT) [8]. By using the inclusion criteria employed by the NLST (age 55–74, 30-pack year smoking history, last smoked within 15 years among former smokers), a large proportion of people at risk may be missed; yet expanding the inclusion criteria may increase false positive rates and the subsequent cost and morbidity of follow-up testing.

Identifying key molecular waypoints in lung cancer would allow early targeting of oncogenic processes. Using SNP microarrays, Nakachi et al. illustrated that in smokers, somatic chromosomal aberrations, such as loss of heterozygosity and allelic copy number variations, are similar in both tumor tissue and high-grade dysplastic lesions relative to normal tissue, demonstrating genomic instability as a precursor event to lung tumorigenesis in smokers [9]. In a small study performing RNA sequencing from normal, dysplastic, and tumor tissue, Ooi et al. characterized a transcriptomal step-wise progression from normal airway epithelium to the development of squamous cell carcinoma (SCC) [10]. Together, these studies illustrate a measurable chain of events in the development of smoking-related lung cancer, which if identified early, could be leveraged to develop premalignancy biomarkers and targeted chemopreventive agents.

Studies utilizing bronchial epithelium have demonstrated the ability to identify diffuse molecular alterations in normal tissue of smokers with and without lung cancer, suggesting that molecular profiles of the host transcriptional response can detect a field of cancerization amongst ever smokers. Kadara et al. characterized both global and tumor adjacent bronchial epithelial gene expression amongst smokers and showed that airway field of cancerization exhibits both site-independent profiles and gradient site-dependent expression patterns with respect to tumor proximity [11]. This study supports the concept of leveraging global bronchial gene expression profiling as a marker of distal disease.

In an attempt to identify genes predictive of future lung cancer development, Blomquist et al. reported 14 genes relating to DNA repair and antioxidant processes in the normal bronchial epithelium that predicted lung cancer development [12]. Using cytologically normal tissue from smokers in an independent validation cohort, they were able to distinguish those with and without cancer. Another study by Spira et al. described an 80 gene bronchial epithelial biomarker capable of distinguishing ever smokers with and without lung cancer [13].

To begin to understand the mechanisms by which some individuals protect themselves from the carcinogenic effects of smoking, Spira et al. characterized the epithelial transcriptome of smokers relative to nonsmokers [14,15]. They noted that the expression level of smoking-induced genes among former smokers began to resemble that of never-smokers after two years of smoking cessation. Genes that reverted to normal within two years of cessation tended to serve metabolizing and antioxidant functions. Several genes, including putative oncogenes and tumor suppressor genes, failed to revert to never-smoker levels years after smoking cessation, perhaps explaining the continued risk for developing lung cancer many years after individuals have stopped smoking.

Efforts to identify readily accessible sampling sites have pushed for extrathoracic lung cancer biomarkers. Genetic abnormalities have been identified in sputum of smokers with lung cancer [16]. Shen et al. recently reported the incorporation of sputum microRNA (short, noncoding RNA strands thought to be relatively stable) analysis to improve CT-based lung cancer diagnosis with an increase in specificity compared to CT alone [17]; however, this has yet to be independently validated. While promising, the developments of these strategies have been challenged due to technical constraints relating to quality of samples and degradation of salivary RNA. Similar strategies have been employed to use profiling of peripheral blood circulating mononuclear cells; however, these are still under development and have yet to be independently validated as a classifier of disease presence.

Future Directions

Since molecular, genetic, and epigenetic abnormalities precede morphological changes in bronchi and alveoli, early detection biomarkers may help select a group of high-risk patients who would benefit most from LDCT screening and chemoprevention strategies. Sridhar et al. studied gene expression profiles in epithelial cells that line the intrathoracic (bronchial) and extrathoracic (buccal and nasal) airways in healthy and current smokers. The authors identified a common set of genes induced by cigarette smoke in buccal, nasal, and bronchial epithelium [18]. Similarly, a study by Boyle et al. using whole-genome gene expression profiling of punch biopsies from the buccal mucosa of 40 healthy smokers and 40 healthy nonsmokers showed a strong relationship between the gene expression responses to smoking in buccal and bronchial epithelium [19]. These studies support the concept that smoking induces a common field of injury throughout the airway and that easily collected buccal and nasal epithelium may provide a biomarker that identifies smokers at high risk for developing lung cancer.

Recently, Silvestri et al. reported the results of two trials where current and former smokers undergoing bronchoscopy for suspected lung cancer were enrolled in two multicenter prospective studies. A 23 gene expression classifier measured in bronchial epithelial cells, combined with a physician pretest probability of lung cancer, was used to assess the probability of lung cancer. The negative predictive value was highest amongst patients with an intermediate pretest probability, offering clinicians a means to decide on need for additional work up in cases with nondiagnostic bronchoscopy [20].

Although not yet applied to clinical trials, a recent study identified a blood-based 13-protein biomarker capable of distinguishing benign from malignant nodules detected on CT scan with a high negative predictive value [21], suggesting a potential role for blood biomarkers in the evaluation of indeterminate pulmonary nodules.

Chemoprevention in high-risk current or former smokers represents one of the most important areas of current research in the prevention of lung cancer. Despite numerous chemoprevention trials, no studies to date have demonstrated a positive outcome in lung cancer mortality, and few studies have demonstrated an effect on intermediate end points such as sputum atypia or genomic markers of epithelial cell damage [22] (reviewed by Kathuria).

Molecular alterations in the field of injury that persist after smoking cessation may help to explain persistent lung cancer risk in former smokers, and risk-related enzymes that metabolize carcinogens in cigarette smoke suggest possible targets for lung cancer chemoprevention. Gustafson et al. observed a significant increase in a genomic signature of phosphatidylinositol 3-kinase (PI3K) pathway activation in cytologically normal bronchial airway of smokers with lung cancer and smokers with dysplastic lesions. The authors reported decreased PI3K activity in the airways of high-risk smokers who had significant regression of dysplasia after treatment with *myo*-inositol, an inhibitor of the PI3K pathway in vitro [23]. The potential impact of chemoprevention is large, but the field awaits the emergence of intermediate markers of cancer risk that must be validated in prospective studies.

CLASSIFICATION AND PROGNOSIS

Histological Classification of Lung Tumors and TNM Staging

Lung tumors are classified as small-cell lung carcinomas (SCLC) or nonsmall-cell carcinomas (NSCLC). Small-cell lung cancers have neuroendocrine features that are identified by immunohistochemistry and histology. Nonsmall-cell carcinomas are subcategorized as adenocarcinomas (the most common), SCCs, and large-cell carcinomas; these are clinically distinct from SCLC. The pathological distinction between SCLC and NSCLC is important since the tumor types are treated differently.

Tumors with the same histological classification behave differently, and morphological classification has not been effective in predicting the aggressiveness of a cancer or how the cancer will respond to therapeutic agents. Yoshizawa et al. studied the impact of a previously reported joint statement from IASLC/ ATS/ERS, which classified adenocarcinomas according to the predominant histological features (lepidic, acinar, papillary, solid and micro papillary). They noted a favorable prognosis for lepidic predominant, intermediate survival for acinar and papillary predominant, and poor prognosis for solid and micropapillary predominant adenocarcinomas [24].

In NSCLC, the anatomic extent or stage of the tumor as defined by TNM descriptors is the most important prognostic variable for predicting survival. In late 2009, the International Association for the Study of Lung Cancer (IASLC) published the seventh revision of the TNM staging system [25] (summarized by Detterbeck) to improve the alignment of the TNM stage with prognosis. Despite these changes and improvements in diagnosis, including both noninvasive methods (CT/PET) and invasive methods (endoscopic biopsies), patients who are diagnosed with similar stages and are treated using similar protocols often respond quite differently and have varying survival rates.

Several combined clinical, histological, and laboratory variables such as age, stage or grade of tumor, and serum protein levels can be used to assess a patient's prognosis with variable accuracy. But these criteria are not able to provide important information about the prognostic diversity within each stage, such as how aggressive a particular subtype will be or how a patient will respond to therapy. By combining clinical variables and histopathology with gene expression profiling, predicting a patient's prognosis may improve.

Molecular Classification and Prognostic Value of Lung Cancer Genomics

Since lung cancer is genetically heterogeneous, morphological tumor classification does not always accurately predict the patient's clinical behavior. Patients with Stage IA lung cancer resected with curative intent still have 30% mortality from local recurrence, distant metastases, and/or new occurrence. Within each clinical stage, there is variability in the presence of specific mutations, deletions of tumor suppressor genes, amplifications of oncogenes, and chromosomal abnormalities. Genomics is a powerful tool for classifying tumor subtypes. Lung cancer patients with biomarkers that predict a poor outcome could be selected for adjuvant chemotherapy, while those that predict a good prognosis may be able to avoid the toxicity and cost of unnecessary chemotherapy.

Microarray studies have identified and validated specific genes whose expression differs between normal and tumor tissue. Using a large-scale meta-analysis approach of five adenocarcinoma gene expression datasets, Chen et al. found 11 genes to be significantly overexpressed in adenocarcinomas compared to normal tissue. Six genes in this signature were specifically overexpressed in adenocarcinoma relative to other NSCLC subtypes [26]. Another study reviewed 14 microRNA expression profiling studies and found miR-210 and miR-21 to be consistently upregulated in lung cancer tissues compared to normal lung tissue [27].

Genomic high-throughput technologies have been used in many studies to identify gene expression signatures that predict patient survival and/or relapse rates. A considerable number of expression profiling studies have been performed on clinical lung cancer specimens [28] (reviewed by Santos). New attempts to compare results from several different studies are being undertaken. Using a cohort of 196 NSCLC, patients with clinical information and long-term follow-up, a prognostic probe set was developed which was then tested in a meta-analysis validation approach of five independent lung cancer cohorts. 14 genes were found to be significantly associated with survival. One of these genes, cell adhesion molecule 1 (CADM1), was confirmed by use of immunohistochemistry on tissue microarrays from two independent NSCLC cohorts. Low CADM1 protein expression was significantly associated with shorter survival, particularly in adenocarcinomas [29]. In a meta-analysis of an independent NSCLC cohort consisting of nine datasets (1382 samples), TAZ expression correlated with decreased survival [30]. Recently, a 4-gene prognostic classifier (*BRCA1, HIF1A, DLC1*, and *XPO1*) that had identified stage 1 lung adenocarcinoma patients at high risk for relapse was validated using a meta-analysis based approach in 12 cohorts (>1000 samples) of stage 1 lung adenocarcinomas patients [31]. This 4-gene classifier, when combined with miR-21 expression and HOXA9 promoter methylation identified high-risk Stage I patients [32].

Future Directions

Currently, lung cancer patients within a given clinical stage and tumor type receive the same treatment despite the genomic heterogeneity that exists between tumors. Although expression profiling of lung cancer has identified prognostic genes and subtypes of adenocarcinomas, until recently those profiles have not been ready to be incorporated into clinical practice. Comparison of array studies has been difficult because the array platforms, sample preparation, and technical factors have been different. In addition, many molecular classification studies do not match the classification based on tumor histology, perhaps because use of whole lungs for microarray studies may not accurately reflect gene expression in the cancer cells due to contamination and differences in abundance of stromal and surrounding normal cells.

To address these issues, a multiinstitutional collaborative study was conducted to generate gene expression profiles from a large number of lung cancer specimens, the results of which will hopefully lead to a gene expression signature that will improve diagnostic options. A pooled analysis of 442 lung adenocarcinomas from multiple institutions established the performance of gene expression signatures across different patient populations and different laboratories. This study identified predictors of survival based on clinical and gene expression microarray data, with better accuracy to predict survival with combined clinical and molecular data [33].

The Cancer Genome Atlas is an additional large-scale initiative aimed at characterizing varying cancers genomic profiles through different platforms.

Redefining tumor classification from strictly morphology-based schemes to molecular-based classifications promises to provide clinically important information on tumor subsets within morphological classes. Prognostic profiles may guide clinical decision making by identifying high-risk lung cancer patients who would benefit from improved diagnostic and treatment options [34,35].

PATHOGENESIS AND TREATMENT OF LUNG CANCER

Molecular Alterations in Lung Cancer

Proto-oncogenes

The *Ras* genes (*Hras, Kras*, and *Nras*) encode GTPase proteins that help transduce survival- and growth-promoting signals. When oncogenic mutations occur, the normal abrogation of RAS signaling is impaired, resulting in persistent signaling (reviewed by Singhal) [36]. Point mutations (found most frequently in codons 12, 13, and 61) are detected in 20–30% of lung adenocarcinomas, and 90% of these mutations are found in *Kras* (Table 9.1).

In 2007, Soda et al. described an EML4-ALK fusion oncogene that generates aberrant signaling in NSCLC [37]. Lung tumors that contain the EML4-ALK fusion oncogene or its variants are associated with specific clinical features, including never or light smoking history, younger age, and adenocarcinoma. ALK gene arrangements are largely mutually exclusive with EGFR or KRAS mutations and occur in 2–7% of NSCLC patients. Crizotinib is a newly approved targeted therapy with activity against the kinases of the products of ALK [38].

Growth Factors and Their Receptors

Lung tumors often express growth factors and their receptors, resulting in regulatory loops that stimulate tumor growth. Specific mutations of the EGFR, also known as ERBB-1, are highly expressed in many epithelial tumors, including 15% of lung adenocarcinomas in the United States [39]. The prevalence of this mutation is higher amongst women, never-smokers, and Asians. From a molecular standpoint, the ligand (EGF or TGF-α) binds to EGFR causing receptor tyrosine kinase activation and a series of downstream signaling events resulting in cellular proliferation, increased cell motility, tumor invasion, antiapoptosis, and resistance to chemotherapy.

Anti-EGFR drugs approved for cancer treatment include the monoclonal antibodies directed against the extracellular domain of the receptor (anti-EGFR Mabs) and small molecule inhibitors of EGFR's tyrosine kinase activity (TKIs). Lee et al. published a comprehensive meta-analysis incorporating results from 23 trials in nearly 15,000 patients with more than 4,000 having molecular analysis. While EGFR TKIs had no impact on overall survival for all-comers (both EGFR mutation positive and negative individuals), it did prolong progression-free survival in individuals harboring EGFR mutations [40]. The failed

TABLE 9.1 Common Molecular Alterations in Lung Cancer

	Type of Mutation	Frequency			Currently available targeted therapies
		Adeno-carcinoma (%)	Squamous Cell Carcinoma (%)	Small Cell Lung Cancer (%)	
Receptor Tyrosine Kinases					
EGFR	Mutation	10	2–3		Erlotinib, Gefitinib, and Afatinib
FGFR1	Amplification		20		
EML4-ALK	Fusion	3–5	<1%		Crizotinib and Ceritinib
MET	Amplification	2–4			Crizotinib
ROS1	Fusion	1–2			Crizotinib
RET	Fusion	1			
HER2	Mutation/ Amplification	2–4			Neratinib, Afatinib, Lapatinib, and Trastuzumab
Signaling					
KRAS	Mutation	15–25	1–2		
BRAF	Mutation	1–6	4–5		
Transcription Factors					
SOX2	Amplification	6	65		
MYC	Amplification	25		30–40	
Cell Cycle					
CDKN2A	Mutation	7	15	>80	
Tumor Suppressor					
P53	Mutation	52	79	75	
LKB1	Mutation	9	2		

impact of TKIs on overall survival lies in the invariable acquisition of resistance amongst patients who initially respond to TKIs. Acquired resistance has been attributed to the development of secondary mutations, most common of which is the T790M gene in 50% of cases or the amplification of the mesenchymal epithelial transition factor (MET) in 20% of cases.

Tumor Suppressor Genes

The role of *p53* is to help maintain genomic integrity after DNA damage. When cells undergo stress, *p53* becomes upregulated and acts as a transcription factor to increase genes such as *p21* (which in turn controls G1/2 cell cycle transition) and induces apoptosis by activating genes such as *BAX*, *PERP*, and others [41] (reviewed by Fong). In lung cancer, missense mutations can occur and cause loss of *p53* function in more than 75% of SCLCs and 50% of NSCLCs.

Decreased expression of p16 by promoter hypermethylation, mutations, or allelic loss occurs in 30–50% of NSCLCs. The *p16^{INK4A}-cyclin D1-CDK4-RB* pathway controls G1/S cell cycle transition, and loss of p16 releases the tumor cells from RB-mediated cell cycle arrest [41]. Alternatively, the *Rb* gene can be inactivated directly by deletions, point mutations, or alternative splicing, and is more commonly found in SCLCs (>90%), than NSCLCs (15–30%).

Oncogenic Pathway Signatures

Oncogenesis in the lung is a complicated process that stems from impairments at all levels of genomic cell cycle regulation. Somatic DNA mutations play a particularly important role as demonstrated by the mortality benefit amongst those patients harboring specific somatic mutations. Mutations and amplifications in many potentially targetable oncogenes have been identified in lung adenocarcinoma, including EGFR, ERBB2, MET, FGFR1, and FGFR2, as well as fusion oncogenes involving ALK, the ROS1 receptor tyrosine kinase, NRG1, NTRK1, and RET [42] (reviewed by Chen).

Epigenetic changes are heritable changes in gene expression that do not alter the primary DNA sequences and are prevalent in all types of cancer. Epigenetic regulation of genes associated with tumorigenesis through DNA methylation, histone modification and micro-RNA associated gene silencing has gained increasing attention [43] (reviewed by Mehta).

Micro-RNAs regulate cell homeostasis and fate through messenger RNA silencing. Using deep sequencing analysis for comprehensive profiling of miR-NAs, Wang et al. described a pattern of "progressive dysregulation" of normal, normal adjacent, and tumor tissue in samples collected from 19 patients with SCC [44]. Functional enrichment of genes associated with cancer development included perturbations in pathways relating to MAPK, proteoglycan, Wnt, PI3K/AKT, and TGFβ pathways. Recently, Edmonds et al. identified distinct miRNA profiles in retrospectively collected samples from those with either relapsing or nonrelapsing adenocarcinoma [45]. Together these studies suggest the potential role miRNAs may play in lung cancer diagnostics, prognostication and therapeutics.

Promoter methylation is an additional cell fate determinant that has been associated with lung cancer [43] and genome-wide scans for aberrant promoter methylation have identified novel targets. A recent study that used Restriction Landmark Genomic Scanning to evaluate promoter DNA methylation status in primary NSCLC tumor samples and matched normal controls determined that 4.8% of all CpG island promoters in a lung cancer genome are targeted for aberrant DNA methylation [46]. Studies looking at methylation patterns in sputum and lung tissue have defined signatures capable of distinguishing cancer and benign samples; however, these have yet to be validated in clinical settings [47].

Gene expression based platforms provide a dynamic platform through which downstream effects of oncogenic pathway dysregulation can be leveraged for diagnostic and therapeutic approaches. Additional applications of gene expression pathway profiling in lung cancer is the repositioning of approved drugs for novel indications.

Molecularly Targeted Therapy to Reduce Lung Cancer Mortality

Biomarkers provide an opportunity to identify subpopulations of patients who are most likely to respond to a given therapy and to identify new targets for drug development. Several studies have now shown the therapeutic efficacy and safety of target-specific inhibitors for specific tumor subtypes. Tyrosine kinase inhibitors prolong progression-free survival amongst end stage adenocarcinoma patients with EGFR mutations. The ALK inhibitor crizotinib improves progression-free survival not only in advanced *ALK*-positive nonsmall cell lung cancer (NSCLC), but also in those with ROS1 chromosomal rearrangements [38,48].

Given the complexity of NSCLC, it is likely that these tumors are dependent on more than one oncogenic signaling pathway. Identifying *KRAS* mutations may become increasingly important with development of new therapies targeting downstream RAS pathways. Recent studies have suggested that targeting *MET, BRAF, AKT1, ERBB2*, and *PIK3CA* mutations and *ROS1* and *RET* fusion oncogenes may be beneficial [42] (reviewed by Chen). Although these strategies are promising, there is insufficient published data on the clinical validity of these methods.

Through next generation sequencing, lung cancers specimens are being analyzed to identify somatic changes in gene copy number, sequence, and expression. Comprehensive molecular profiling of 230 lung adenocarcinomas showed that 62% (143/230) of tumors harbored activating mutations in known driver oncogenes, including KRAS (32%), EGFR (11%), and BRAF (7%). RTK/RAS/RAF pathway was activated in 76% of these cases. On the other hand, MET and ERBB2 amplifications as well as NF1 and RIT1 mutations were enriched in samples otherwise lacking an activated oncogene, suggesting a driver role for these events in certain tumors [35].

In lung SCC, DDR2, FGFR1, FGFR2, FGFR3, and genes in the PI3K pathway seem to be more commonly mutated [34]. Recently, 110 small cell lung cancers (SCLC) have been sequenced and biallelic inactivation of TP53 and RB1 has been found in almost all specimens. In addition, 25% of SCLC samples had inactivating mutations in NOTCH family genes, identifying candidate therapeutic targets [49].

It is likely that combination-targeted therapy directed at multiple oncogenic pathways may not only prove more effective than single agents alone, but may also prevent or delay secondary resistance. Gene expression signatures from tumor biopsy specimens have been developed that can predict sensitivity to individual chemotherapeutics. Furthermore, these chemotherapy response signatures are being integrated with signatures of oncogenic pathway deregulation to potentially identify new therapeutic strategies. Results from a Phase II clinical trial program, BATTLE (Biomarker-integrated Approaches of Targeted Therapy for Lung Cancer Elimination), suggest that patients prescribed existing drugs based on their tumor biomarkers benefit more than patients whose treatments are not based on their tumor biomarkers [50]. The ongoing BATTLE-2 trial, which aims to analyze biomarkers and therapies involving several pathways in NSCLC, including dual targeting of two different combinations, demonstrates the considerable potential of using gene expression profiling of tumors to define causal molecular pathways and potential therapeutic targets for individuals.

CONCLUSION

To date, significant advances in lung cancer treatment have come from advances in our understanding of EGFR activating mutations, ALK fusion genes, ROS-1 rearrangements, and ongoing research in targeted therapies for K-RAS and MET. Lung cancer is being subdivided into molecular subtypes with dedicated targeted and chemotherapeutic strategies. Yet lung cancer mortality rates remain higher than the next three major cancers combined.

A multitargeted approach focusing on prevention, early detection, and molecularly targeted therapy is needed to optimize lung cancer diagnosis and treatment. Tobacco use must be the first target, since prevention is the most effective way to reduce lung cancer mortality. Early detection must be the second target, using LDCT scanning combined with molecular profiling to identify therapeutically exploitable differences between normal, precancerous, and cancer cells. These approaches must be combined with genomic and genetic biomarkers that identify current and former smokers at highest risk for developing lung cancer, who may well benefit from one of the many chemopreventive medications now being tested in clinical trials.

The final target must be better treatment. Better understanding of cancer heterogeneity, including variability in prognosis and patient response to therapy, requires stronger working relationships and collaborations between bench

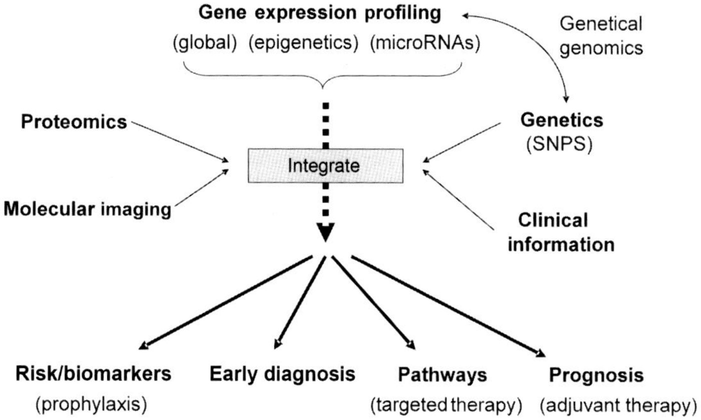

FIGURE 9.1 Diagram depicting contributions of gene expression profiling which, combined with genetics, clinical information, proteomics, and imaging studies, can be applied to developing (1) risk-assessment tools to identify current and former smokers at highest risk for developing lung cancer and who might benefit from chemopreventive therapy; (2) tools for early diagnosis of current and former smokers with Stage I, potentially resectable lung cancer; (3) tools to define molecular pathways that have led to individual lung cancers and to predict what pharmacological approaches might be used to define best approaches to treatment of individual lung cancers; and (4) prognosis of resected cancers, defining which patients have high probability of recurrence or metastases and should therefore receive adjuvant therapy.

scientists and their clinical counterparts. The implementation of next-generation sequencing-based methods and their application to clinical medicine is growing rapidly. Pooling data sets of lung cancer specimens, including gene expression array data and sequencing data have already provided insights into the mutation spectra of lung cancer. The contributions of modern genomic technologies, particularly those that provide measures of global gene expression, are depicted in Fig. 9.1. Combined with genome-wide SNP screens that hold the promise of determining heritable predispositions to developing lung cancer and defining pathway-based pharmacogenomics, there is hope that lung cancer may eventually no longer be the number one cause of cancer death in the world.

GLOSSARY TERMS

Chemoprevention the use of drugs, vitamins, or other agents to try to reduce the risk of, or delay the development or recurrence of cancer

Field cancerization the constellation of locoregional changes triggered by long-term or repeated exposure of a field of tissue to a carcinogenic insult (e.g., tobacco or alcohol); field cancerization may induce dysplasia, carcinoma in situ (CIS), or carcinoma.

REFERENCES

[1] Lin H, Huang Y-S, Yan H-H, Yang X-N, Zhong W-X, Ye H-W, et al. A family history of can-cer and lung cancer risk in never-smokers: A clinic-based case–control study. Lung Cancer 2015;89(2):94–8.

[2] Dela Cruz CS, Tanoue LT, Matthay RA. Lung cancer: epidemiology, etiology, and prevention. Clin Chest Med 2011;32(4):605–44.

[3] Fagan P, Moolchan ET, Pokhrel P, Herzog T, Cassel KD, Pagano I, et al. Biomarkers of tobacco smoke exposure in racial/ethnic groups at high risk for lung cancer. Am J Public Health 2015;105(6):1237–45.

[4] Sun S, Schiller JH, Gazdar AF. Lung cancer in never smokers – a different disease. Nat Rev Cancer 2007;7(10):778–90.

[5] Bell DW, Gore I, Okimoto RA, Godin-Heymann N, Sordella R, Mulloy R, et al. Inherited susceptibility to lung cancer may be associated with the T790M drug resistance mutation in EGFR. Nat Genet 2005;37(12):1315–6.

[6] Truong T, Sauter W, McKay JD, Hosgood HD, Gallagher C, Amos CI, et al. International Lung Cancer Consortium: coordinated association study of 10 potential lung cancer suscep-tibility variants. Carcinogenesis 2010;31(4):625–33.

[7] Anthonisen NR, Skeans MA, Wise RA, Manfreda J, Kanner RE, Connet JE. The effects of a smoking cessation intervention on 14.5-year mortality: a randomized clinical trial. Ann Intern Med 2005;142(4):233.

[8] National Lung Screening Trial Research Team Aberle DR, Adams AM, Berg CD, Black WC, Clapp JD, et al. Reduced lung-cancer mortality with low-dose computed tomographic screen-ing. N Engl J Med 2011;365(5):395–409.

[9] Nakachi I, Rice JL, Coldren CD, Edwards MG, Stearman RS, Glidewell SC, et al. Application of SNP microarrays to the genome-wide analysis of chromosomal instability in premalignant airway lesions. Cancer Prev Res (Phila) 2014;7(2):255–65.

[10] Ooi AT, Gower AC, Zhang KX, Vick JL, Hong L, Nagao B, et al. Molecular profiling of premalignant lesions in lung squamous cell carcinomas identifies mechanisms involved in stepwise carcinogenesis. Cancer Prev Res (Phila) 2014;7(5):487–95.

[11] Kadara H, Fujimoto J, Yoo S-Y, Maki Y, Gower AC, Kabbout M, et al. Transcriptomic archi-tecture of the adjacent airway field cancerization in non-small cell lung cancer. J Natl Cancer Inst 2014;106(3) dju004–dju004.

[12] Blomquist T, Crawford EL, Mullins D, Yoon Y, Hernandez D-A, Khuder S, et al. Pattern of antioxidant and DNA repair gene expression in normal airway epithelium associated with lung cancer diagnosis. Cancer Res 2009;69(22):8629–35.

[13] Spira A, Beane JE, Shah V, Steiling K, Liu G, Schembri F, et al. Airway epithelial gene expression in the diagnostic evaluation of smokers with suspect lung cancer. Nat Med 2007;13(3):361–6.

[14] Beane J, Sebastiani P, Liu G, Brody JS, Lenburg ME, Spira A. Reversible and permanent effects of tobacco smoke exposure on airway epithelial gene expression. Genome Biol 2007;8(9):R201.

[15] Spira A, Beane J, Shah V, Liu G, Schembri F, Yang X, et al. Effects of cigarette smoke on the human airway epithelial cell transcriptome. Proc Natl Acad Sci U S A 2004;101(27):10143–8.

[16] Hassanein M, Callison JC, Callaway-Lane C, Aldrich MC, Grogan EL, Massion PP. The state of molecular biomarkers for the early detection of lung cancer. Cancer Prev Res (Phila) 2012;5(8):992–1006.

[17] Shen J, Liao J, Guarnera MA, Fang H, Cai L, Stass SA, et al. Analysis of microRNAs in sputum to improve computed tomography for lung cancer diagnosis. J Thorac Oncol 2014;9(1):33–40.

[18] Sridhar S, Schembri F, Zeskind J, Shah V, Gustafson AM, Steiling K, et al. Smoking-induced gene expression changes in the bronchial airway are reflected in nasal and buccal epithelium. BMC Genomics 2008;9:259.

[19] Boyle JO, Gümüs ZH, Kacker A, Choksi VL, Bocker JM, Zhou XK, et al. Effects of cigarette smoke on the human oral mucosal transcriptome. Cancer Prev Res (Phila) 2010;3(3):266–78.

[20] Silvestri GA, Vachani A, Whitney D, Elashoff M, Porta Smith K, Ferguson JS, et al. AEGIS Study Team. A bronchial genomic classifier for the diagnostic evaluation of lung cancer. N Engl J Med 2015;373(3):243–51.

[21] Li X, Hayward C, Fong PY, Dominguez M, Hunsucker SW, Lee LW, et al. A blood-based proteomic classifier for the molecular characterization of pulmonary nodules. Sci Transl Med 2013;5(207) 207ra142.

[22] Kathuria H, Gesthalter Y, Spira A, Brody J, Steiling K. Updates and controversies in the rapidly evolving field of lung cancer screening, early detection, and chemoprevention. Cancers 2014;6(2):1157–79.

[23] Gustafson AM, Soldi R, Anderlind C, Scholand MB, Qian J, Zhang X, et al. Airway PI3K pathway activation is an early and reversible event in lung cancer development. Sci Transl Med 2010;2(26) 26ra25.

[24] Yoshizawa A, Motoi N, Riely GJ, Sima CS, Gerald WL, Kris MG, et al. Impact of proposed IASLC/ATS/ERS classification of lung adenocarcinoma: prognostic subgroups and implications for further revision of staging based on analysis of 514 stage I cases. Mod Pathol 2011;24(5):653–64.

[25] Detterbeck FC, Boffa DJ, Tanoue LT. The new lung cancer staging system. Chest 2009;136(1):260–71.

[26] Chen R, Khatri P, Mazur PK, Polin M, Zheng Y, Vaka D, et al. A meta-analysis of lung cancer gene expression identifies PTK7 as a survival gene in lung adenocarcinoma. Cancer Res 2014;74(10):2892–902.

[27] Guan P, Yin Z, Li X, Wu W, Zhou B. Meta-analysis of human lung cancer microRNA expression profiling studies comparing cancer tissues with normal tissues. J Exp Clin Cancer Res 2012;31(1):54.

[28] Santos ES, Blaya M, Raez LE. Gene expression profiling and non-small-cell lung cancer: where are we now? Clin Lung Cancer 2009;10(3):168–73.

[29] Botling J, Edlund K, Lohr M, Hellwig B, Holmberg L, Lambe M, et al. Biomarker discovery in non-small cell lung cancer: integrating gene expression profiling, meta-analysis, and tissue microarray validation. Clin Cancer Res 2013;19(1):194–204.

[30] Noguchi S, Saito A, Horie M, Mikami Y, Suzuki HI, Morishita Y, et al. An integrative analysis of the tumorigenic role of TAZ in human non-small cell lung cancer. Clin Cancer Res 2014;20(17):4660–72.

[31] Okayama H, Schetter AJ, Ishigame T, Robles AI, Kohno T, Yokota J, et al. The expression of four genes as a prognostic classifier for stage I lung adenocarcinoma in 12 independent cohorts. Cancer Epidemiol Biomarkers Prev 2014;23(12):2884–94.

[32] Robles AI, Arai E, Mathé EA, Okayama H, Schetter AJ, Brown D, et al. An integrated prognostic classifier for stage I lung adenocarcinoma based on mRNA, microRNA, and DNA methylation biomarkers. J Thorac Oncol 2015;10(7):1037–48.

[33] Director's Challenge Consortium for the Molecular Classification of Lung Adenocarcinoma Shedden K, Taylor JM, Enkemann SA, Tsao MS, Yeatman TJ, et al. Gene expression-based

survival prediction in lung adenocarcinomas: a multi-site, blinded validation study. Nat Med 2008;14(8):822–7.

[34] Cancer Genome Atlas Research Network Comprehensive genomic characterization of squamous cell lung cancers. Nature 2012;489(7417):519–25.

[35] Collisson EA, Campbell JD, Brooks AN, Berger AH, Lee W, Chmielecki J, et al. Comprehensive molecular profiling of lung adenocarcinoma. Nature 2014;511(7511):543–50.

[36] Singhal S, Vachani A, Antin-Ozerkis D, Kaiser LR, Albelda SM. Prognostic implications of cell cycle, apoptosis, and angiogenesis biomarkers in non-small cell lung cancer: a review. Clin Cancer Res 2005;11(11):3974–86.

[37] Soda M, Choi YL, Enomoto M, Takada S, Yamashita Y, Ishikawa S, et al. Identification of the transforming EML4-ALK fusion gene in non-small-cell lung cancer. Nature 2007;448(7153):561–6.

[38] Shaw AT, Kim D-W, Nakagawa K, Seto T, Crinó L, Ahn M-J, et al. Crizotinib versus chemotherapy in advanced *ALK*-positive lung cancer. N Engl J Med 2013;368(25):2385–94.

[39] Reissmann PT, Koga H, Figlin RA, Holmes EC, Slamon DJ. Amplification and overexpression of the cyclin D1 and epidermal growth factor receptor genes in non-small-cell lung cancer. Lung Cancer Study Group. J Cancer Res Clin Oncol 1999;125(2):61–70.

[40] Lee CK, Brown C, Gralla RJ, Hirsh V, Thongprasert S, Tsai C-M, et al. Impact of EGFR inhibitor in non-small cell lung cancer on progression-free and overall survival: a meta-analysis. J Natl Cancer Inst 2013;105(9):595–605.

[41] Fong KM, Sekido Y, Gazdar AF, Minna JD. Lung cancer. 9: molecular biology of lung cancer: clinical implications. Thorax 2003;58(10):892–900.

[42] Chen Z, Fillmore CM, Hammerman PS, Kim CF, Wong K-K. Non-small-cell lung cancers: a heterogeneous set of diseases. Nat Rev Cancer 2014;14(8):535–46.

[43] Mehta A, Dobersch S, Romero-Olmedo AJ, Barreto G. Epigenetics in lung cancer diagnosis and therapy. Cancer Metastasis Rev 2015;34(2):229–41.

[44] Wang J, Li Z, Ge Q, Wu W, Zhu Q, Luo J, et al. Characterization of microRNA transcriptome in tumor, adjacent, and normal tissues of lung squamous cell carcinoma. J Thorac Cardiovasc Surg 2015;149(5) 1404–1414.e4.

[45] Edmonds MD, Eischen CM. Differences in miRNA expression in early stage lung adenocarcinomas that did and did not relapse. PLoS ONE 2014;9(7):e101802.

[46] Risch A, Plass C. Lung cancer epigenetics and genetics. Int J Cancer 2008;123(1):1–7.

[47] Brothers JF, Hijazi K, Mascaux C, El-Zein RA, Spitz MR, Spira A. Bridging the clinical gaps: genetic, epigenetic and transcriptomic biomarkers for the early detection of lung cancer in the post-National Lung Screening Trial era. BMC Med 2013;11:168.

[48] Shaw AT, Ou S-HI, Bang Y-J, Camidge DR, Solomon BJ, Salgia R, et al. Crizotinib in *ROS1*-rearranged non-small-cell lung cancer. N Engl J Med 2014;371(21):1963–71.

[49] George J, Lim JS, Jang SJ, Cun Y, Ozretić L, Kong G, et al. Nature 2015;524(7563):47–53.

[50] Kim ES, Herbst RS, Wistuba II, Lee JJ, Blumenschein Jr GR, Tsao A, et al. The BATTLE trial: personalizing therapy for lung cancer. Cancer Discov 2011;1(1):44–53.

Chapter 10

Breast Cancer

Paul K. Marcom
Duke University School of Medicine, Durham, NC, United States

Chapter Outline

INTRODUCTION

Developments in genetics and genomics have had a profound impact on breast cancer management. Given the prevalence of breast cancer worldwide, and a clear connection with increasing levels of affluence, breast cancer will continue to increase as a worldwide public health problem. Advances in genomics technologies have allowed a much richer investigation of breast cancer biology. The resulting knowledge has been translated into an increasingly sophisticated approach to breast cancer risk assessment and treatment, allowing more personalized and precise care. These advances promise to continue reducing breast cancer mortality while also minimizing the toxicities of treatments.

The challenges for developing personalized breast cancer care based on genomics have also become more manifest. Initial successes with targeted therapies against HER-2 positive breast cancer have not been as easily achieved in other breast cancer subtypes. Validating and quantitative the performance of molecular assays in guiding therapy can be a slow process. The ability to

Genomic and Precision Medicine. DOI: http://dx.doi.org/10.1016/B978-0-12-800685-6.00010-2

acquire massive amounts of data regarding gene mutations has outstripped the ability to understand and apply the findings from those data [1,2]. Nevertheless, ongoing successes continue to support developing personalized approaches for improving breast cancer outcomes.

GERMLINE GENETIC PREDISPOSITION

While the inherited contribution to breast cancer predisposition has been appreciated for over a century, only in the last two decades has this knowledge been developed for clinical use. After an initial time when genetic evaluation was predominantly limited to BRCA1/2 testing, since 2013, there has been a marked increase and expansion in genetic testing for inherited breast cancer predisposition [1]. Many events have contributed to this phenomenon. The Supreme Court decision striking down patents on BRCA1/2 opened the commercial space for testing, and many more labs began offering the service. High-profile celebrity disclosures of breast-cancer-related genetic testing results have increased the general public and medical communities' awareness about the issue [3]. Finally, next-generation sequencing technologies have made it possible to test larger panels of cancer predisposing genes.

The availability of next-generation sequencing is driving fundamental changes in the models for genetic counseling, testing, and use of germline predisposition information. The expanded list of genes assessing breast cancer genetic predisposition can be extensive, with upwards of 40 genes tested, making more comprehensive assessments that are less restricted by the need to consider one cancer predisposing syndrome at a time. However, the expanded list of genes tested often includes genes with less well-defined penetrance data, and minimal information on clinical actionability. Larger gene lists also lead to a higher probability of finding variants of uncertain significance, a potential source of confusion for patients, and clinicians [4]. BRCA1 and BRCA2 are still the more important genes sequenced, but many of the additional genes associated with breast cancer risk are also known to play a role in sensing and repairing DNA damage. Table 10.1 is a nonexhaustive list of genes with known mutations that increase lifetime breast cancer risk to over 20%. Mutation status has been primarily used for guiding screening recommendations and prophylactic surgeries. For the listed genes, screening breast MRI is recommended, with varying data to support discussion of risk reducing mastectomies (RRM) and prophylactic bilateral salpingo-oophorectomy (BSO) [5]. Importantly, the clinical actionability of germline information is maturing quickly [6]. For example, data are now available supporting the predictive function of germline testing for selecting specific chemotherapies and targeted therapies [7–9].

Many important aspects of breast cancer genetic predisposition need further elucidation. For example, the penetrance of these genes can be highly variable and influenced by other environmental andgenetic factors and family history context [17]. Frequencies of specific genes in various populations can vary, as can the penetrance. The contribution of many other candidate genes needs to be better

TABLE 10.1 Genes Associated with >20% Risk of Breast Cancer When Mutated

Genes	Function	Syndrome	Mutation Frequency	RRM?	BSO?
BRCA1 and BRCA2	DNA repair: homologous recombination	Hereditary Breast/Ovarian Cancer (Fanconi Anemia Type D1 if homozygous for BRCA2 mutation)	1:400 (Higher for certain ethnicities: Ashkenazi Jewish 1:40; Icelanders 1:160)	↑	↑
TP53	Pleomorphic: cell cycle regulation, DNA damage sensing and repair	Li-Fraumeni syndrome	1:5000 to 1:20,000	↑	–
PTEN	Regulation of cell growth signaling pathways	PTEN hamartoma tumor syndrome (Cowden)	1:200,000 (likely underestimate)	↑	–
PALB2	DNA repair: homologous recombination; partners BRCA1 and 2	Breast/Pancreatic cancer syndrome (Fanconi Anemia Type N if homozygous for PALB2 mutation)	Not established	↑	–
CDH1	Cell–cell adhesion and invasion suppression	Hereditary diffuse gastric cancer syndrome; Lobular breast cancer only	<0.1:100,000	↑	–
ATM	Double strand DNA break detection	Ataxia telangiectasia mutated (ATM syndrome if homozygous ATM mutation)	1:200	–	–
CHEK2	DNA repair: mediates cell cycle arrest in response to double strand DNA breaks	Breast and colorectal cancer (previously called Li-Fraumeni-like)	1:140 (in Northern/Western Europeans; lower in other ethnicities)	–	–
STK11	Regulation of cell metabolism and energy homeostasis	Peutz–Jeghers syndrome	1:60,000 to 300,000	–	–

RRM = risk reducing mastectomy; BSO = bilateral salpingo-oophorectomy [5,10–16].

defined. For example, other DNA repair genes such as RAD51C, NBN, BARD1, and PMS2 are associated with an as yet poorly defined risk. Lastly, the majority of the general population risk is likely secondary to accumulation of lower risk single nucleotide polymorphism (SNP) markers. Commercial assays for testing breast cancer predisposition SNPs are available, and claim to refine breast cancer risk estimates by summing up the presence of a variety of SNPs. Recent validation studies show great promise for including panels of SNPs for refining breast cancer risk estimates, particularly in the context of family history [18]. However, to date, no guidelines have included these as recommended for patient risk assessment.

EARLY-STAGE BREAST CANCER MANAGEMENT

Development of Molecular Guided Management

The most robust contribution of genomic approaches to breast cancer management has been in defining specific molecular subtypes. The dissection of specific molecular subtypes has been instrumental to refining prognostic estimates, individualizing therapy, predicting response to therapy, and identifying new therapeutic targets. The vast majority of patients are diagnosed with breast cancer at an early stage, when the goal of treatment is to improve the chances of long-term disease-free status, while minimizing the toxicities of treatment. Genomic insights have fundamentally changed the approach to adjuvant therapy.

Prior to the genomic era, breast cancer was classified by standard histopathology, and estimates for developing distant, fatal, metastatic recurrences were based on conventional staging using primary tumor size and lymph node status. Tumor grade as assessed by a variety of systems was also known to be an important prognostic factor. Randomized adjuvant chemotherapy trials were designed to show proportional reduction in risk of recurrence (ROR) with the administration of a given course of therapy. Estimates of the benefit to an individual patient for receiving chemotherapy in absolute terms were made by applying proportional risk reduction estimates from adjuvant trials to the estimated ROR based on tumor size and lymph node status. While some low-risk histologic subtypes such as tubular cancers were identified, the estimates of the benefit of chemotherapy were generally applied uniformly without too much consideration of underlying biology. This "systemic model," acknowledging the risk for micrometastatic spread throughout the body at the time of presentation with early stage disease, was a step forward in breast cancer management [19]. However, it likely led to an indiscriminant use of chemotherapy.

Personalization of breast cancer care began to develop through the 1980–90s with the development of predictive biomarkers for guiding treatment. The first step toward using molecular information to personalize cancer care was the testing of estrogen and progesterone receptors, and demonstrating that such expression predicted for benefit from antiestrogen endocrine therapy. This advance took almost three decades to fully develop and apply in clinical practice, but was one of the first and most powerful demonstrations of the use of a biologic marker for personalizing cancer therapy [20]. The next advance was the discovery of the HER-2 amplicon, and the recognition of an "oncogene addicted" subtype. This led to the development of a specific target therapy, the monoclonal antibody trastuzumab, over the course of a decade, and marked a revolution in oncology care [21]. Together, the combination of standard histopathologic assessment combined with hormone and HER2 receptor assessment still provides the core approach for personalizing breast cancer treatment. However, refining prognostic estimates, particularly in the heterogeneous and most common hormone receptor (HR) positive subtype, requires more sophisticated genomic tools.

Defining the Intrinsic Breast Cancer Subtypes

The current organizational paradigm for studying breast cancer was developed by Sorlie and Perou in the early 2000s, and defined the intrinsic breast cancer subtypes based on gene expression profiling (GEP). Using an unsupervised hierarchical clustering analysis of full transcriptome data, four core subtypes were identified and are uniformly accepted: Luminal A, luminal B, HER2 over-expressed, and basal-like [22,23]. These subtypes are consistent with those defined by conventional immunohistochemistry staining for ER, PR, and HER-2, but provided greater insight into their origins [24–26]. The basal-like subtype is characterized by lacking expression for ER, PR, and HER-2 receptors. This subtype is almost uniformly high grade and has been demonstrated to have a worse prognosis in general, but a greater sensitivity to chemotherapy. Intriguing epidemiologic associations have also been made for this subtype, such as the more common occurrence in young women with African ancestry. The HER-2 positive cancers define a second subtype, having a distinct gene expression profile that is associated with HER-2 gene amplification. This subtype also has a worse prognosis that is abrogated by treatment with anti-HER-2 therapy. The luminal A and B subtypes comprise the majority of ER expressing cancers. Luminal A cancers have a better prognosis and greater sensitivity to endocrine therapy with little sensitivity to chemotherapy. Luminal B cancers, on the other hand, demonstrate a greater tendency towards resistance to endocrine therapy, and greater sensitivity to chemotherapy. The classification scheme has been validated by a number of groups, and has come to guide much of the fundamental molecular study of breast cancer [27,28].

The GEP-based intrinsic subtypes do not entirely overlap with clinical classification, however, and should be acknowledged as predominantly useful as a research guiding structure. For example, one study correlating immunohistochemistry markers and RT-PCR classification of intrinsic subtypes, only 75% of ER-negative tumors were classified into the HER-2 enriched or basal-like subtypes [24]. Clinically, HER-2 positive cancers can be clustered into many categories including the luminal B and basal-like categories [29]. Using immunohistochemistry markers as surrogates has the best correlation for determining basal-like versus nonbasal-like breast cancers [30]. Despite these limitations, the intrinsic subtype classification system has been applied successfully to banked samples for conducting important outcomes research [27].

Comprehensive Molecular Analysis: The Cancer Genome Anatomy Project

A comprehensive landmark molecular analysis of early-stage breast cancer is being performed by The Cancer Genome Atlas network. The initial network findings were published in 2012 and included analysis of genomic DNA copy number arrays, DNA methylation, exome sequencing, messenger RNA arrays, micro RNA sequencing, and reverse phase protein arrays. Analysis of data across all five platforms confirmed the four main intrinsic subtypes as described above; the claudin-low type and normal type were rare. Within each main class there was additional significant molecular heterogeneity. Only three genes had somatic mutations at a greater than 10% frequency: TP53, PIK3CA, and GATA3. However, mutational patterns were seen in specific subtypes. PIK3CA mutations were most common in the luminal A subtype at 45%, and least common in basal-like at 9%, with an overall mutation frequency across all types of 36%. Previously defined mutation patterns were also confirmed, such as a high frequency of p53 mutations, particularly in the basal-like and HER-2 positive subtypes. With continued addition of cases, the TCGA dataset will continue to provide additional insights into breast cancer biology [31].

Gene Expression Profiles for Clinical Management

A number of genomic assays are available for guiding breast cancer management. Most use some form of gene expression measurement to develop a score that predicts for outcome, but were developed using various strategies and platforms as well as validation approaches. An ASCO guideline has also been published that addresses the clinical utility of these assays for guiding therapy [32].

The oncotype Dx assay was the first multianalyte genomic assay marketed in the United States for use in guiding early-stage breast cancer management. This test was specifically developed to guide the management of HR-positive breast cancer without lymph node involvement. The development approach used a candidate gene strategy to identify a set of 21 genes for the assay, 16 that interrogate biology and five that are used for normalization of gene expression. A recurrence score (RS) is generated that estimates the risk of distant disease recurrence over a 10-year interval with adjuvant tamoxifen, classifying cases into low, moderate, and high-risk groups. This test garnered relatively rapid acceptance in the United States because validation studies could be conducted using a "retrospective/prospective" using banked paraffin blocks from randomized clinical trials. The first validation used the NSABP B-14 study of tamoxifen versus placebo, and confirmed that the RS predicted prognosis with tamoxifen therapy in node-negative disease. The work was then extended to the NSABP B-20 study comparing chemotherapy followed by tamoxifen versus tamoxifen alone HR-positive and -negative cancer, and indicated that the benefit of chemotherapy was confined to the cancers with RSs over 31, an observation that shows the proportional reduction in recurrence risk achieved with chemotherapy is not evenly applied to the anatomic staging prognostic estimates.

A meta-analysis incorporating pathologic variables (grade, tumor size) indicates that these variables do further refine prognostic estimates provide by the RS, but do not improve prediction of chemotherapy benefit [33]. A prospective trial validating the RS (TAILORx) has been fully accrued, and the results for the low RS cohort (scores <11) confirm the excellent prognosis of this group when treated with endocrine therapy only; [34] results for the intermediate- and high-risk groups are still pending, and a trial in node positive disease is actively recruiting. The combination of an easily ordered approachable molecular assay based on familiar biology, and validation in the context of familiar pivotal clinical trials, has led to the widespread adoption of the Oncotype assay in the United States; it has been endorsed by both ASCO and NCCN guidelines for guiding adjuvant chemotherapy use in HR-positive, HER2 negative, node-negative early-stage breast cancer [32,35].

The Mammaprint assay was developed by investigators at the Netherlands Cancer Institute using a collection of early-stage breast cancers not treated with medical therapy, and analyzing microarray-derived gene expression data with an unsupervised clustering approach. A panel of 70 genes was thus derived to develop a prognostic signature. Importantly, the development set included both HR-positive and negative samples, and cancers involving 1–3 lymph nodes as well as node-negative disease. The test has received FDA clearance for analytic validity, and can also provide a report of intrinsic subtype. A prospective trial, the MINDACT study, assigned treatment based on the concordance of the 70 gene prediction and standard clinical-pathologic assessment using the Adjuvant! Online algorithm; patients concordant for low-risk did not receive chemotherapy, concordant for high-risk received chemotherapy, and discordant for prognosis were randomized to chemotherapy or no chemotherapy. The published results of this study are pending and may be practice changing. The assay is not currently recommended by the ASCO or NCCN guidelines, though is recommended by the St. Gallen's, ESMO, and Japanese Medical Society.

The Prosigna ROR score was developed by combining a 50 gene classifier for the intrinsic subtypes (PAM-50) with pathologic features [26]. The original classifier used microarray data, but has been re-developed on the Nanostring platform, a technology using fluorophore labeled probes that does not require PCR amplification of extracted RNA [36]. The assay has been compared to the Oncotype Dx assay in the transATAC cohort and shown provide more prognostic information in HR-positive, node-negative disease [37]. Based on this data, the Prosigna assay is recommended by ASCO for use in HR-negative, HER2-negative node-negative disease [32].

The proliferation of molecular assays for guiding early-stage breast cancer management is a testament to the strong interest in personalizing breast cancer management and avoiding toxic therapy. Fig. 10.1 lists the genes assayed in the above tests, as well as the EndoPredict, Breast Cancer Index (BCI), and Mammostrat assays, clustered by purported gene function. While the overlap in genes is modest, the figure demonstrates the heterogeneity of breast cancer biology, as well as reflecting the variation in techniques for developing these

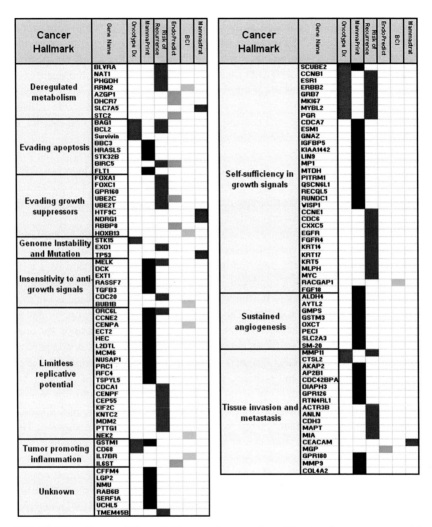

FIGURE 10.1 Commercially available genomic assays for guiding breast cancer management. The genes measured in each assay are shown, clustered by the approximate "Cancer Hallmark" they may contribute [38,39].

tests. GEP-based tests also have promise for specific clinical situations beyond chemotherapy use such as duration of endocrine therapy. Studies of the BCI provide clinical validity for predicting late recurrences, and some prediction of benefit from extended aromatase inhibitor therapy [40].

A synthesis of current clinical practice approach to early-stage breast cancer is presented in Fig. 10.2. HER2 status now takes the primary role in guiding therapy, since almost all invasive HER2 positive disease has a recurrence risk

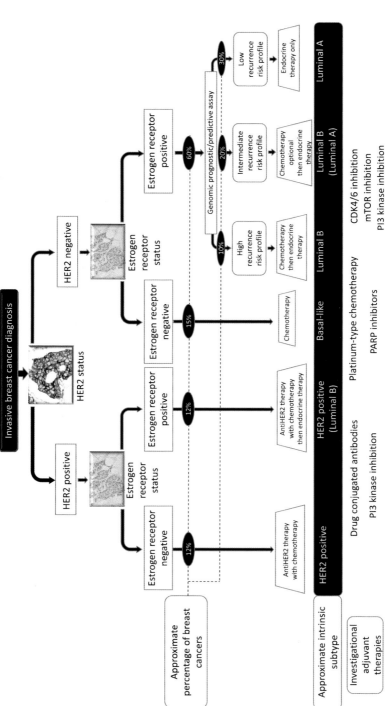

FIGURE 10.2 Current paradigm for treating early stage breast cancer. The approximate percent of breast cancers in each subtype is shown, as well as the intrinsic subtype, and some investigational therapies being developed for that subtype.

justifying consideration of anti-HER2 monoclonal antibody therapy in conjunction with chemotherapy. While endocrine therapy is recommended for all HR-positive disease, molecular prognostic/predictive assays are used to guide use of chemotherapy.

ADVANCED METASTATIC DISEASE MANAGEMENT

Metastatic breast cancer is still considered an incurable process, but the division into molecular subtypes has led to improvements in care and facilitated a personalized management approach. The conventional markers of HER-2, ER, and PR also guide management in the metastatic setting. The identification of HER-2 positive breast cancer has led to the most dramatic change, with prolonged disease control in many patients either presenting with or developing metastatic HER-2 positive breast cancer. Consensus guidelines recommend repeating these conventional markers on a biopsy of suspected metastatic disease, since numerous studies have documented the potential for changes in the conventional markers on metastasis. Meta-analyses of these studies suggest that ER status can change from positive to negative in 5.7% to 9.5% of cases, and negative to positive in 3% to 8.8% of cases. HER-2 can change in either direction in approximately 5.5% of cases. While the benefit of reassessing markers is unproven, the consensus recommendation is to guide treatment based on the metastatic cancer [41]. The broader point is that assessment of metastatic biology is much more complex, but important for optimal personalized disease management.

The successful application of conventional biomarkers has led to incorporation of biomarker development with new agents. The FDA has given guidance recommendations to include biomarker companion diagnostics with all new drugs, a challenging task [42]. The PIK3CA gene is often mutated in HR-positive breast cancer. Clinical trials developing PTEN inhibitors and PIK3CA inhibitors have looked at the performance of mutation status for predicting benefit, supported by preclinical cell line data. For example, in the BOLERO01 trial everolimus was shown to improve progression free survival (PFS) in combination with exemestane, but overall mutation status did not predict benefit [43,44]. However, there was a trend in benefit when looking at specific mutations. PIK3CA exon-9 mutations were associated with more benefit from everolimus than exon-20 mutations [44]. A similar story was seen in the PALOMA-1 trial studying the effectiveness of palbociclib, an inhibitor of the cell cycle regulating CDK 4/6 kinase, in combination with aromatase inhibition. While the overall study showed a doubling of PFS with combination treatment, a rationally designed cohort selecting for Cyclin D1 amplification or p16 loss failed to show a superior outcome. In this case, the activity seen in the overall group made it statistically challenging to improve outcomes using biomarker selection [45]. Both of these studies demonstrate the challenge of developing predictive markers for personalized therapy, particularly in the metastatic setting where the cancer process is more heterogeneous, standards for sample selection and marker testing are not well established, and statistical validation is hard to achieve.

Early clinical trial efforts to use molecular lesions to assign therapy for metastatic breast cancer have likewise been challenging and had mixed results. In a small pilot study using a variety of molecular profiling techniques, patients with refractory metastatic breast cancer had treatment assigned based on a fresh metastatic cancer sample. A novel measure of clinical benefit called the growth modulation index (GMI), determined by the ratio of time to progression (TTP) for the molecular guided treatment in comparison to the TTP with the immediately preceding treatment, was assessed. In an evaluable cohort of 25 patients, 44% had a GMI greater ≥ 1.3, suggesting an improvement in clinical outcome with molecularly selected therapy [46]. In the more disappointing SHIVA trial, patients with metastatic solid tumors were randomized to molecularly targeted therapy based on tumor molecular profile versus conventional provider selected therapy; 22% were breast cancer patients. In the experimental arm, treatment was selected based on classification of mutation status into one of three pathways: HR-positive, PI3K-pathway mutated, or RAF/MEK mutated. The patients in the molecularly guided arm had no greater improvement in PFS [47].

The source material for marker testing to guide personalized therapy is also important. For the trials above, somatic testing on tumor-associated DNA or protein was generally used. Two examples of alternative marker information have been shown to have greater predictability for guiding therapy. Germline mutations in BRCA1 and BRCA2 can predict benefit from inhibitors of poly (adenosine diphosphate [ADP]–ribose) polymerase (PARP), an enzyme involved in compensatory DNA repair when the repair associated with these genes is disrupted [8]. Circulating tumor DNA (ctDNA) may also be a more reliable predictor of treatment benefit. In a recently presented trial of a PIK3CA kinase inhibitor in combination with fulvestrant, ctDNA PIK3CA mutation status was able to predict efficacy [48].

CONCLUSION

Personalized approaches to breast cancer management have developed rapidly over the past two decades. The explosion in technical ability to perform multiple biologic assays from multiple sources presents great opportunities and challenges. Demonstrating that measurement of a marker has analytic validity, clinical validity, and clinical utility requires careful development of assays and clinical investigations. Nevertheless, the record so far clearly demonstrates that personalizes breast cancer care will result in improved outcomes with less toxicity overall, and will benefit the patients afflicted with this disease.

REFERENCES

[1] Stadler ZK, Schrader KA, Vijai J, Robson ME, Offit K. Cancer genomics and inherited risk. J Clin Oncol 2014;32(7):687–98.

[2] Tripathy D, Harnden K, Blackwell K, Robson M. Next generation sequencing and tumor mutation profiling: are we ready for routine use in the oncology clinic? BMC Med 2014;12:140.

[3] Burstein HJ. Lou gehrig, angelina jolie, and cancer genetics. J Nat Compr Cancer Network 2013;11(6):631–2.

[4] Slavin TP, Niell-Swiller M, Solomon I, Nehoray B, Rybak C, Blazer KR, et al. Clinical application of multigene panels: challenges of next-generation counseling and cancer risk management. Front Oncol 2015;5:208.

[5] Daly MB, Pilarski R, Axilbund JE, Berry M, Buys SS, Crawford B, et al. Genetic/familial high-risk assessment: breast and ovarian, Version 2.2015. J Nat Compr Cancer Network 2016;14(2):153–62.

[6] Robson ME, Bradbury AR, Arun B, Domchek SM, Ford JM, Hampel HL, et al. American society of clinical oncology policy statement update: genetic and genomic testing for cancer susceptibility. J Clin Oncol 2015;33(31):3660–7.

[7] Farmer H, McCabe N, Lord CJ, ANJ Tutt, Johnson DA, Richardson TB, et al. Targeting the DNA repair defect in BRCA mutant cells as a therapeutic strategy. Nature 2005;434(7035):917–21.

[8] Fong PC, Boss DS, Yap TA, Tutt A, Wu P, Mergui-Roelvink M, et al. Inhibition of poly(ADP-ribose) polymerase in tumors from BRCA mutation carriers. N Engl J Med 2009;361(2):123–34.

[9] Byrski T, Huzarski T, Dent R, Marczyk E, Jasiowka M, Gronwald J, et al. Pathologic complete response to neoadjuvant cisplatin in BRCA1-positive breast cancer patients. Breast Cancer Res Treat 2014;147(2):401–5.

[10] Petrucelli N DM, Feldman GL. BRCA1 and BRCA2 hereditary breast and ovarian cancer. Available from: <http://www.ncbi.nlm.nih.gov/books/NBK1247/>.

[11] Schneider K ZK, Nichols KE. Li–Fraumeni syndrome. Available from: <http://www.ncbi.nlm.nih.gov/books/NBK1311/>.

[12] Eng. PTEN hamartoma tumor syndrome (PHTS). Available from: <http://www.ncbi.nlm.nih.gov/books/NBK1488/>.

[13] Oliveira C, Seruca R, Hoogerbrugge N, Ligtenberg M, Carneiro F. Clinical utility gene card for: hereditary diffuse gastric cancer (HDGC). Eur J Hum Genet 2013;21(8). Available from: <http://www.nature.com/ejhg/journal/v21/n8/full/ejhg2012247a.html>.

[14] Partner and Localizer of BRCA2; palb2. Mim 610355. Available from: <http://omim.org/entry/610355?search=palb2&highlight=palb2>. Accessed 02/04/2016.

[15] Renwick A, Thompson D, Seal S, Kelly P, Chagtai T, Ahmed M, et al. ATM mutations that cause ataxia-telangiectasia are breast cancer susceptibility alleles. Nat Genet 2006;38(8):873–5.

[16] Thompson D, Seal S, Schutte M, McGuffog L, Barfoot R, Renwick A, et al. A multicenter study of cancer incidence in CHEK2 1100delC mutation carriers. Cancer Epidemiol Biomarkers Prev 2006;15(12):2542–5.

[17] Peterlongo P, Chang-Claude J, Moysich KB, Rudolph A, Schmutzler RK, Simard J, et al. Candidate genetic modifiers for breast and ovarian cancer risk in BRCA1 and BRCA2 mutation carriers. Cancer Epidemiol Biomarkers Prev 2015;24(1):308–16.

[18] Mavaddat N, Pharoah PDP, Michailidou K, Tyrer J, Brook MN, Bolla MK, et al. Prediction of breast cancer risk based on profiling with common genetic variants. JNCI J Nat Cancer Inst 2015;107(5):djv036.

[19] Early Breast Cancer Trialists' Collaborative G. Polychemotherapy for early breast cancer: an overview of the randomised trials. Lancet 1998;352(9132):930–42.

[20] Early Breast Cancer Trialists' Collaborative G. Effects of chemotherapy and hormonal therapy for early breast cancer on recurrence and 15-year survival: an overview of the randomised trials. Lancet 2005;365(9472):1687–717.

[21] Hudis CA. Trastuzumab—mechanism of action and use in clinical practice. N Engl J Med 2007;357(1):39–51.

[22] Perou CM, Sorlie T, Eisen MB, van de Rijn M, Jeffrey SS, Rees CA, et al. Molecular portraits of human breast tumours. Nature 2000;406(6797):747–52.

[23] Sorlie T, Perou C, Tibshirani R, Aas T, Geisler S, Johnsen H, et al. Gene expression patterns of breast carcinomas distinguish tumor subclasses with clinical implications. Proc Natl Acad Sci U S A 2001;98:10869–74.

[24] Bastien RR, Rodríguez-Lescure Á, Ebbert MT, Prat A, Munárriz B, Rowe L, et al. PAM50 breast cancer subtyping by RT-qPCR and concordance with standard clinical molecular markers. BMC Med Genomics 2012;5(1):1–12.

[25] Carey LA, Perou CM, Livasy CA, Dressler LG, Cowan D, Conway K, et al. Race, breast cancer subtypes, and survival in the carolina breast cancer study. JAMA 2006;295(21):2492–502.

[26] Parker JS, Mullins M, Cheang MC, Leung S, Voduc D, Vickery T, et al. Supervised risk predictor of breast cancer based on intrinsic subtypes. J Clin Oncol 2009;27(8):1160–7.

[27] Nielsen TO, Perou CM. CCR 20th anniversary commentary: the development of breast cancer molecular subtyping. Clin Cancer Res 2015;21(8):1779–81.

[28] Fan C, Oh DS, Wessels L, Weigelt B, Nuyten DS, Nobel AB, et al. Concordance among gene-expression-based predictors for breast cancer. N Engl J Med 2006;355(6):560–9.

[29] Carey LA, Berry DA, Cirrincione CT, Barry WT, Pitcher BN, Harris LN, et al. Molecular heterogeneity and response to neoadjuvant human epidermal growth factor receptor 2 targeting in CALGB 40601, a randomized phase III trial of paclitaxel plus trastuzumab with or without lapatinib. J Clin Oncol 2016;34(6):542–9.

[30] Allott EH, Cohen SM, Geradts J, Sun X, Khoury T, Bshara W, et al. Performance of three-biomarker immunohistochemistry for intrinsic breast cancer subtyping in the AMBER consortium. Cancer Epidemiol Biomarkers Prev 2016;25(3):470–8.

[31] Cancer Genome Atlas Network. Comprehensive molecular portraits of human breast tumours. Nature 2012;490(7418):61–70.

[32] Harris LN, Ismaila N, McShane LM, Andre F, Collyar DE, Gonzalez-Angulo AM, et al. Use of biomarkers to guide decisions on adjuvant systemic therapy for women with early-stage invasive breast cancer: American Society of Clinical Oncology clinical practice guideline. J Clin Oncol 2016;34(10):1134–50.

[33] Tang G, Cuzick J, Costantino JP, Dowsett M, Forbes JF, Crager M, et al. Risk of recurrence and chemotherapy benefit for patients with node-negative, estrogen receptor–positive breast cancer: recurrence score alone and integrated with pathologic and clinical factors. J Clin Oncol 2011;29(33):4365–72.

[34] Sparano JA, Gray RJ, Makower DF, Pritchard KI, Albain KS, Hayes DF, et al. Prospective validation of a 21-gene expression assay in breast cancer. N Engl J Med 2015;373(21):2005–14.

[35] Gradishar WJ, Anderson BO, Balassanian R, Blair SL, Burstein HJ, Cyr A, et al. Breast cancer version 2.2015. J Nat Compr Cancer Network 2015;13(4):448–75.

[36] Wallden B, Storhoff J, Nielsen T, Dowidar N, Schaper C, Ferree S, et al. Development and verification of the PAM50-based Prosigna breast cancer gene signature assay. BMC Med Genomics 2015;8(1):1–14.

[37] Dowsett M, Sestak I, Lopez-Knowles E, Sidhu K, Dunbier AK, Cowens JW, et al. Comparison of PAM50 risk of recurrence score with oncotype DX and IHC4 for predicting risk of distant recurrence after endocrine therapy. J Clin Oncol 2013;31(22):2783–90.

[38] Hanahan D, Weinberg Robert A. Hallmarks of cancer: the next generation. Cell 2011;144(5):646–74.

[39] Tian S, Roepman P, van't Veer LJ, Bernards R, de Snoo F, Glas AM. Biological functions of the genes in the mammaprint breast cancer profile reflect the hallmarks of cancer. Biomark Insights 2010;5:129–38.

[40] Bell DW, Lynch TJ, Haserlat SM, Harris PL, Okimoto RA, Brannigan BW, et al. Epidermal growth factor receptor mutations and gene amplification in non-small-cell lung cancer: molecular analysis of the IDEAL/INTACT gefitinib trials. J Clin Oncol 2005;23(31):8081–92.

[41] Van Poznak C, Somerfield MR, Bast RC, Cristofanilli M, Goetz MP, Gonzalez-Angulo AM, et al. Use of biomarkers to guide decisions on systemic therapy for women with metastatic breast cancer: American Society of Clinical Oncology clinical practice guideline. J Clin Oncol 2015;33(24):2695–704.

[42] FDA. Available from: <http://www.fda.gov/downloads/MedicalDevices/DeviceRegulation andGuidance/GuidanceDocuments/UCM262327.pdf>; 2014.

[43] Baselga J, Campone M, Piccart M, Burris HAI, Rugo HS, Sahmoud T, et al. Everolimus in postmenopausal hormone-receptor–positive advanced breast cancer. N Engl J Med 2012;366(6):520–9.

[44] Hortobagyi GN, Chen D, Piccart M, Rugo HS, Burris HA, Pritchard KI, et al. Correlative analysis of genetic alterations and everolimus benefit in hormone receptor–positive, human epidermal growth factor receptor 2–negative advanced breast cancer: results from BOLERO-2. J Clin Oncol 2016;34(5):419–26.

[45] Finn RS, Crown JP, Lang I, Boer K, Bondarenko IM, Kulyk SO, et al. The cyclin-dependent kinase 4/6 inhibitor palbociclib in combination with letrozole versus letrozole alone as first-line treatment of oestrogen receptor-positive, HER2-negative, advanced breast cancer (PALOMA-1/TRIO-18): a randomised phase 2 study. Lancet Oncol 2015;16(1):25–35.

[46] Jameson GS, Petricoin EF, Sachdev J, Liotta LA, Loesch DM, Anthony SP, et al. A pilot study utilizing multi-omic molecular profiling to find potential targets and select individualized treatments for patients with previously treated metastatic breast cancer. Breast Cancer Res Treat 2014;147(3):579–88.

[47] Le Tourneau C, Delord J-P, Gonçalves A, Gavoille C, Dubot C, Isambert N, et al. Molecularly targeted therapy based on tumour molecular profiling versus conventional therapy for advanced cancer (SHIVA): a multicentre, open-label, proof-of-concept, randomised, controlled phase 2 trial. Lancet Oncol 2015;16(13):1324–34.

[48] Baselga J, Im S-A, Iwata H, Clemons M, Ito Y, Awada A, et al. Abstract S6-01: PIK3CA status in circulating tumor DNA (ctDNA) predicts efficacy of buparlisib (BUP) plus fulvestrant (FULV) in postmenopausal women with endocrine-resistant HR + /HER2− advanced breast cancer (BC): first results from the randomized, phase III BELLE-2 trial. Cancer Res 2016;76(4 Supplement):S6-01.

Chapter 11

Colorectal Cancer

Roland P. Kuiper[1,2], Robbert D.A. Weren[1] and Ad Geurts van Kessel[1]

[1]*Radboud University Medical Centre, Nijmegen, The Netherlands*
[2]*Princess Máxima Center for Pediatric Onolocy, Nijmegen, The Netherlands*

Chapter Outline

ABBREVIATIONS

AFAP attenuated familial adenomatous polyposis
CIMP CpG island methylator phenotype
CIN chromosome instability
CRC colorectal cancer
FAP familial adenomatous polyposis
HNPCC hereditary nonlypolysis colorectal cancer
GWAS genome-wide association studies
MAP MUTYH-associated polyposis
MMR mismatch repair
MSI microsatellite instability
NAP NTHL1-associated polyposis
PPAP polymerase proofreading-associated polyposis
SAC spindle assembly checkpoint

Genomic and Precision Medicine. DOI: http://dx.doi.org/10.1016/B978-0-12-800685-6.00011-4

INTRODUCTION

Colorectal cancer (CRC) is the third most commonly diagnosed cancer in males and the second in females, with worldwide over 1.4 million new cases and nearly 700,000 deaths reported in 2012 [1]. The incidence rates of CRC are higher in Western countries compared to those in Africa and Asia. An increasing incidence of CRC has been reported in several Asian and Eastern European countries, which may reflect the adoption of Western lifestyles and other risk factors in these countries, like unhealthy diet, obesity, and smoking [2]. In most parts of the world, CRC rates are higher in men than women, and there is a strong age dependency, that is, increasing from the age of 40 and peaking after the age of 50 [3].

About four decades ago, the concept was introduced that CRC develops from benign precursor lesions in the gastrointestinal tract, referred to as polyps [4]. Clinically, these polyps can differ in their localization (proximal or distal), histopathology (e.g., conventional adenomas or serrated adenomas), and their potential to develop into invasive and metastatic carcinomas. CRC is a heterogeneous disease, and at least three distinct pathways have been reported to be involved in CRC development, namely the chromosomal instability (CIN) pathway, the serrated neoplasia pathway, and the microsatellite instability (MSI) pathway.

GENETICS OF COLORECTAL CANCER

The CIN pathway was the first CRC pathway that was characterized at the molecular level by the pioneering work of Fearon and Vogelstein [5]. They postulated that CRC development acts as a multistep process that is accompanied by the sequential accumulation of genetic mutations, often occurring over many years. Based on genetic studies of early adenomatous lesions, it was found that inactivating mutations in the tumor-suppressor gene *APC* represent an initiating event of the CIN pathway [6]. These adenomas have the potential to evolve into an in situ carcinoma by the accumulation of additional somatic mutations in oncogenes (e.g., *KRAS*) and tumor-suppressor genes (e.g., *TP53*), resulting in genetic instability (Fig. 11.1). CIN tumors account for approximately 80% of the CRCs and exhibit a microsatellite stable but aneuploid phenotype [7].

During the last decade, it has become clear that also other types of polyps, that is, serrated and hyperplastic polyps, have a malignant potential [8]. The development of CRC from serrated polyps differs from the adenoma–carcinoma sequence and accounts for approximately 15% of all CRCs [8]. The most pronounced (epi) genetic alterations observed in these tumors include activating *BRAF* mutations and widespread hypermethylation of CpG islands, referred to as CpG island methylator phenotype (CIMP) [8]. One of the genes that may be silenced by hypermethylation is the mismatch repair (MMR) gene *MLH1*, which results in sporadic MSI [9]. MSI may, however, also be observed in the absence of CIMP in tumors derived from patients with Lynch syndrome, who carry germ line mutations in *MLH1* or any of the other MMR genes (described in detail below). This latter syndrome underlies the onset of CRC in approximately 5% of the cases.

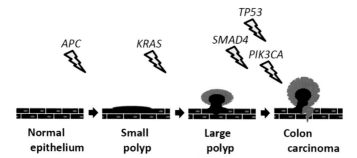

| Normal epithelium | Small polyp | Large polyp | Colon carcinoma |

FIGURE 11.1 A genetic model of colonic tumorigenesis according to the adenoma-to-carcinoma sequence based on an accumulation of mutations in tumor-suppressor genes and oncogenes, as proposed by Fearon and Vogelstein [5]. Loss of *APC* is considered the initiating event of tumor development, resulting in aberrant proliferation of the epithelial cells. Subsequently, an activating mutation in *KRAS* will accelerate the growth of a colonic polyp. Additional mutations in, for example, *SMAD4*, *TP53*, and *PIK3CA* will result in the formation of an in situ carcinoma with a CIN phenotype.

With the advent of next-generation sequencing technologies, the CRC genome has been extensively explored, which has resulted in the identification of several additional genes that may drive the multistep process of tumorigenesis. The anomalies encountered in these genes involve both sequence mutations (point mutations and indels) and structural alterations (copy number variants and chromosomal translocations) [10]. Later, these genomic characterizations were refined by the integration of proteomics data, which revealed five distinct proteomic signatures. Two of these signatures were found to overlap with the serrated/CIMP pathway, but with distinct mutation, methylation and protein expression patterns, and a different clinical outcome [11]. Recently, it was reported that human intestinal stem cells, in which four of the most commonly mutated genes in CRC were inactivated, showed an in vitro growth factor-independent growth and a highly efficient in vivo tumor-forming potential when xenotransplanted in immunodeficient mice [12].

GENETICS OF COLORECTAL-CANCER-ASSOCIATED SYNDROMES

Although most CRCs present as sporadic cancers, hereditary factors may affect an individuals' life-time risk to develop cancer (Fig. 11.2). This increased risk may be subtle, as is mostly the case, or may be substantial. A highly increased risk to develop CRC is observed in Mendelian syndromes, which are caused by single-gene mutations in the germ line [13]. During the last few decades, we have witnessed the elucidation of the genetic basis of several inherited CRC-associated syndromes, together representing ~5% of all CRC cases. These CRC-associated syndromes may be caused by mutations in at least 15 different genes (listed in Table 11.1) and can be subdivided into two major groups encompassing nonpolyposis or polyposis cases, respectively [14–16]. The identification

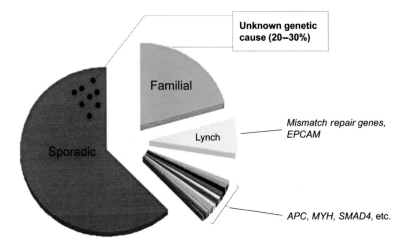

FIGURE 11.2   Colorectal cancers can be divided into sporadic, familial, and hereditary. In total, 35% of the colorectal cancers are assumed to result from a genetic defect but, as yet, only ~5% of them has been associated with a high-penetrant dominant or recessive inherited syndrome, such as Lynch syndrome. In 20–30% of the familial and de novo early-onset colorectal cancer patients (the latter usually classified as sporadic; black dots), the genetic defect remains to be resolved. *Adapted from Lynch and Lynch, Dis Markers 2004;20:181–98.*

of the genes underlying these syndromes has resulted in immediate benefits for the patients and their families through genetic testing, presymptomatic screening, genetic counseling, and prevention and, in addition, has provided important insights into the molecular mechanisms underlying cancer susceptibility.

Nonpolyposis Colorectal Cancer Syndromes

Lynch syndrome (also known as hereditary nonpolyposis CRC) is an autosomal-dominant syndrome caused by heterozygous germ line mutations in either one of the MMR genes (*MLH1, MSH2, MSH6,* and *PMS2*) [17], or by epigenetic silencing of the *MSH2* gene due to germ line deletions affecting the 3′ exon of the *EPCAM* gene [18]. Lynch syndrome patients are at a high risk to develop early-onset CRC (60–90%), endometrial cancer (20–60%), and, to a lesser extent, cancer of the bowel, stomach, ovary, urinary tract, and hepatobiliary tract [17]. The DNA MMR pathway prevents the accumulation of single base pair mismatches and small nucleotide insertions and deletions (indels), which are formed during replication of small tandem repeat sequences [17]. Tumors of Lynch syndrome patients exhibit a functional loss of the remaining wild-type MMR allele, resulting in inability to recognize and correct single base pair mismatches and small nucleotide insertions and deletions. As a consequence, these tumors show a hypermutation signature, which may affect important CRC driver genes such as *APC* and *TGFBR2* [17]. Subtypes of Lynch syndrome include Muir–Torre syndrome [19] and Turcot syndrome [20].

TABLE 11.1 High-Penetrant Polyposis or Colorectal Cancer Predisposing Syndromes and Genes

Gene	Cancer Syndrome	Mode of Inheritance[a]	Year of Discovery	Polyposis
APC	Familial adenomatous polyposis	AD	1991	Yes
AXIN2	Oligodentia-colorectal cancer syndrome	AD	2004	Yes
MUTYH	MUTYH-associated polyposis	AR	2002	Yes
NTHL1	NTHL1-associated polyposis	AR	2015	Yes
POLD1	Polymerase proofreading-associated polyposis	AD	2013	Yes
POLE	Polymerase proofreading-associated polyposis	AD	2013	Yes
BMPR1A	Juvenile polyposis syndrome	AD	2001	Yes
PTEN	PTEN hamartoma tumor syndrome	AD	1997	Yes
SMAD4	Juvenile polyposis syndrome	AD	1998	Yes
STK11	Peutz–Jeghers syndrome	AD	1998	Yes
GREM1	Hereditary mixed polyposis syndrome	AD	2012	Yes
MLH1	Lynch syndrome	AD	1994	No
MSH2	Lynch syndrome	AD	1993	No
MSH6	Lynch syndrome	AD	1997	No
PMS2	Lynch syndrome	AD	1994	No

[a]AD, autosomal dominant; AR, autosomal recessive.

Autosomal-Dominant Polyposis-Associated Syndromes

The most common high-penetrant Mendelian polyposis syndrome is familial adenomatous polyposis (FAP), which is caused by monoallelic germ line mutations in the *APC* gene and shows an autosomal-dominant inheritance pattern [21]. An inactivating somatic mutation in the remaining wild-type allele results in loss of APC activity, which is one of the initiating events of the adenoma–carcinoma sequence (Fig. 11.1). The high proliferation rate in the colorectum leads to a substantial number of cells that accrue somatic inactivation of the second allele, which explains why FAP patients usually develop hundreds to thousands of adenomatous polyps and, if untreated, inevitably develop CRC. In the majority of FAP patients (~95%), a truncating nonsense or frameshift germ line mutation in *APC* is observed [21].

The clinical manifestation of FAP is related to the genomic position of the *APC* gene mutation. The most severe phenotype, also known as the profuse phenotype (thousands of colorectal polyps), is caused by *APC* mutations located between codons 1250 and 1464. In contrast, the much milder attenuated FAP phenotype (AFAP; <100 colorectal polyps and a later age-of-onset) is caused by mutations in the 5′ and 3′ regions of the gene [21].

Recently, a novel adenomatous polyposis predisposing syndrome was identified, for which the name polymerase proofreading-associated polyposis (PPAP) has been proposed [15]. PPAP is an autosomal-dominant syndrome caused by monoallelic germ line mutations in the exonuclease (proofreading) domains of *POLE* and *POLD1* [15]. The exact incidence of PPAP in the population has not been established yet, but one of the common germ line variants in *POLE*, the pathogenic p. L424V variant, has been encountered with variable frequencies (1:67 to 1:858) in different cohorts of unexplained polyposis and/or early-onset and familial CRC cases [15,22,23]. *POLE* and *POLD1* encode the polymerases ε and δ, respectively, and accurate proofreading via their exonuclease domains is required to correct mispaired bases inserted during DNA replication. All mutation carriers identified thus far have properly functioning polymerase domains but impaired proofreading activity. As a consequence, these patients accumulate base substitutions during their life, which eventually results in the development of hypermutated tumors [15]. Next to several large PPAP families that have been reported, also de novo mutation cases have been identified [22,23]. The prevalence of these de novo mutations remains to be determined. The exact clinical phenotype of PPAP has not been established yet, but available data strongly indicate that PPAP results in a high-penetrant predisposition to develop polyposis (usually <100 polyps) and early-onset CRC [15,22,23].

Several autosomal-dominant hamartomatous polyposis predisposing syndromes have also been described, of which the incidences are relatively low. Examples are PTEN hamartoma tumor syndrome, juvenile polyposis syndrome, and Peutz–Jeghers syndrome. PTEN hamartoma tumor syndrome, caused by germ line mutations in the *PTEN* gene, is associated with the development of breast, endometrium, and thyroid cancers [24]. The risk to develop CRC in this syndrome is, as yet, not well established. Germ line aberrations in the *SMAD4* and *BMPR1A* genes, both involved in the transforming growth factor-beta (TGF-β)-signaling pathway, predispose to the development of juvenile polyposis syndrome, which is associated with an highly increased risk to develop CRC [25,26]. Patients with Peutz–Jeghers syndrome carry germ line mutations in the *STK11* gene and are at a high risk to develop gastrointestinal cancers, including pancreatic cancer and CRC [27,28].

Hereditary mixed polyposis syndrome is an autosomal-dominant syndrome associated with multiple types of colorectal polyps and an increased risk to develop CRC. Recently, genome-wide screening assays have led to the identification of a 40-kb germ line duplication upstream of the *GREM1* locus, which increases the expression of GREM1 and concomitantly initiates tumorigenesis [29,30]. This duplication was found to be recurrent in hereditary mixed polyposis patients of Ashkenazi descent [31].

Autosomal Recessive Polyposis-Associated Syndromes

MUTYH-associated polyposis (MAP) is considered to be the second most common high-penetrant Mendelian cancer syndrome associated with adenomatous polyposis, with an estimated incidence of 1:5000–40,000 in Europe [32]. In contrast to FAP and PPAP, MAP is inherited in an autosomal recessive manner due to biallelic germ line mutations in the base excision repair gene *MUTYH* [33]. Lack of expression of functional MUTYH results in an inability to recognize and correct mismatches between an oxidation-damaged guanine base and an adenine base [34]. Consequently, this will lead to an increased incidence of G:C to T:A transversions, and an enrichment for this specific base substitution is, indeed, observed in tumors derived from MAP patients [33]. Both truncating and missense germ line variants are frequently encountered in MAP patients [32], but the observed mutations strongly differ between different ethnicities [35]. Compared to FAP, MAP patients develop a milder phenotype and the majority of MAP patients are diagnosed with less than 100 polyps [32]. MAP patients have an approximately 28-fold increased life-time risk to develop CRC compared to the general population, and the reported penetrance at the age of 60 ranges from 43 to 100% [32,35].

Very recently, a novel recessive adenomatous polyposis syndrome was discovered, which is caused by a biallelic germ line truncating mutation in another base excision repair gene, *NTHL1* [16]. The current knowledge of the *NTHL1*-associated polyposis (NAP) syndrome is still limited, with only seven affected individuals identified in three families. Since all seven homozygous carriers had adenomatous polyps, six of them had (multiple) malignancies including CRC, and none of the heterozygous or wild-type family members were affected, the penetrance is likely to be high. Similar to what has been observed in MAP, the adenomas and CRCs in NAP showed a specific mutation bias, in this case toward G:C to A:T transitions. The reason for this difference in mutation spectra between NAP and MAP is still unclear but may be explained by differences in substrate specificity of the two DNA glycosylases involved. Whereas MUTYH specifically removes mismatched adenine bases opposite oxidation-damaged guanine bases (8-oxoG), NTHL1 lacks this activity and, instead, targets oxidized pyrimidines and ring-opened purines [36].

OTHER CRC-ASSOCIATED SYNDROMES

Besides the above-mentioned Mendelian CRC-associated syndromes, CRC predisposition has also been observed in patients affected by various developmental syndromes. An example is Bloom syndrome, which is caused by biallelic mutations in the RECQL helicase gene *BLM*. This syndrome is characterized by a severe pre- and postnatal growth retardation, erythematas, and an increased risk to develop CRC [37]. Recently, we found indications that monoallelic carriers of pathogenic *BLM* mutations are also at an increased risk to develop CRC, albeit with a moderate-to-low penetrance [38]. Also oligodontia-CRC syndrome, which is caused by deficiencies in the *AXIN2* gene, features next to teeth malformations CRC predisposition [39]. In addition, an increased CRC risk has been encountered in patients with Li–Fraumeni syndrome, which is caused by mutations in the *TP53* gene [40].

EPIMUTATIONS AND HEREDITARY COLORECTAL CANCER

Studies on CRC families have revealed that not only genetic mechanisms but also epigenetic mechanisms may underlie CRC predisposition. These so-called constitutional epimutations result in aberrant gene expression in multiple somatic tissues due to an epigenetic DNA modification rather than a DNA sequence mutation [41]. Epimutations can be primary, when no underlying genetic change is involved, or secondary in case the epigenotype is due to an in cis or an in trans genetic change [42].

Several examples of constitutional epimutations have been found in Lynch syndrome cases. Gazzoli et al. [43], for example, identified a person with early-onset CRC who did not carry a pathogenic germ line variant in one of the MMR genes but, instead, showed constitutional hemi-allelic hypermethylation of the *MLH1* gene. The tumor of this patient showed MSI and a complete loss of *MLH1* expression due to somatic deletion of the second, unmethylated allele. Since then, many similar cases with constitutional *MLH1* epimutations and classical Lynch-like features have been identified, often without a pronounced family history for cancer [41]. It has been estimated that *MLH1* epimutations occur in 1–10% of the patients that meet the criteria for Lynch syndrome and have MLH1-deficient tumors in the absence of a germ line mutation [44]. In a limited number of cases, the *MLH1* epimutation was transmitted to multiple family members and over generations, and could be linked to a haplotype bearing a c.−27C > A promoter variant as the probable cause of the epimutation [41]. These cases are, thus, due to a secondary epimutation.

Another example of a heritable secondary epimutation in Lynch syndrome involves 3′ end *EPCAM* deletions, which induce transcriptional read-through into the neighboring *MSH2* gene leading to in cis hypermethylation of the *MSH2* promoter and, thereby, allele-specific silencing in tissues expressing *EPCAM* [18]. The *EPCAM* deletions appear to be variable in size and location but invariably encompass the last two exons of the *EPCAM* gene, including its polyadenylation signals [45]. In addition, it was found that carriers of these deletions have a high risk to develop CRC, very similar to that noted in carriers of *MLH1* or *MSH2* gene mutations. In contrast, they were found to exhibit a strikingly low risk to develop endometrial cancer. Patients with deletions extending close to the *MSH2* promoter, however, appeared to have an unaltered risk to develop endometrial cancer [46]. Although the rationale for this remarkable clinical difference remains to be established, the identification and characterization of *EPCAM* deletions by copy number analysis has provided a basis for an optimized protocol for the recognition and targeted prevention of cancer in *EPCAM* deletion carriers, which has now been implemented in routine clinical diagnostics for suspected Lynch syndrome families [47].

GENOME-WIDE ASSOCIATION STUDIES

Together, the above-mentioned syndromes are responsible for ~5% of all CRCs, whereas twin studies have suggested that up to 35% of all CRCs may be ascribed to a genetic susceptibility (Fig. 11.2) [14]. A majority of this risk is attributed to low-to-moderate susceptibility genetic variants, which can be identified by unbiased genome-wide genotype–phenotype association studies. Since 2007, multiple genome-wide association studies (GWAS) have been conducted, eventually resulting in combined meta-analyses of up to 10,000 patients and controls, and the identification of 26 susceptibility loci thus far [48–50]. Some of the variants associated with an increased CRC risk are in or near genes implicated in important functional pathways, including the bone morphogenetic protein signaling pathway (*SMAD7, GREM1, BMP4,* and *BMP2*) [51,52]. As yet, however, a large fraction of the variants cannot be placed within any functional context related to the coding part of the genome. Due to the small effect sizes of the loci identified in GWAS, the clinical utility of the outcome of these studies has been limited [48]. It has been estimated that individuals carrying multiple low-penetrance risk loci may have an up to twofold increased lifetime CRC risk, underscoring its limited feasibility for clinical risk prediction. Together, these common variants may explain 5–10% of the heritability of CRC [48], indicating that another 10% of its heritability may still be missing.

NOVEL COLORECTAL CANCER PREDISPOSING GENES

At least part of the still missing heritability in CRC may be explained by rare moderate-to-high-penetrant risk factors in a variety of genes not previously associated with CRC predisposition. In search for these genes, genomic profiling studies, like genome-wide copy number analysis and whole-exome and whole-genome sequencing, have been applied to genetically suspected CRC patients. Although this work has revealed several novel candidate CRC predisposing genes, interpretation of the findings of these studies appears challenging. One explanation may be that our knowledge of normal (benign) variation, particularly of rare single nucleotide variants, is still far from complete. Furthermore, the expected incomplete penetrance of many of the still to be discovered risk factors seriously complicates the functional validation process, particularly in smaller sized studies. Therefore, a promising strategy to identify true candidates might be to focus not only on recurrently affected genes, but also on recurrent downstream effects like, for example, increased WNT signaling or genomic instability. With respect to the latter, several interesting candidates have already been proposed. For example, in a recent exome sequencing study, germ line nonsense and missense variants in the Fanconi anemia-associated nuclease 1 (FAN1) gene *FAN1* were identified in five independent CRC families [53]. FAN1 is involved in the repair of DNA interstrand crosslinks by interacting with the mono-ubiquitylated FANCI-FANCD2 protein complex [54], which provides a first link between CRC predisposition and the Fanconi anemia DNA repair pathway [53]. In another study, genome-wide and targeted copy number and mutation analyses in a cohort of patients with nonpolyposis early-onset CRC revealed a potential predisposing role of heterozygous germ line aberrations in the spindle assembly checkpoint (SAC) genes *BUB1* and *BUB3* [55]. Disruption of this SAC complex may result in chromosome segregation artifacts resulting in aneuploidies and structural anomalies, hallmarks of cancer. Cytogenetic analyses in several of these patients revealed mosaic-variegated aneuploidies in multiple tissues. In addition, mild craniofacial dysmorphisms were observed. These findings suggest that chromosome segregation defects caused by germ line mutations in SAC components may underlie a novel class of syndromic forms of CRC. Clearly, large case-control validation studies are required to confirm the association of all these rare variants with CRC predisposition.

PERSPECTIVES

Predictive clinical genetic testing is currently available for the various known Mendelian CRC-associated syndromes (Table 11.1), and patients carrying mutations in either one of the underlying high-penetrance genes may immediately be subjected to surveillance according to the latest recommendations [56]. At present, next-generation targeted sequencing strategies are being tested to enable the screening of larger numbers of patients using less stringent selection criteria [57,58]. It is generally anticipated that such approaches, once implemented in a routine clinical settings, will increase the number of predisposed patients and provide more insight into the entire phenotypic spectrum associated with each of the CRC predisposing genes. On the other hand, this approach may also result in the identification of more unclassified variants which complicates the interpretation.

Genetic testing and counseling still remain a challenge in families with an obvious strong suspicion for a genetic predisposition, but without an overt pathogenic mutation in any of the known high-penetrance genes. For these families, the utility of exome sequencing in a clinical diagnostic setting was recently assessed, but due to the extensive heterogeneity of CRC predisposition, the results appeared difficult to interpret [59]. This genetic heterogeneity may also explain why novel CRC syndromes, despite the extensive application of next-generation sequencing to unexplained families, have resulted in only two new Mendelian CRC syndromes: PPAP and NAP [15,16]. Based on the relatively low incidence of these new syndromes, novel CRC syndromes still to be discovered in the future will likely be very rare as well, which makes genotype–phenotype correlation studies, including CRC risk estimates, challenging.

We expect that a comprehensive genomic characterization (i.e., identification of mutation spectra) of polyps and tumors derived from familial CRC patients may provide new insight into the genetic heterogeneity of CRC predisposition. Such studies may reveal the presence of specific tumor subtypes or aberrant signaling pathways and will uncover the presence of genomic instability or mutation signatures that point toward specific subsets of mutated genes in the germ line, including those with a lower penetrance.

Functional validation of germ line variants, both in established and novel candidate CRC predisposing genes, is another essential step toward unraveling the heterogeneity of CRC predisposition. The recent application of organoids has provided a potentially limitless source of patient-derived colon tissue [60]. Using organoids, cells that are involved in the development of polyps and carcinomas can be studied in a situation that is strongly reminiscent of the in vivo cellular environment. As such, organoids may allow us to study the function of a gene, including its putative CRC predisposing role in a patient- and tissue-dependent context.

Together, these studies are expected to provide further insight into the etiology of CRC and to open up new avenues for improved diagnosis, prognosis, and surveillance of CRC patients and their family members. Ultimately, these studies may lead to the development of new therapeutic intervention strategies for CRC.

ACKNOWLEDGMENTS

The authors thank R. de Voer, M.M. Hahn, M.J.L. Ligtenberg, and N. Hoogerbrugge for their contributions. This work was supported by the Dutch Cancer Society (KWF) and the Netherlands Organization for Health Research and Development (ZonMW).

GLOSSARY TERMS

Polyposis the development of multiple (tens to hundreds) polyps in the colon or rectum

Microsatellite instability the condition of genetic hypermutability caused by impaired DNA mismatch repair, resulting in variability in the length of DNA sequence repeats called microsatellites

Constitutional epimutations aberrant changes in gene expression in multiple somatic tissues due to epigenetic DNA modifications

REFERENCES

[1] Torre LA, Bray F, Siegel RL, Ferlay J, Lortet-Tieulent J, Jemal A. Global cancer statistics, 2012. CA Cancer J Clin 2015;65(2):87–108.

[2] Center MM, Jemal A, Ward E. International trends in colorectal cancer incidence rates. Cancer Epidemiol Biomarkers Prev 2009;18(6):1688–94.

[3] Haggar FA, Boushey RP. Colorectal cancer epidemiology: incidence, mortality, survival, and risk factors. Clin Colon Rectal Sur 2009;22(4):191–7.

[4] Muto T, Bussey HJ, Morson BC. The evolution of cancer of the colon and rectum. Cancer 1975;36(6):2251–70.

[5] Fearon ER, Vogelstein B. A genetic model for colorectal tumorigenesis. Cell 1990;61(5): 759–67.

[6] Powell SM, Zilz N, Beazer-Barclay Y, Bryan TM, Hamilton SR, Thibodeau SN, et al. APC mutations occur early during colorectal tumorigenesis. Nature 1992;359(6392):235–7.

[7] Vogelstein B, Kinzler KW. Cancer genes and the pathways they control. Nat Med 2004; 10(8):789–99.

[8] IJspeert JE, Vermeulen L, Meijer GA, Dekker E. Serrated neoplasia-role in colorectal carcinogenesis and clinical implications. Nat Rev Gastroenterol Hepatol 2015;12(7):401–9.

[9] Cunningham JM, Christensen ER, Tester DJ, Kim CY, Roche PC, Burgart LJ, et al. Hypermethylation of the hMLH1 promoter in colon cancer with microsatellite instability. Cancer Res 1998;58(15):3455–60.

[10] Cancer Genome Atlas N. Comprehensive molecular characterization of human colon and rectal cancer. Nature 2012;487(7407):330–7.

[11] Zhang B, Wang J, Wang X, Zhu J, Liu Q, Shi Z, et al. Proteogenomic characterization of human colon and rectal cancer. Nature 2014;513(7518):382–7.

[12] Drost J, van Jaarsveld RH, Ponsioen B, Zimberlin C, van Boxtel R, Buijs A, et al. Sequential cancer mutations in cultured human intestinal stem cells. Nature 2015;521(7550): 43–7.

[13] Jasperson KW, Tuohy TM, Neklason DW, Burt RW. Hereditary and familial colon cancer. Gastroenterology 2010;138(6):2044–58.

[14] de la Chapelle A. Genetic predisposition to colorectal cancer. Nat Rev Cancer 2004;4(10): 769–80.

[15] Palles C, Cazier JB, Howarth KM, Domingo E, Jones AM, Broderick P, et al. Germ line mutations affecting the proofreading domains of POLE and POLD1 predispose to colorectal adenomas and carcinomas. Nat Genet 2013;45(2):136–44.

[16] Weren RD, Ligtenberg MJ, Kets CM, de Voer RM, Verwiel ET, Spruijt L, et al. A germ line homozygous mutation in the base-excision repair gene NTHL1 causes adenomatous polyposis and colorectal cancer. Nat Genet 2015;47(6):668–71.

[17] Lynch HT, Snyder CL, Shaw TG, Heinen CD, Hitchins MP. Milestones of Lynch syndrome: 1895–2015. Nat Rev Cancer 2015;15(3):181–94.

[18] Ligtenberg MJ, Kuiper RP, Chan TL, Goossens M, Hebeda KM, Voorendt M, et al. Heritable somatic methylation and inactivation of MSH2 in families with Lynch syndrome due to deletion of the 3′ exons of TACSTD1. Nat Genet 2009;41(1):112–7.

[19] Kruse R, Rutten A, Lamberti C, Hosseiny-Malayeri HR, Wang Y, Ruelfs C, et al. Muir–Torre phenotype has a frequency of DNA mismatch-repair-gene mutations similar to that in hereditary nonpolyposis colorectal cancer families defined by the Amsterdam criteria. Am J Hum Genet 1998;63(1):63–70.

[20] Hamilton SR, Liu B, Parsons RE, Papadopoulos N, Jen J, Powell SM, et al. The molecular basis of Turcot's syndrome. N Engl J Med 1995;332(13):839–47.

[21] Nieuwenhuis MH, Vasen HF. Correlations between mutation site in APC and phenotype of familial adenomatous polyposis (FAP): a review of the literature. Crit Rev Oncol Hematol 2007;61(2):153–61.

[22] Valle L, Hernandez-Illan E, Bellido F, Aiza G, Castillejo A, Castillejo MI, et al. New insights into POLE and POLD1 germ line mutations in familial colorectal cancer and polyposis. Hum Mol Genet 2014;23(13):3506–12.

[23] Elsayed FA, Kets CM, Ruano D, van den Akker B, Mensenkamp AR, Schrumpf M, et al. Germ line variants in POLE are associated with early onset mismatch repair deficient colorectal cancer. Eur J Hum Genet 2015;23(8):1080–4.

[24] Pilarski R, Stephens JA, Noss R, Fisher JL, Prior TW. Predicting PTEN mutations: an evaluation of Cowden syndrome and Bannayan–Riley–Ruvalcaba syndrome clinical features. J Med Genet 2011;48(8):505–12.

[25] Houlston R, Bevan S, Williams A, Young J, Dunlop M, Rozen P, et al. Mutations in DPC4 (SMAD4) cause juvenile polyposis syndrome, but only account for a minority of cases. Hum Mol Genet 1998;7(12):1907–12.

[26] Howe JR, Roth S, Ringold JC, Summers RW, Jarvinen HJ, Sistonen P, et al. Mutations in the SMAD4/DPC4 gene in juvenile polyposis. Science 1998;280(5366):1086–8.

[27] Hemminki A, Markie D, Tomlinson I, Avizienyte E, Roth S, Loukola A, et al. A serine/threonine kinase gene defective in Peutz–Jeghers syndrome. Nature 1998;391(6663):184–7.

[28] Jenne DE, Reimann H, Nezu J, Friedel W, Loff S, Jeschke R, et al. Peutz–Jeghers syndrome is caused by mutations in a novel serine threonine kinase. Nat Genet 1998;18(1):38–43.

[29] Jaeger E, Leedham S, Lewis A, Segditsas S, Becker M, Cuadrado PR, et al. Hereditary mixed polyposis syndrome is caused by a 40-kb upstream duplication that leads to increased and ectopic expression of the BMP antagonist GREM1. Nat Genet 2012;44(6):699–703.

[30] Lewis A, Freeman-Mills L, de la Calle-Mustienes E, Giraldez-Perez RM, Davis H, Jaeger E, et al. A polymorphic enhancer near GREM1 influences bowel cancer risk through differential CDX2 and TCF7L2 binding. Cell Rep 2014;8(4):983–90.

[31] Laitman Y, Jaeger E, Katz L, Tomlinson I, Friedman E. GREM1 germ line mutation screening in Ashkenazi Jewish patients with familial colorectal cancer. Genet Res 2015;97:e11.

[32] Aretz S, Genuardi M, Hes FJ. Clinical utility gene card for: MUTYH-associated polyposis (MAP), autosomal recessive colorectal adenomatous polyposis, multiple colorectal adenomas, multiple adenomatous polyps (MAP)-update 2012. Eur J Hum Genet 2013;21(1), Available from: http://dx.doi.org/10.1038/ejhg.2010.77.

[33] Al-Tassan N, Chmiel NH, Maynard J, Fleming N, Livingston AL, Williams GT, et al. Inherited variants of MYH associated with somatic G:C-- > T:A mutations in colorectal tumors. Nat Genet 2002;30(2):227–32.

[34] Slupska MM, Luther WM, Chiang JH, Yang H, Miller JH. Functional expression of hMYH, a human homolog of the *Escherichia coli* MutY protein. J Bacteriol 1999;181(19):6210–3.

[35] Nielsen M, Morreau H, Vasen HF, Hes FJ. MUTYH-associated polyposis (MAP). Crit Rev Oncol Hematol 2011;79(1):1–16.

[36] Krokan HE, Bjoras M. Base excision repair. Cold Spring Harb Perspect Biol 2013;5(4): a012583.

[37] German J, Sanz MM, Ciocci S, Ye TZ, Ellis NZ. Syndrome-causing mutations of the BLM gene in persons in the Bloom's Syndrome Registry. Hum Mutat 2007;28:743–53.

[38] de Voer RM, Hahn MM, Mensenkamp AR, Hoischen A, Gilissen C, Henkes A, et al. Deleterious germ line BLM mutations and the risk for early-onset colorectal cancer. Sci Rep 2015;5:14060.

[39] Lammi L, Arte S, Somer M, Jarvinen H, Lahermo P, Thesleff I, et al. Mutations in AXIN2 cause familial tooth agenesis and predispose to colorectal cancer. Am J Hum Genet 2004;74(5): 1043–50.

[40] Wong P, Verselis SJ, Garber JE, Schneider K, DiGianni L, Stockwell DH, et al. Prevalence of early onset colorectal cancer in 397 patients with classic Li–Fraumeni syndrome. Gastroenterology 2006;130(1):73–9.

[41] Hitchins MP. Constitutional epimutation as a mechanism for cancer causality and heritability? Nat Rev Cancer 2015;15(10):625–34.

[42] Horsthemke B. Epimutations in human disease. Curr Top Microbiol Immunol 2006;310:45–59.

[43] Gazzoli I, Loda M, Garber J, Syngal S, Kolodner RD. A hereditary nonpolyposis colorectal carcinoma case associated with hypermethylation of the MLH1 gene in normal tissue and loss of heterozygosity of the unmethylated allele in the resulting microsatellite instability-high tumor. Cancer Res 2002;62(14):3925–8.

[44] Hitchins MP, Lynch HT. Dawning of the epigenetic era in hereditary cancer. Clin Genet 2014;85(5):413–6.

[45] Kuiper RP, Vissers LE, Venkatachalam R, Bodmer D, Hoenselaar E, Goossens M, et al. Recurrence and variability of germ line EPCAM deletions in Lynch syndrome. Hum Mutat 2011;32(4):407–14.

[46] Kempers MJ, Kuiper RP, Ockeloen CW, Chappuis PO, Hutter P, Rahner N, et al. Risk of colorectal and endometrial cancers in EPCAM deletion-positive Lynch syndrome: a cohort study. Lancet Oncol 2011;12(1):49–55.

[47] Vasen HF, Blanco I, Aktan-Collan K, Gopie JP, Alonso A, Aretz S, et al. Revised guidelines for the clinical management of Lynch syndrome (HNPCC): recommendations by a group of European experts. Gut 2013;62(6):812–23.

[48] Dunlop MG, Tenesa A, Farrington SM, Ballereau S, Brewster DH, Koessler T, et al. Cumulative impact of common genetic variants and other risk factors on colorectal cancer risk in 42,103 individuals. Gut 2013;62(6):871–81.

[49] Whiffin N, Hosking FJ, Farrington SM, Palles C, Dobbins SE, Zgaga L, et al. Identification of susceptibility loci for colorectal cancer in a genome-wide meta-analysis. Hum Mol Genet 2014;23(17):4729–37.

[50] Al-Tassan NA, Whiffin N, Hosking FJ, Palles C, Farrington SM, Dobbins SE, et al. A new GWAS and meta-analysis with 1000Genomes imputation identifies novel risk variants for colorectal cancer. Sci Rep 2015;5:10442.

[51] Broderick P, Carvajal-Carmona L, Pittman AM, Webb E, Howarth K, Rowan A, et al. A genome-wide association study shows that common alleles of SMAD7 influence colorectal cancer risk. Nat Genet 2007;39(11):1315–7.

[52] Tomlinson IP, Carvajal-Carmona LG, Dobbins SE, Tenesa A, Jones AM, Howarth K, et al. Multiple common susceptibility variants near BMP pathway loci GREM1, BMP4, and BMP2 explain part of the missing heritability of colorectal cancer. PLoS Genet 2011;7(6):e1002105.

[53] Segui N, Mina LB, Lazaro C, Sanz-Pamplona R, Pons T,Navarro M. et al., Germ line mutations in FAN1 cause hereditary colorectal cancer by impairing DNA repair. Gastroenterology 2015;149(3):563–6.

[54] Liu T, Ghosal G, Yuan J, Chen J, Huang J. FAN1 acts with FANCI-FANCD2 to promote DNA interstrand cross-link repair. Science 2010;329(5992):693–6.

[55] de Voer RM, Geurts van Kessel A, Weren RD, Ligtenberg MJ, Smeets D, Fu L, et al. Germ line mutations in the spindle assembly checkpoint genes BUB1 and BUB3 are risk factors for colorectal cancer. Gastroenterology 2013;145(3):544–7.

[56] Vasen HF, Tomlinson I, Castells A. Clinical management of hereditary colorectal cancer syndromes. Nat Rev Gastroenterol Hepatol 2015;12(2):88–97.

[57] Pritchard CC, Smith C, Salipante SJ, Lee MK, Thornton AM, Nord AS, et al. ColoSeq provides comprehensive Lynch and polyposis syndrome mutational analysis using massively parallel sequencing. J Mol Diagn 2012;14(4):357–66.

[58] Yurgelun MB, Allen B, Kaldate RR, Bowles KR, Judkins T, Kaushik P, et al. Identification of a variety of mutations in cancer predisposition genes in patients with suspected Lynch syndrome. Gastroenterology 2015;149(3) 604–613 e20.

[59] Neveling K, Feenstra I, Gilissen C, Hoefsloot LH, Kamsteeg EJ, Mensenkamp AR, et al. A post-hoc comparison of the utility of sanger sequencing and exome sequencing for the diagnosis of heterogeneous diseases. Hum Mutat 2013;34(12):1721–6.

[60] Sato T, Vries RG, Snippert HJ, van de Wetering M, Barker N, Stange DE, et al. Single Lgr5 stem cells build crypt-villus structures in vitro without a mesenchymal niche. Nature 2009;459(7244):262–5.

Chapter 12

Prostate Cancer

Wennuan Liu, Rong Na, Carly Conran and Jianfeng Xu
NorthShore University HealthSystem, Evanston, IL, United States

Chapter Outline

INTRODUCTION

In personalized medicine (PM), the aim is to provide individual risk assessment for medical conditions, or to predict the efficacy of measures intended to monitor, prevent, or treat these conditions (http://www.personalizedmedicinecoalition.org). The approaches of PM could be important in addressing clinical and public health issues involved in a variety of diseases, including cancers that are detected via population-level screening. This is particularly relevant to prostate cancer (PCa), where concerns have been raised regarding prostate-specific antigen (PSA) screening, subsequent overdiagnosis of low-grade diseases, and ultimately overtreatment of many indolent cancers. These interrelated issues have prompted a significant effort to identify markers that can effectively differentiate individuals who have different risks for PCa onset or progression. Improved risk estimation may help to address this major public health problem,

Genomic and Precision Medicine. DOI: http://dx.doi.org/10.1016/B978-0-12-800685-6.00012-6
211

as the prostate is the most common site of cancer diagnosis, accounting for approximately 26% of all new cancer diagnoses and 9% of cancer deaths in US men. This translates to an estimated 220,800 PCa diagnoses and 27,540 deaths in US men each year [1].

Findings in several genetics/genomics-based fields have provided new insights regarding the development and progression of PCa and have offered great potential to be translated into clinically useful biomarkers. There is now a great opportunity to utilize these findings in order to address some of the most vexing issues in the healthcare of men affected by PCa. In this chapter, we will discuss promising discoveries from germline genetics, somatic genetics, and epigenetics, as well as their potential implementation as clinical tools.

GERMLINE GENETICS OF PROSTATE CANCER

A variety of risk factors have been proposed as underlying PCa risk and mortality, but the only confirmed risk factors to-date are age, race, family history (FH), and inherited genetic variants [2]. It is arguable that genetic inheritance plays a significant role in the pathogenesis of PCa, as demonstrated by the observation that the risk of PCa increases with additional affected first-degree relatives [3], and by the findings from a large twin study determining that the heritability of PCa is 42%, being notably the greatest among all common cancers [4]. These findings have solicited the need to identify and study the role of inherited (germline) genetic variants in association with altered risks for PCa onset or progression.

Discoveries of Inherited Variants of Prostate Cancer

Rare High-penetrant PCa Susceptibility Genes

Genome-wide linkage studies were commonly used in the past three decades to identify high-penetrant PCa susceptibility genes in hereditary PCa (HPC) families, typically defined as three or more first-degree affected relatives. These linkage results have led to the reports of several high-penetrant PCa susceptibility genes in the early 2000s, including *HPC1* [5], *RNASEL* [6], and *MSR1* [7]. The molecular mechanisms of these three candidate PCa susceptibility genes are largely unknown, and confirmations of these genes in other series of HPC families are inconsistent.

In 2012, by combining linkage analysis and targeted next-generation sequencing (NGS), a rare, but recurrent, mutation (G84E) in the *HOXB13* gene, a homeobox transcription factor gene that is important in prostate development was found to cosegregate with PCa in four HPC families [8]. All 18 men with PCa and available DNA in these four families carried the mutation. In addition, the carrier rate of the G84E mutation was increased by a factor of approximately 20 among 5083 unrelated PCa patients of European descent, with the mutation found in 72 case subjects (1.4%), as compared with 1 in 1401 control subjects

(0.1%), $p = 8.5 \times 10^{-7}$. The mutation was significantly more common in men with early-onset, familial PCa (3.1%) than in those with late-onset, nonfamilial PCa (0.6%), $p = 2.0 \times 10^{-6}$.

Efforts were also made to examine roles of several established high-penetrant genes for other types of cancer in PCa [9]. Several population-based studies examined PCa risk among mutation carriers of *BRCA1* and *BRCA2*, DNA damage repair pathway genes known to confer a significantly increased risk of breast and ovarian cancer. These studies show that carriers of *BRCA2* mutations confer a higher risk of PCa, ranging from 2.5 to 8.6 times, as compared to the general population [10]. *BRCA2* mutation carriers were statistically significantly more likely to have disease that was Gleason ≥8, T3/T4 stage, nodal involvement, and metastases at the time of diagnoses when compared to noncarriers. Moderately increased risk (3.5fold) for PCa was also found for carriers of *BRCA1* mutations [11].

The magnitude and importance of DNA repair genes in the development of metastatic PCa were highlighted in a recent study of 150 men with metastatic castrate resistant PCa (mCRPC) where 12 patients (8%) were found to have loss-of-function mutations in several DNA repair genes such as *BRCA2*, *ATM*, and *BRCA1* [12]. Similar findings were reported in other studies [13].

Common PCa Risk-associated Genetic Variants

Since 2007, genome-wide association studies (GWASs) have used large sample populations to identify common genetic variants that are associated with specific diseases. These variants are known as single nucleotide polymorphisms, or "SNPs." Approximately, 100 SNPs have been associated with PCa to-date. Though the effects of individual SNPs are very modest as they relate to one's potential risk of developing PCa, the cumulative effect of PCa risk-associated SNPs can confer significant risk of developing the disease. In general, the more PCa risk-associated SNPs an individual harbors, the more likely he is to develop PCa.

Many of the PCa-associated SNPs identified by GWAS are in introns or intergenic regions (i.e., "gene-desert" regions) of the genome; thus, the molecular mechanisms underlying these SNPs remain largely unknown. Nonetheless, the consistent replications have in part led to an intensive effort to evaluate these risk variants for clinical applications, particularly in prediction of disease risk for unaffected men. Although the risk-associated alleles of these SNPs are common (usually >5%) in the general population, each has a small individual effect on disease risk, with odds ratios (ORs) generally in the range of 1.1–1.3 [14]. However, larger ORs are observed when multiple SNPs are combined in a risk prediction model to assess cumulative lifetime risk. For example, when the first five PCa risk-associated SNPs identified from GWAS and FH were examined, men who have five or more of these risk factors have an OR of 9.46 for PCa, as compared with men without any of the risk factors [15]. In a later study that examined the first 14 PCa risk-associated SNPs identified from GWAS and family history, it was found that 55-year-old men with a FH and 14 risk alleles have a 52% risk of being diagnosed with PCa in the next 20 years. In comparison, without knowledge of genotype and family history, these men had an average population absolute risk of 13% [16].

Translational and Clinical Research

Clinical Validity of Inherited Genetic Variants

The clinical validity of high-penetrance mutations for risk prediction in PCa has been well established. Specifically, loss-of-function mutations in high-penetrance genes such as *BRCA2* and *ATM* have been shown to be strongly associated with PCa [12,17–21]. To establish the clinical validity of mutations of high-penetrance genes for their abilities to predict PCa diagnosis or prognosis, it is necessary to consider impact of mutations on gene function, associations of the particular gene with PCa in multiple populations, and biallelic alterations in germline and/or somatic DNA.

For most studies of PCa risk-associated SNPs, cumulative SNP effect is assessed using a quantitative measure commonly known as a "Genetic Risk Score" or "GRS." The clinical validity of PCa risk-associated SNPs, specifically in the form of GRSs, has been well established among large samples of independent populations [14,15,22]. There are three commonly used methods for calculating a GRS: (1) a simple count of risk alleles; (2) a weighted count of risk alleles, which incorporates the effect size of each SNP based on OR; and (3) population-standardization by incorporating risk allele frequency, which provides a value one's genetic risk relative to their same demographic population. Using the population-standardized method, GRS values center around a value of 1.0, no matter how many SNPs are included, with values greater and less than 1.0 conferring risk above or below the average population risk of developing PCa, respectively.

Various studies of these three GRS methods, particularly the weighted risk allele count and population-standardized methods, have demonstrated the superior ability of GRS to assess risk of developing PCa compared to the currently used methods in biopsy cohorts, clinical trial populations, and large case–control studies [14,22,23]. At present, FH of PCa is the most commonly used variable by clinicians, in addition to age and race, for assessing a man's risk of developing PCa. However, only ~7–17% of men in the general population are estimated to have a FH of PCa [2], and many of those men are unaware of their comprehensive PCa FH information. A recent study analyzing data from the Prostate Cancer Prevention Trial demonstrated the ability of GRS to identify twice as many men at higher risk of developing PCa when GRS was combined with FH compared to when FH is used to identify high-risk men alone [24].

Though GRSs show great promise as a tool for allowing clinicians to more accurately assess one's risk of developing PCa, some studies have shown that

GRS may not yet be ready for clinical use as a PCa risk stratification tool [25]. However, utilizing the data from Reduction by Dutasteride of Prostate Cancer Events where four sets of sequentially discovered SNP were analyzed, we found GRS values calculated from each of the four sequential sets of PCa risk-associated SNPs were significantly associated with PCa risk and had a better performance in discriminating PCa from non-PCa than FH. Although there was variability in GRS values for individual patients when using these four sequential sets of risk-associated SNPs, they were highly correlated. More importantly, multiple GRS values from evolving SNP sets actually provide a valuable tool for refining risk for all subjects. Risk reclassification effectively captures men with GRS values in a gray zone that are at intermediate risk, and men who have consistently lower or higher GRS values from multiple SNP sets are further assured of their low or high genetic risk, respectively ([24], submitted).

Clinical Utility of Inherited Genetic Variants

Testing for high-penetrance mutations (HPMs) and GRSs based on PCa risk-associated SNPs can be used for all men in early adulthood to identify those at higher genetic risk for lethal PCa. HPMs may also be used to determine which men are at significantly higher risk for PCa, as more men in the general population may harbor HPMs in genes such as *BRCA2* than previously thought. A recent study by Robinson et al. found that 8% of men with mCRPC harbored clinically actionable high-penetrance mutations [12]. Similarly, approximately 20% of men in the general population are expected to have an elevated risk of PCa [24]. Once a man is determined to be at high risk for PCa, based on HPMs, GRS, or FH, he should undergo regular PSA screening for early diagnosis [26,27]. If an individual has an elevated PSA test, germline DNA testing can also be used to guide biopsy decision making [23]. Once PCa is diagnosed, germline HPMs (e.g., in *BRCA2*) can predict prognosis and guide drug choices in early-stage disease [19,28–31].

SOMATIC GENETICS OF PROSTATE CANCER

As in many other human cancers, substantial genomic changes, including base substitutions (mutations), insertion/deletions, translocations, inversions, gene fusions, and copy number alterations (CNAs) have been identified in prostate tumors. Some of these acquired genetic alterations are found to be frequent across prostatic tumor samples and are thus demonstrated to play a crucial role in the tumorigenesis and progression of PCa. Notably, PCa is a heterogeneous disease with various combinations of somatic-level genomic alterations identified among different tumors or even from different cancer cells within the same tumor tissue [12,32–35]. This genetic heterogeneity is considered the primary obstacle for distinguishing aggressive from indolent forms of PCa and for treating the disease effectively. In this section, we will highlight several significant discoveries in this field and discuss the potential implications of these findings for clinical management of PCa.

Major Types of Somatic Genetic Alterations

Copy Number Alterations

CNAs refer to changes in the number of copies of DNA in specific regions of the somatic genome that range in size from 1 kb to a complete chromosome arm and mainly involve copy number losses (deletions) or gains (amplifications). Genome-wide analyses of PCa have revealed that CNAs are a major component of the landscape of the PCa tumor genome. The most frequently detected CNAs in primary PCa tumors include amplifications of genomic regions containing oncogenic *MYC* (8q24.21, 10–40%), and deletions of DNA sequences containing tumor suppressor genes such as *NKX3-1* (8p21.2, 40–70%), *PTEN* (10q23.31, 10–40%), *CDKN1B* (12p13.1, 20–30%), *RB1* (13q14.2, 30–50%), and *TP53* (17p13.1, 20–30%) (Fig. 12.1).

It is notable that the frequencies of these recurrent CNAs usually vary greatly among different tumor samples, depending on the compositions of cohorts, tumor grade, and degree of tumor cell heterogeneity, reflecting the complex roles of these CNAs in PCa. For example, high-grade, metastatic prostatic tumors typically contain more frequent CNAs than clinically localized PCa [36,37]. The majority of PCa-specific CNAs resulting in deletions of tumor suppressor genes involve the loss of only one copy of the genes. It is speculated that either these genes are haploinsufficient or these hemizygous deletions are complemented by additional genetic and epigenetic alterations on the other copy of the gene. For example, the combination of a hemizygous germline frameshift in one copy and an acquired somatic deletion of the other copy was found to inactivate *BRCA2* [12,38]. The combination of a somatic point mutation in one copy and a hemizygous deletion of another was reported to abolish *PTEN* and *TP53* [12,39]. Homozygous deletions causing the loss of both copies of the sequence have been observed in many genes. The most common homozygously deleted genes are *PTEN* (~15%) and *CHD1* (~10%) in PCa [40]. In addition, it has been reported that each patient with metastatic PCa typically harbors a unique signature of CNAs that is shared among tumors from different metastatic sites, and that each of these tumors also notably accumulate a pattern of CNAs that is distinctive from each other, suggesting a clonally evolving nature of metastatic tumors [41–43].

Gene Fusions

Unlike most other solid tumors, PCa contains a high frequency of fusion genes that typically result from chromosomal rearrangements such as translocations and interstitial deletions. For example, the archetypal *TMPRSS2-ERG* fusion gene, formed as a consequence of translocation or deletion, has been found in 40–70% of PCa tumors [44,45] (Fig. 12.2). To date, more than 10 such fusion genes have been identified in PCa tumor genomes. One interesting, common feature of these fusion genes is that they often involve combinations of the 5′ promoter/untranslated regions of androgen-regulated genes such as *TMPRSS2*,

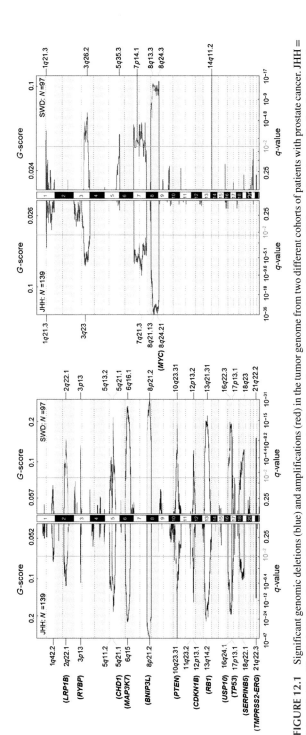

FIGURE 12.1 Significant genomic deletions (blue) and amplifications (red) in the tumor genome from two different cohorts of patients with prostate cancer. JHH = Johns Hopkins Hospital, SWD = Sweden. G-score is a measure of the significance of a copy number alteration event, which considers the amplitude of the aberration as well as the frequency of its occurrence across samples. q-value is a measure of the False Discovery Rate for the aberrant regions.

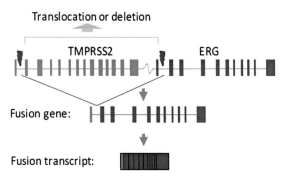

FIGURE 12.2   *TMPRESS2-ERG* gene fusion caused by DNA rearrangements.

SLC45A3, *KLK2*, *ACSL2*, and others, and 3′-coding regions of E-twenty-six (ETS) transcription factors such as *ERG*, *ETV1*, *ETV4*, and *ETV5* [44]. Although these potentially oncogenic ETS transcription factors remain silent in normal prostatic epithelial cells, gene fusions cause androgen-induced overexpression of these ETS proteins. Although the functions of these fusion genes are yet to be sufficiently defined, current evidence supports the notion that these fusions may represent an early, crucial genetic lesion which in collaboration with other genetic/molecular alterations drives the whole oncogenic process of PCa.

Point Mutations

Point mutations, also called somatic single nucleotide variations (SNVs), refer to changes in the sequence of DNA bases and include substitutions, insertions, and deletions of one or a few bases. Although PCa harbors the lowest point mutation rate (~0.33/Mb) among the major human cancers [46], relatively high frequencies of point mutations at cancer-related genes (oncogenes or tumor suppressor genes) *TP53* (12%), *PTEN* (7%), *SPOP* (7%), *KRAS* (4%), *FOXA1* (3%), *KMT2C* (3%), *EGFR* (3%), and *CTNNB1* (3%) have been documented in prostate adenocarcinoma (http://www.sanger.ac.uk/genetics/CGP/ cosmic, as January 31, 2016). Mutation frequencies in each of the genes vary greatly among different PCa tumor samples, which possibly reflect the genetic heterogeneity of PCa. Systematic analyses of SNVs across the tumor genome using exome and whole genome NGS have revealed a comprehensive mutation landscape of clinically localized PCa. Although the majority of these mutations are observed in a relatively small proportion of tumors, many genes, including *SPOP, FOXA1, TP53, PTEN, CDKN1B, MED12, THSD7B, SCN11A, NIPA2, PIK3CA*, ZNF595, *C14orf49, CDC27, MLL3, KDM6A*, and *KIF5A*, are considered to be significantly mutated. *SPOP* and *TP53* are the most frequently mutated genes with 10–15% across different cohorts in primary PCa [35,47,48]. Defects in DNA repair are apparently the initial cause of genome-wide alterations and subsequent genomic instability, with higher frequencies in advanced metastatic PCa. For example, in addition to enriched *TP53* mutation and *PTEN* homozygous deletion, biallelic *BRCA2* inactivation is reported in more than 10% of mCRPC, with ~5% damaging germline mutations [12,49]. Although point mutation frequencies are relatively low compared with other cancer types, the mutations at these genes may still contribute significantly to the pathogenesis of PCa, possibly by complementing the alterations at other genes in their corresponding pathways to favor the oncogenesis of PCa.

Translational and Clinical Research

Promising progress has been achieved in the translational research of these PCa-associated somatic genomic alterations, which investigates whether they can be used as biomarkers, alone or in combination with already-established clinicopathologic parameters, to improve screening, diagnosis, staging, and risk prediction of PCa [50]. For example, loss of *PTEN* has been shown to be a late genetic event and is predictive of a shorter time to biochemical recurrence of PCa. The relative copy number gain of *MYC* is reportedly associated with worse PCa-specific survival. By combining the cancer-specific *TMPRSS2-ERG* score with levels of PSA and PCA3 (the product of PCa gene 3), Tomlins et al. demonstrated a more accurate stratification of men at high risk for developing clinically significant PCa [51]. While promising, application in the clinical setting is still years away. In spite of lagging behind expression-based markers in clinical application, DNA-based markers hold some advantages such as resistance to degradation and remaining constant despite physiological and environmental fluctuations for diagnosis and prognosis of PCa.

Determination of the profiles of somatic genetics in individual PCa patients can also be used to direct anticancer therapy for the disease. There are currently many *Food and Drug Administration* (FDA)-approved drugs for use in oncology that have pharmacogenomic information in their labels; a strong majority of these target somatic genetic alterations [52]. For example, dysregulation of the PI3K–PTEN–AKT pathway usually caused by deletion/mutation of *PTEN*, amplification/mutation-induced upregulation of *PIK3CA* and AKT genes, and others has been closely implicated in prostate carcinogenesis and progression, and thus is the target of several pharmaceutical drugs that have been developed or are under development. In addition, the poly[adenosine diphosphate (ADP)-ribose] polymerase inhibitor, olaparib [19], and a platinum-based chemotherapy [29] have been demonstrated as an effective treatment for patients with CRPC harboring aberrations in DNA repair genes. Given the general interpatient genetic heterogeneity, the clinical success of these therapeutic drugs can be maximized by prospectively identifying patients harboring molecular abnormalities in these pathways, who may have a higher likelihood of responding. Furthermore, because cancer cells are able to evolve genetically, especially under pressure of therapeutic drugs, genome-wide analysis of further acquired alterations of DNA in response to cancer treatments will shed light on prognosis, new drug targets, and selection of strategies for the effective management of PCa for PM.

EPIGENETICS OF PROSTATE CANCER

The NIH has defined epigenetics as "an emerging field of science that studies heritable changes caused by the activation and deactivation of genes without any change in the underlying DNA sequence of the organism." (http://www. genome.gov/Glossary/index.cfm?id=528&textonly=true). Epigenetics consists of a variety of molecular mechanisms including DNA methylation, histone modifications, and noncoding RNAs. A common feature of these mechanisms is the potential to modify gene expression and alter the growth state of cells, thus contributing to carcinogenesis and tumor progression. As regulation of genes by epigenetic mechanisms is considered reversible, it may be possible to treat PCa using molecular therapies that selectively correct the epigenetic processes going awry during tumorigenesis.

Epigenetic Mechanisms

DNA Methylation

DNA methylation is a well-characterized epigenetic modification that plays a key role in gene regulation, with alterations contributing to tumorigenesis in PCa. This involves the covalent addition of a methyl group at the 5-position carbon of a cytosine ring in mammals. Methylation usually occurs at CG dinucleotides, where a cytosine (C) is joined by a phosphate (p) to a guanine (G), and is frequently abbreviated as CpG. When a cluster of CpGs occur within a DNA sequence, they may be referred to as a CpG island, and thus a likely location for methylation. DNA methylation patterns are established and maintained by DNA methyltransferases (DNMTs). The activity of these enzymes can be modified by several small-molecule chemicals such as vidaza (5-azacytidine) and decitabine (5-aza-2'-deoxycytidine), whose chemotherapeutic potentials for PCa are being explored. Methylated DNA can be actively demethylated by 10–11 translocation family proteins encoded by TET1, TET2, and TET3. The majority of DNA methylation in normal human cells is observed in noncoding regions of the genome, with 34% of intragenic CpG islands methylated, as opposed to only 2% of CpG islands in promoters [53]. Throughout the process of PCa initiation and progression, global hypomethylation is observed, particularly in intragenic regions [54,55]. Nonetheless, promoter hypermethylation was reported for more than 60 genes in PCa, which may be independent of the global hypomethylation status in PCa [56]. For example, promoter hypermethylation of *GSTP1*, which is the most frequently studied DNA-methylated gene in PCa, is observed in >90% of prostate tumors and results in reduced expression of this detoxification enzyme, presumably creating an environment that is more permissive to DNA damage and ultimately tumorigenesis and progression [57]. Changes in DNA methylation can be associated with high grade/stage or progression of PCa [58–61]. In addition, DNA hypomethylation associated with neuroendocrine and metastatic lethal PCa has been reported [62,63].

Histone Modifications

A variety of chemical modifications have been identified on histone proteins, including methylation, acetylation, phosphorylation, ubiquitination, sumoylation, and ADP-ribosylation. These modifications could change the structure and dynamics of chromatin where the modified histones reside and could potentially influence the expression of genes located nearby. Among these, the reversible methylation and acetylation of histones, that is, the attachment or removal of methyl and acetyl groups on histone proteins via the actions of histone methyltransferases, acetyl transferases, deacetylases (HDACs), and their coregulators, have been studied the most to date and have an established role in regulation of gene expressions in mammalian cells. In general, transcription of genes within a chromosomal region is reduced when specific lysine residues of histones are deacetylated and/or methylated, on the other hand, becomes more permissive or enhanced when acetylated and/or demethylated. Due to these regulatory roles in controlling gene expression, dysregulation of histone methylation and acetylation has been widely implicated in various human cancers, including PCa. In addition, compared with normal tissues, prostate tumors have been observed to harbor histone acetylation and demethylation in specific patterns that are permissive to transcription, and these modifications are linked to risk of PCa recurrence and metastasis [64,65].

Noncoding RNAs

A significant proportion of genomic DNA is transcribed into RNAs that are not translated into proteins and are thus termed "noncoding RNAs." The functions of these RNAs were not well understood until recent years, when growing evidence revealed that they may be involved in regulating diverse cellular processes such as splicing, transcription, translation, genomic imprinting, telomere maintenance, etc., and are implicated in many human diseases, including cancers. The best-understood of these noncoding RNAs are microRNAs, consisting of approximately 22 nucleotides that can base-pair with complementary sequences of messenger RNA (mRNA), causing mRNA degradation or preventing its translation into a protein. MicroRNAs have been suggested to be important players in the tumorigenesis and progression of PCa. For example, microRNA-205, which prevents cell growth and migration via apoptosis and cell-cycle arrest, has been observed to be downregulated in prostate tumors compared to normal tissues [66–69]. In addition, expression of certain microRNAs has been shown to be also associated with metastatic PCa [70,71]. A number of long noncoding RNAs (lncRNAs), including *PCA3, SChLAP1, PCAT1, PCGEM1, CTBP1-AS, PCAT29,* and *Linc00963,* have been identified in PCa [72], some of which are associated with aggressive or progressing PCa [73,74]. It should be noted, however, that while the area is promising, overall results have remained inconsistent. Various technical hurdles remain, including study design, sample handling, and assay design. As standards are developed, perhaps clearer patterns will emerge.

Translational and Clinical Research

A better understanding of DNA methylation, histone modifications, and non-coding RNAs may allow for the development of clinical applications that target elements of these molecular mechanisms. Disease screening based on DNA methylation status is promising. Many studies have examined the measurement of *GSTP1* promoter methylation as a predictive test. A meta-analysis observed high specificity (0.89) for *GSTP1* promoter methylation in a variety of body fluids, suggesting that this measure may be useful as complement to PSA screening as it can help address the relatively low specificity of PSA [75]. DNA methylation status in *APC*, *GSTP1*, and *RASSF1* has been identified as a promising utility to reduce unnecessary repeat biopsy in PCa [76,77]. Disease staging represents another potential application of epigenetic assays. For example, patterns of histone alteration could be measured across the genome to predict clinical staging and outcomes, and thus could also be useful in guiding treatment decisions.

Molecular guided therapies represent another promising direction. DNMT inhibitors (vidaza and decitabine) are already FDA-approved for treating myelodysplastic syndrome. There is a body of research evaluating HDAC inhibitors acting via mechanisms such as cell-cycle arrest and triggering of host immune responses as a means to induce apoptosis in tumors [78,79]. Another exciting area in the future will be epigenetic approaches to cancer prevention, such as a trial of compounds found in cruciferous vegetables that may prevent PCa development by inhibiting HDAC activity (http://clinicaltrials.gov/ct2/show/NCT01 265953?term=HDAC+prostate&rank=1).

One of the most widely used lncRNA biomarkers for detecting PCa is *PCA3* in urinary samples [80]. This biomarker has been shown to outperform PSA for predicting PCa; however, the ability of *PCA3* for predicting high-grade PCa is limited [72]. *SChLAP1*, which can be tested in urinary and tissue samples, has been shown to predict lethal PCa [80]. MicroRNAs, highly stable in both body fluids and tissue specimens, may be used as diagnostic or prognostic biomarkers [81,82].

GENOMIC PROFILING OF RNA IN PROSTATE CANCER

Genetic alterations and epigenetic modifications are known to affect gene expression. Altered gene expression leads to biochemical/biological changes at the cellular level, thereby causing phenotypic pathogenesis at the tissue level. The information of both genetic variations and epigenetic regulations can be effectively captured by genomic profiling of RNA. PCa prognosis can be assessed using several RNA tests for clinical use to predict PCa progression.

Translational and Clinical Research

Based on recent, unprecedented genomic findings, several RNA profiling panels for predicting PCa progression have been developed [83]. Most RNA profiling begins by designing a retrospective cohort or case–control study and analyzing the differential expression of RNA in tumor samples via transcriptome analysis. The next step involves selecting the genes with the highest and most significant differential expression between aggressive PCa and indolent diseases, as well as those with important biological functions. The final step is to combine the candidate and housekeeping genes into one panel for validation in an independent population. At present, three RNA profiling panels are commercially available and have all been approved for clinical use by FDA: *Oncotype Dx*, *Prolaris*, and *Decipher*.

Oncotype Dx

Oncotype Dx PCa Assay is a multigene expression assay based on real-time polymerase chain reaction (RT-PCR) technique, developed by Genomic Health, Inc., Redwood City, CA, USA. The assay measures the expression of 17 genes using approximately 1 mm of fixed paraffin-embedded prostate biopsy tissue, and the Genomic Prostate Score will be calculated thereafter. Previous studies suggest that this assay can predict PCa recurrence after radical prostatectomy (RP) or PCa progression in an active surveillance (AS) population, which may help guide decision making for further treatment [84,85].

Prolaris

Cancer cells, which actively proliferate at faster rates, especially in aggressive disease, transcribe more cell-cycle progression (CCP) genes than normal cells. Therefore, CCP gene expression is considered to reflect tumor biology, which may be useful for predicting cancer outcomes. The Prolaris PCa test (by Myriad Genetics Inc., Salt Lake City, UT, USA) was designed on the basis of this theory to test the expressions of 31 CCP genes and 15 housekeeping genes using quantitative RT-PCR. After testing, a CCP score is calculated as the average expression of the 31 genes for predicting cancer recurrence, metastases, and PCa-specific mortality in RP patients, and PCa progression in AS patients [86–90].

Decipher

Decipher, developed by GenomeDx Biosciences Inc., Vancouver, BC, Canada, is a genetic classifier (GC) that uses a profiling panel of 22 markers. A GC score is calculated. Previous studies suggest that GC score can predict early metastasis and cancer-specific mortality after RP, which has the potential to indicate that early intervention should be used in these patients [91–95].

Limitations and Future Directions

Although these RNA profiling panels have shown optimistic results of their clinical utilities, several limitations are notable and should be addressed in future studies: (1) the studies supporting these tests were retrospective with relatively small sample sizes, so larger scale, prospective randomized trials are necessary for validation; (2) RNA qualities vary among panels and a more robust method is preferred; and (3) high cost of RNA testing limits the use of the panels, and further evaluation of their cost-effectiveness is needed.

THE PYRAMID MODEL FOR PERSONALIZED CANCER CARE

The Pyramid Model (Fig. 12.3) describes an approach to cancer care in which PM is offered even before the onset of disease (i.e., to guide prevention and cancer screening practices). This model was developed to broaden the use of PM to more than late-stage disease, which is often the case [96]. The need for and feasibility of engaging this Pyramid Model of personalized cancer care can be best illustrated in PCa because of its current clinical challenges and availability of genomic findings at each stage of PCa. In the bottom tier (*prevention and screening*), which is earliest in chronology, the use of germline genomic tests is suggested for all men starting in young adulthood to assess their underlying, inherited risk for developing PCa. Higher risk men should undergo regular PSA testing, whereas men at lower risk can forgo PSA testing. The second tier (*diagnosis*) suggests the use of PCa-specific biomarkers such as Prostate Health Index [97], PCA3 [98], TMPRSS2-ERG [99], and 4-K Score [100], to better determine who should undergo biopsy for diagnosis of PCa. The third tier (*early-stage PCa*) is to be used along with pathological characteristics of the disease (stage

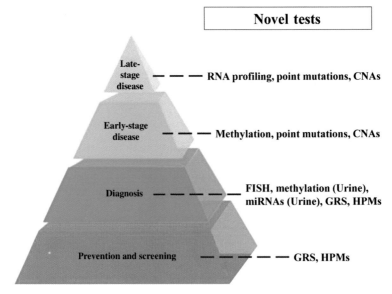

FIGURE 12.3   Novel Tests.

and grade) and involves testing of somatic (tumor) DNA as well as RNA to determine disease prognosis. Patients with prostate tumors exhibiting RNA and DNA profiles that are associated with a poorer prognosis should be advised to receive more aggressive treatment, possibly including adjuvant therapy, whereas patients without these risk-associated alterations may consider AS. The fourth tier (*late-stage PCa*) utilizes analyses of genomic information in one's germline as well as tumor(s) to develop individualized treatment strategies for metastatic disease.

CONCLUSIONS

The burgeoning field of germline genetics, somatic genetics, and epigenetics has dramatically reshaped studies of PCa, one of the major public healthcare problems. The various germline and somatic, genetic and epigenetic variations or alterations identified to be associated with increased risk for PCa onset and progression have not only widened our view and deepened our understanding of the complex mechanisms underlying the pathogenesis of PCa but also offer great potential to be translated into clinically useful biomarkers or novel drug targets for PM of this disease.

REFERENCES

[1]   Siegel RL, Miller KD, Jemal A. Cancer statistics, 2015. CA Cancer J Clin 2015;65:5–29.

[2]   Xu J, Sun J, Zheng SL. Prostate cancer risk-associated genetic markers and their potential clinical utility. Asian J Androl 2013;15:314–22.

[3]   Johns LE, Houlston RS. A systematic review and meta-analysis of familial prostate cancer risk. BJU Int 2003;91:789–94.

[4]   Lichtenstein P, Holm NV, Verkasalo PK, Iliadou A, Kaprio J, Koskenvuo M, et al. Environmental and heritable factors in the causation of cancer—analyses of cohorts of twins from Sweden, Denmark, and Finland. N Engl J Med 2000;343:78–85.

[5]   Tavtigian SV, Simard J, Teng DH, Abtin V, Baumgard M, Beck A, et al. A candidate prostate cancer susceptibility gene at chromosome. Nat Genet 2001;27:172–80.

[6]   Carpten J, Nupponen N, Isaacs S, Sood R, Robbins C, Xu J, et al. Germline mutations in the ribonuclease L gene in families showing linkage with HPC1. Nat Genet 2002;30:181–4.

[7]   Xu J, Zheng SL, Komiya A, Mychaleckyj JC, Isaacs SD, Hu JJ, et al. Germline mutations and sequence variants of the macrophage scavenger receptor 1 gene are associated with prostate cancer risk. Nat Genet 2002;32:321–5.

[8]   Ewing CM, Ray AM, Lange EM, Zuhlke KA, Robbins CM, Tembe WD, et al. Germline mutations in HOXB13 and prostate-cancer risk. N Engl J Med 2012;366:141–9.

[9]   Pilie PG, Giri VN, Cooney KA. HOXB13 and other high penetrant genes for prostate cancer. Asian J Androl 2016;18:530–2.

[10] Mersch J, Jackson MA, Park M, Nebgen D, Peterson SK, Singletary C, et al. Cancers associated with BRCA1 and BRCA2 mutations other than breast and ovarian. Cancer 2015;121:269–75.

[11] Castro E, Goh C, Olmos D, Saunders E, Leongamornlert D, Tymrakiewicz M, et al. Germline BRCA mutations are associated with higher risk of nodal involvement, distant metastasis, and poor survival outcomes in prostate cancer. J Clin Oncol 2013;31:1748–57.

[12] Robinson D, Van Allen EM, Wu YM, Schultz N, Lonigro RJ, Mosquera JM, et al. Integrative clinical genomics of advanced prostate cancer. Cell 2015;161:1215–28.

[13] Hart SN, Ellingson MS, Schahl K, Vedell PT, Carlson RE, Sinnwell JP, et al. Determining the frequency of pathogenic germline variants from exome sequencing in patients with castrate-resistant prostate cancer. BMJ Open 2016;6:e010332.

[14] Al Olama AA, Kote-Jarai Z, Berndt SI, Conti DV, Schumacher F, Han Y, et al. A meta-analysis of 87,040 individuals identifies 23 new susceptibility loci for prostate cancer. Nat Genet 2014;46:1103–9.

[15] Zheng SL, Sun J, Wiklund F, Smith S, Stattin P, Li G, et al. Cumulative association of five genetic variants with prostate cancer. N Engl J Med 2008;358:910–9.

[16] Xu J, Sun J, Kader AK, Lindstrom S, Wiklund F, Hsu FC, et al. Estimation of absolute risk for prostate cancer using genetic markers and family history. Prostate 2009;69:1565–72.

[17] Akbari MR, Wallis CJ, Toi A, Trachtenberg J, Sun P, Narod SA, et al. The impact of a BRCA2 mutation on mortality from screen-detected prostate cancer. Br J Cancer 2014;111:1238–40.

[18] Decker B, Karyadi DM, Davis BW, Karlins E, Tillmans LS, Stanford JL, et al. Biallelic BRCA2 mutations shape the somatic mutational landscape of aggressive prostate tumors. Am J Hum Genet 2016;98:818–29.

[19] Mateo J, Carreira S, Sandhu S, Miranda S, Mossop H, Perez-Lopez R, et al. DNA-repair defects and olaparib in metastatic prostate cancer. N Engl J Med 2015;373:1697–708.

[20] Thorne H, Willems AJ, Niedermayr E, Hoh IM, Li J, Clouston D, et al. Decreased prostate cancer-specific survival of men with BRCA2 mutations from multiple breast cancer families. Cancer Prev Res 2011;4:1002–10.

[21] Tryggvadottir L, Vidarsdottir L, Thorgeirsson T, Jonasson JG, Olafsdottir EJ, Olafsdottir GH, et al. Prostate cancer progression and survival in BRCA2 mutation carriers. J Natl Cancer Inst 2007;99:929–35.

[22] Hoffmann TJ, Van Den Eeden SK, Sakoda LC, Jorgenson E, Habel LA, Graff RE, et al. A large multiethnic genome-wide association study of prostate cancer identifies novel risk variants and substantial ethnic differences. Cancer Discov 2015;5:878–91.

[23] Kader AK, Sun J, Reck BH, Newcombe PJ, Kim ST, Hsu FC, et al. Potential impact of adding genetic markers to clinical parameters in predicting prostate biopsy outcomes in men following an initial negative biopsy: findings from the REDUCE trial. Eur Urol 2012;62:953–61.

[24] Chen H, Liu X, Brendler CB, Ankerst DP, Leach RJ, Goodman PJ, et al. Adding genetic risk score to family history identifies twice as many high-risk men for prostate cancer: results from the Prostate Cancer Prevention Trial. Prostate 2016;76:1120–9.

[25] Krier J, Barfield R, Green RC, Kraft P. Reclassification of genetic-based risk predictions as GWAS data accumulate. Genome Med 2016;8:20.

[26] Gronberg H, Adolfsson J, Aly M, Nordstrom T, Wiklund P, Brandberg Y, et al. Prostate cancer screening in men aged 50–69 years (STHLM3): a prospective population-based diagnostic study. Lancet Oncol 2015;16:1667–76.

[27] Bancroft EK, Page EC, Castro E, Lilja H, Vickers A, Sjoberg D, et al. Targeted prostate cancer screening in BRCA1 and BRCA2 mutation carriers: results from the initial screening round of the IMPACT study. Eur Urol 2014;66:489–99.

[28] Castro E, Goh C, Leongamornlert D, Saunders E, Tymrakiewicz M, Dadaev T, et al. Effect of BRCA mutations on metastatic relapse and cause-specific survival after radical treatment for localised prostate cancer. Eur Urol 2015;68:186–93.

[29] Cheng HH, Pritchard CC, Boyd T, Nelson PS, Montgomery B. Biallelic inactivation of BRCA2 in platinum-sensitive metastatic castration-resistant prostate cancer. Eur Urol 2015;69:992–5.

[30] Kaufman B, Shapira-Frommer R, Schmutzler RK, Audeh MW, Friedlander M, Balmana J, et al. Olaparib monotherapy in patients with advanced cancer and a germline BRCA1/2 mutation. J Clin Oncol 2015;33:244–50.

[31] Sandhu SK, Omlin A, Hylands L, Miranda S, Barber LJ, Riisnaes R, et al. Poly(ADP-ribose) polymerase (PARP) inhibitors for the treatment of advanced germline BRCA2 mutant prostate cancer. Ann Oncol 2013;24:1416–8.

[32] Berger MF, Lawrence MS, Demichelis F, Drier Y, Cibulskis K, Sivachenko AY, et al. The genomic complexity of primary human prostate cancer. Nature 2011;470:214–20.

[33] Grasso CS, Wu YM, Robinson DR, Cao X, Dhanasekaran SM, Khan AP, et al. The mutational landscape of lethal castration-resistant prostate cancer. Nature 2012;487:239–43.

[34] Baca SC, Prandi D, Lawrence MS, Mosquera JM, Romanel A, Drier Y, et al. Punctuated evolution of prostate cancer genomes. Cell 2013;153:666–77.

[35] Barbieri CE, Baca SC, Lawrence MS, Demichelis F, Blattner M, Theurillat JP, et al. Exome sequencing identifies recurrent SPOP, FOXA1 and MED12 mutations in prostate cancer. Nat Genet 2012;44:685–9.

[36] Sun JS, Liu WN, Adams TS, Sun JL, Li XN, Turner AR, et al. DNA copy number alterations in prostate cancers: a combined analysis of published CGH studies. Prostate 2007;67:692–700.

[37] Taylor BS, Schultz N, Hieronymus H, Gopalan A, Xiao Y, Carver BS, et al. Integrative genomic profiling of human prostate cancer. Cancer Cell 2010;18:11–22.

[38] Hong MK, Macintyre G, Wedge DC, Van Loo P, Patel K, Lunke S, et al. Tracking the origins and drivers of subclonal metastatic expansion in prostate cancer. Nat Commun 2015;6:6605.

[39] Carreira S, Romanel A, Goodall J, Grist E, Ferraldeschi R, Miranda S, et al. Tumor clone dynamics in lethal prostate cancer. Sci Transl Med 2014;6:254ra125.

[40] Liu W, Lindberg J, Sui G, Luo J, Egevad L, Li T, et al. Identification of novel CHD1-associated collaborative alterations of genomic structure and functional assessment of CHD1 in prostate cancer. Oncogene 2012;31:3939–48.

[41] Gundem G, Van Loo P, Kremeyer B, Alexandrov LB, Tubio JM, Papaemmanuil E, et al. The evolutionary history of lethal metastatic prostate cancer. Nature 2015;520:353–7.

[42] Kumar A, Coleman I, Morrissey C, Zhang X, True LD, Gulati R, et al. Substantial interindividual and limited intraindividual genomic diversity among tumors from men with metastatic prostate cancer. Nat Med 2016;22:369–78.

[43] Liu W, Laitinen S, Khan S, Vihinen M, Kowalski J, Yu G, et al. Copy number analysis indicates monoclonal origin of lethal metastatic prostate cancer. Nat Med 2009;15:559–65.

[44] Kumar-Sinha C, Tomlins SA, Chinnaiyan AM. Recurrent gene fusions in prostate cancer. Nat Rev Cancer 2008;8:497–511.

[45] Tomlins SA, Rhodes DR, Perner S, Dhanasekaran SM, Mehra R, Sun XW, et al. Recurrent fusion of TMPRSS2 and ETS transcription factor genes in prostate cancer. Science 2005;310:644–8.

[46] Kan Z, Jaiswal BS, Stinson J, Janakiraman V, Bhatt D, Stern HM, et al. Diverse somatic mutation patterns and pathway alterations in human cancers. Nature 2010;466:869–73.

[47] Kumar A, White TA, MacKenzie AP, Clegg N, Lee C, Dumpit RF, et al. Exome sequencing identifies a spectrum of mutation frequencies in advanced and lethal prostate cancers. Proc Natl Acad Sci U S A 2011;108:17087–92.

[48] Lindberg J, Mills IG, Klevebring D, Liu W, Neiman M, Xu J, et al. The mitochondrial and autosomal mutation landscapes of prostate cancer. Eur Urol 2013;63:702–8.

[49] Beltran H, Yelensky R, Frampton GM, Park K, Downing SR, Macdonald TY, et al. Targeted next-generation sequencing of advanced prostate cancer identifies potential therapeutic targets and disease heterogeneity. Eur Urol 2013;63:920–6.

[50] Liu W. DNA alterations in the tumor genome and their associations with clinical outcome in prostate cancer. Asian J Androl 2016;18:533–42.

[51] Tomlins SA, Aubin SM, Siddiqui J, Lonigro RJ, Sefton-Miller L, Miick S, et al. Urine TMPRSS2:ERG fusion transcript stratifies prostate cancer risk in men with elevated serum PSA. Sci Transl Med 2011;3:94ra72.

[52] US FDA. Table of pharmacogenomic biomarkers in drug labels. Available from: <http://www.fda.gov/drugs/scienceresearch/researchareas/pharmacogenetics/ucm083378.htm>; 2015.

[53] Maunakea AK, Nagarajan RP, Bilenky M, Ballinger TJ, D'Souza C, Fouse SD, et al. Conserved role of intragenic DNA methylation in regulating alternative promoters. Nature 2010;466:253–7.

[54] Feinberg AP, Vogelstein B. Hypomethylation distinguishes genes of some human cancers from their normal counterparts. Nature 1983;301:89–92.

[55] Kim JH, Dhanasekaran SM, Prensner JR, Cao X, Robinson D, Kalyana-Sundaram S, et al. Deep sequencing reveals distinct patterns of DNA methylation in prostate cancer. Genome Res 2011;21:1028–41.

[56] Ongenaert M, Van Neste L, De Meyer T, Menschaert G, Bekaert S, Van Criekinge W. PubMeth: a cancer methylation database combining text-mining and expert annotation. Nucleic Acids Res 2008;36:D842–6.

[57] Brooks JD, Weinstein M, Lin X, Sun Y, Pin SS, Bova GS, et al. CG island methylation changes near the GSTP1 gene in prostatic intraepithelial neoplasia. Cancer Epidemiol Biomarkers Prev 1998;7:531–6.

[58] Bhasin JM, Lee BH, Matkin L, Taylor MG, Hu B, Xu Y, et al. Methylome-wide sequencing detects DNA hypermethylation distinguishing indolent from aggressive prostate cancer. Cell Rep 2015;13:2135–46.

[59] Angulo JC, Andres G, Ashour N, Sanchez-Chapado M, Lopez JI, Ropero S. Development of castration resistant prostate cancer can be predicted by a DNA hypermethylation profile. J Urol 2016;195:619–26.

[60] Stott-Miller M, Zhao S, Wright JL, Kolb S, Bibikova M, Klotzle B, et al. Validation study of genes with hypermethylated promoter regions associated with prostate cancer recurrence. Cancer Epidemiol Biomarkers Prev 2014;23:1331–9.

[61] Yegnasubramanian S, Kowalski J, Gonzalgo ML, Zahurak M, Piantadosi S, Walsh PC, et al. Hypermethylation of CpG islands in primary and metastatic human prostate cancer. Cancer Res 2004;64:1975–86.

[62] Aryee MJ, Liu W, Engelmann JC, Nuhn P, Gurel M, Haffner MC, et al. DNA methylation alterations exhibit intraindividual stability and interindividual heterogeneity in prostate cancer metastases. Sci Transl Med 2013;5:169ra110.

[63] Beltran H, Prandi D, Mosquera JM, Benelli M, Puca L, Cyrta J, et al. Divergent clonal evolution of castration-resistant neuroendocrine prostate cancer. Nat Med 2016;22:298–305.

[64] Seligson DB, Horvath S, Shi T, Yu H, Tze S, Grunstein M, et al. Global histone modification patterns predict risk of prostate cancer recurrence. Nature 2005;435:1262–6.

[65] Yu J, Yu J, Rhodes DR, Tomlins SA, Cao X, Chen G, et al. A polycomb repression signature in metastatic prostate cancer predicts cancer outcome. Cancer Res 2007;67:10657–63.

[66] Ambs S, Prueitt RL, Yi M, Hudson RS, Howe TM, Petrocca F, et al. Genomic profiling of microRNA and messenger RNA reveals deregulated microRNA expression in prostate cancer. Cancer Res 2008;68:6162–70.

[67] Gandellini P, Folini M, Longoni N, Pennati M, Binda M, Colecchia M, et al. miR-205 Exerts tumor-suppressive functions in human prostate through down-regulation of protein kinase Cepsilon. Cancer Res 2009;69:2287–95.

[68] Porkka KP, Pfeiffer MJ, Waltering KK, Vessella RL, Tammela TL, Visakorpi T. MicroRNA expression profiling in prostate cancer. Cancer Res 2007;67:6130–5.

[69] Tong AW, Fulgham P, Jay C, Chen P, Khalil I, Liu S, et al. MicroRNA profile analysis of human prostate cancers. Cancer Gene Ther 2009;16:206–16.

[70] Thieu W, Tilki D, deVere White RW, Evans CP. The role of microRNA in castration-resistant prostate cancer. Urol Oncol 2014;32:517–23.

[71] Watahiki A, Wang Y, Morris J, Dennis K, O'Dwyer HM, Gleave M, et al. MicroRNAs associated with metastatic prostate cancer. PLoS ONE 2011;6:e24950.

[72] Malik B, Feng FY. Long noncoding RNAs in prostate cancer: overview and clinical implications. Asian J Androl 2016;18:568–74.

[73] Prensner JR, Iyer MK, Sahu A, Asangani IA, Cao Q, Patel L, et al. The long noncoding RNA SChLAP1 promotes aggressive prostate cancer and antagonizes the SWI/SNF complex. Nat Genet 2013;45:1392–8.

[74] Wang L, Han S, Jin G, Zhou X, Li M, Ying X, et al. Linc00963: a novel, long non-coding RNA involved in the transition of prostate cancer from androgen-dependence to androgen-independence. Int J Oncol 2014;44:2041–9.

[75] Wu T, Giovannucci E, Welge J, Mallick P, Tang WY, Ho SM. Measurement of GSTP1 promoter methylation in body fluids may complement PSA screening: a meta-analysis. Br J Cancer 2011;105:65–73.

[76] Partin AW, Van Neste L, Klein EA, Marks LS, Gee JR, Troyer DA, et al. Clinical validation of an epigenetic assay to predict negative histopathological results in repeat prostate biopsies. J Urol 2014;192:1081–7.

[77] Trock BJ, Brotzman MJ, Mangold LA, Bigley JW, Epstein JI, McLeod D, et al. Evaluation of GSTP1 and APC methylation as indicators for repeat biopsy in a high-risk cohort of men with negative initial prostate biopsies. BJU Int 2012;110:56–62.

[78] Bolden JE, Peart MJ, Johnstone RW. Anticancer activities of histone deacetylase inhibitors. Nat Rev Drug Discov 2006;5:769–84.

[79] Dokmanovic M, Marks PA. Prospects: histone deacetylase inhibitors. J Cell Biochem 2005;96:293–304.

[80] Groskopf J, Aubin SM, Deras IL, Blase A, Bodrug S, Clark C, et al. APTIMA PCA3 molecular urine test: development of a method to aid in the diagnosis of prostate cancer. Clin Chem 2006;52:1089–95.

[81] Hunter MP, Ismail N, Zhang X, Aguda BD, Lee EJ, Yu L, et al. Detection of microRNA expression in human peripheral blood microvesicles. PLoS ONE 2008;3:e3694.

[82] Kumar B, Lupold SE. MicroRNA expression and function in prostate cancer: a review of current knowledge and opportunities for discovery. Asian J Androl 2016;18:559–67.

[83] Na R, Wu Y, Ding Q, Xu J. Clinically available RNA profiling tests of prostate tumors: utility and comparison. Asian J Androl 2016;18:575–9.

[84] Cullen J, Rosner IL, Brand TC, Zhang N, Tsiatis AC, Moncur J, et al. A Biopsy-based 17-gene genomic prostate score predicts recurrence after radical prostatectomy and adverse surgical pathology in a racially diverse population of men with clinically low- and intermediate-risk prostate cancer. Eur Urol 2015;68:123–31.

[85] Klein EA, Cooperberg MR, Magi-Galluzzi C, Simko JP, Falzarano SM, Maddala T, et al. A 17-gene assay to predict prostate cancer aggressiveness in the context of Gleason grade heterogeneity, tumor multifocality, and biopsy undersampling. Eur Urol 2014;66:550–60.

[86] Bishoff JT, Freedland SJ, Gerber L, Tennstedt P, Reid J, Welbourn W, et al. Prognostic utility of the cell cycle progression score generated from biopsy in men treated with prostatectomy. J Urol 2014;192:409–14.

[87] Cooperberg MR, Simko JP, Cowan JE, Reid JE, Djalilvand A, Bhatnagar S, et al. Validation of a cell-cycle progression gene panel to improve risk stratification in a contemporary prostatectomy cohort. J Clin Oncol 2013;31:1428–34.

[88] Cuzick J, Berney DM, Fisher G, Mesher D, Moller H, Reid JE, et al. Prognostic value of a cell cycle progression signature for prostate cancer death in a conservatively managed needle biopsy cohort. Br J Cancer 2012;106:1095–9.

[89] Cuzick J, Swanson GP, Fisher G, Brothman AR, Berney DM, Reid JE, et al. Prognostic value of an RNA expression signature derived from cell cycle proliferation genes in patients with prostate cancer: a retrospective study. Lancet Oncol 2011;12:245–55.

[90] Freedland SJ, Gerber L, Reid J, Welbourn W, Tikishvili E, Park J, et al. Prognostic utility of cell cycle progression score in men with prostate cancer after primary external beam radiation therapy. Int J Radiat Oncol Biol Phys 2013;86:848–53.

[91] Cooperberg MR, Davicioni E, Crisan A, Jenkins RB, Ghadessi M, Karnes RJ. Combined value of validated clinical and genomic risk stratification tools for predicting prostate cancer mortality in a high-risk prostatectomy cohort. Eur Urol 2015;67:326–33.

[92] Den RB, Feng FY, Showalter TN, Mishra MV, Trabulsi EJ, Lallas CD, et al. Genomic prostate cancer classifier predicts biochemical failure and metastases in patients after postoperative radiation therapy. Int J Radiat Oncol Biol Phys 2014;89:1038–46.

[93] Erho N, Crisan A, Vergara IA, Mitra AP, Ghadessi M, Buerki C, et al. Discovery and validation of a prostate cancer genomic classifier that predicts early metastasis following radical prostatectomy. PLoS ONE 2013;8:e66855.

[94] Klein EA, Yousefi K, Haddad Z, Choeurng V, Buerki C, Stephenson AJ, et al. A genomic classifier improves prediction of metastatic disease within 5 years after surgery in node-negative high-risk prostate cancer patients managed by radical prostatectomy without adjuvant therapy. Eur Urol 2015;67:778–86.

[95] Ross AE, Johnson MH, Yousefi K, Davicioni E, Netto GJ, Marchionni L, et al. Tissue-based genomics augments post-prostatectomy risk stratification in a natural history cohort of intermediate- and high-risk men. Eur Urol 2016;69:157–65.

[96] Conran CA, Brendler CB, Xu J. Personalized prostate cancer care: from screening to treatment. Asian J Androl 2016;18:505–8.

[97] Loeb S, Sokoll LJ, Broyles DL, Bangma CH, van Schaik RH, Klee GG, et al. Prospective multicenter evaluation of the Beckman Coulter Prostate Health Index using WHO calibration. J Urol 2013;189:1702–6.

[98] Durand X, Moutereau S, Xylinas E, de la Taille A. Progensa PCA3 test for prostate cancer. Expert Rev Mol Diagn 2011;11:137–44.

[99] Leyten GH, Hessels D, Jannink SA, Smit FP, de Jong H, Cornel EB, et al. Prospective multicentre evaluation of PCA3 and TMPRSS2-ERG gene fusions as diagnostic and prognostic urinary biomarkers for prostate cancer. Eur Urol 2014;65:534–42.

[100] Carlsson S, Maschino A, Schroder F, Bangma C, Steyerberg EW, van der Kwast T, et al. Predictive value of four kallikrein markers for pathologically insignificant compared with aggressive prostate cancer in radical prostatectomy specimens: results from the European Randomized Study of Screening for Prostate Cancer section Rotterdam. Eur Urol 2013;64:693–9.

Chapter 13

Asthma

Michael J. Mcgeachie, Damien C. Croteau-Chonka and
Scott T. Weiss
Harvard Medical School, Boston, MA, United States

Chapter Outline

INTRODUCTION

Asthma is a complex disease affecting over 300 million individuals in the developed world [1]. Most asthma cases, including asthma in adults, have their origins in childhood. The increases in asthma prevalence and hospitalization rates [2] that occurred between 1980 and 2010, when the self-reported prevalence of asthma in the United States of America increased from 3% to 8.2% [3], seem to have leveled off and stabilized. However, this increase has resulted in health care utilization and mortality that is currently an estimated 50 billion dollars [4]. Despite the availability of several classes of therapeutic agents for asthma, it has been estimated that as many as one-half of asthmatic patients do not respond to treatment with β2-agonists, leukotriene (LT) antagonists, or inhaled corticosteroids [5–7]. Asthma is an important and prevalent disease, with some effective therapies available, and a disease in which genetics and genomics has had an impact on all aspects of asthma understanding.

ASTHMA: BASIC PATHOBIOLOGY

Asthma is a clinical syndrome of unknown etiology characterized by reversible episodes of airflow obstruction, airway hyperresponsiveness (AHR), and

Genomic and Precision Medicine. DOI: http://dx.doi.org/10.1016/B978-0-12-800685-6.00013-8
231

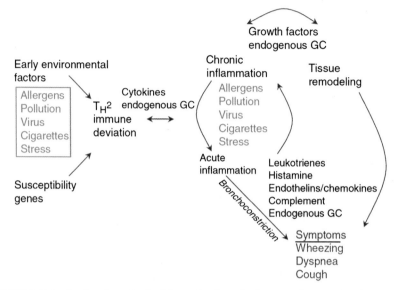

FIGURE 13.1   Depicts the pathophysiology of asthma development. Genes and environmental exposures lead to TH₂ inflammation that can be both acute and chronic (Adapted from reference 8).

a chronic inflammatory process of the airways of which mast cells, eosinophils, T-lymphocytes, epithelial cells, and airway smooth muscle (ASM) cells play a prominent role [8]. Fig. 13.1 shows important pathobiologic features of asthma. CD4+ lymphocytes produce IL-3, IL-4, IL-5, IL-13, and Granulocyte Macrophage–Colony Stimulating Factor (GM-CSF) and thereby promote the synthesis of Immunoglobulin E (IgE), an important allergic effector molecule. Chemokines, such as eotaxin, Regulated on Activation Normal T-cell Expressed and Secreted (RANTES), and IL-8 produced by epithelial and inflammatory cells, serve to amplify and perpetuate the inflammatory events. Several bronchoactive mediators, such as histamine, LTs, and neuropeptides are released into the airways and precipitate an asthma attack by causing ASM constriction, mucus secretion, and edema. Over time, there is smooth muscle growth and the deposition of subepithelial connective tissue, a process referred to as airway remodeling. Clinically, asthmatics have difficulty exhaling air because of an increase in airway resistance that is a consequence of smooth muscle contraction, inflammation, and remodeling (Fig. 13.2). Physiological impairment is quantified most commonly by the forced expiratory volume in one second (FEV 1, Fig. 13.2). FEV 1 is the volume of air a person can "blow out" in 1 s and is very useful as a measurement of lung function because it is easily obtained, reproducible, and correlated with asthma severity and therapeutic responses [9]. The above description of events, while true, is likely to only represent a small percentage of the total pathobiology that will be ultimately elucidated by genomic methods.

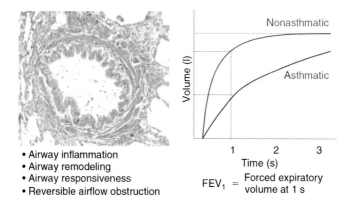

- Airway inflammation
- Airway remodeling
- Airway responsiveness
- Reversible airflow obstruction

FEV_1 = Forced expiratory volume at 1 s

FIGURE 13.2 Depicts an asthmatic airway on the left with inflammation and remodeling. On the right, the physiologic consequences of airway narrowing are depicted, showing decreased Forced Expiratory Flow in one second (FEV_1), in asthma compared to normal.

PREDISPOSITION (GENETIC AND NONGENETIC) TO ASTHMA

Estimates of the heritability of asthma in several twin studies conducted around the world have ranged from 36% to 79% [10–14], with the highest values coming from studies that had a more comprehensive phenotypic assessment of asthma [15]. Modern estimates of the heritability of asthma due to additive genetic effects from common variants are between 61.5% ± 16% [16]. These studies indicate that asthma is a significantly genetic disease, and that has been borne out by the genetic, genomic, and pharmacogenomic studies performed since.

GENOME-WIDE ASSOCIATION STUDIES OF ASTHMA

Earlier candidate gene and linkage studies have been largely eclipsed by genome-wide association studies (GWAS), although through candidate gene studies, variants in nine genes (IL4, IL4RA, IL13, ADRB2, CD14, TNFA, HLA-DRB1, HLA-DQB1, and FCERIB) have been associated with asthma phenotypes in at least ten populations [17]. GWAS are designed to sample the variability at nearly all locations in the genome simultaneously, without bias or hypothesis. These studies have been successful, with sufficient sample sizes of cases and controls, at identifying common genomic variants with typically very small-to-small effect sizes on a variety of common diseases, including asthma. In particular, there have been two very large GWAS metastudies of asthma, composed of many smaller GWAS, which have identified several variants near genes that have then been repeatedly replicated.

GABRIEL [18] is a large cohort of European-ancestry asthmatics including 10,365 people, and 16,110 controls. The GABRIEL study identified asthma susceptibility genes IL1RL1/IL18R1, HLA-DQ, IL33, SMAD3, IL2RB, and found that the ORMDL3/GSDMB locus was only associated with childhood-onset asthma. EVE [19] is an American meta-analysis of asthma GWAS including asthmatics from three major racial ancestry groups: European, African, and Hispanic. EVE included 5416 asthma cases with replication in another 12,649 subjects, finding five significant replications: ORMDL3/GSDMB, IL1RL1, TSLP, IL33, and also identifying PYHIN1 as a risk factor for asthma in African descent individuals only.

Other well-powered GWAS have identified additional genes, including an association of HAS2 with adult asthma in a Japanese population [20]; PSORS1C1 in American Latinos [21]; PDE4D in a combination of cohorts [22]; GAB1, 10p14, and 12q13 [23]; DENND1B [24]; SLC30A8 [25]; RORA [26]; CDHR3 and RAD50 [27]; and IL6R and LRRC32 [28]. Other GWAS have established links between other genes and specific subtypes of asthma, in particular allergic asthma cooccurring with allergic rhinitis: ZBTB10, CLEC16A, WDR36, and GSDMA were associated [29]. Others have also looked at allergic asthma, finding associations with TLR1 [30]. Some studies have tried to establish a particular difference between asthma, allergic asthma, and atopy: finding PERLD1 associated with just allergic asthma but not the others [31]. Some work has looked at severe asthma specifically, finding additional associations with ORMDL3/GSDMB and IL1RL1/IL18R1 [32]. GWAS in children have established some associations with childhood asthma, including ATPAF1 [33], and the 3p26.2 region which was associated with earlier onset of asthma in children [34]. Studies of occupational or diisocyanate-induced asthma have additionally identified GWAS-significant associations with HERC2, ODZ3, and CDH17 [35] that mediate the incidence of isocyanate-induced asthma. Finally, some GWAS have looked at the incidence of asthma-control indicators, such as asthma exacerbations requiring hospitalization or other doctor's visit and additional high-dose systemic steroids for asthma control, identifying CTNNA3, and SEMA3D as genes associated with worse asthma exacerbation [36].

GWAS are ideal for identifying associations with commonly occurring variants, but will not find associations with rarer or spontaneous mutations that may dispose to asthma incidence. The next generation of DNA sequencing technology, beyond microarray GWA chips, is allowing a person's whole genome to be sequenced, and many times over to assure correctness and redundancy. Although only a recent development, whole-genome sequencing of asthmatics has identified a deletion near NEDD4L associated with asthma [37]. Whole-genome sequencing can be expected to advance the knowledge of asthma genetics considerably in the next 3–5 years.

ASTHMA GENOMICS

Complementing these whole genome genetic surveys are studies examining global gene expression patterns in asthma and allergic inflammation. Variability in gene expression is an important driver of variability in complex biological phenotypes, such as human disease [38].

While asthma is a disease of chronic inflammation localized to the airways, blood is the most easily obtained tissue for scientific study and is an important source of known systemic inflammatory biomarkers that can also reflect issues in the lungs. Profiles of controlled and therapy-resistant childhood asthma in peripheral blood leukocytes have been shown to be distinct from each other [39]. Pathway-enrichment analyses indicated that the controlled profiles were characterized by increased expression of natural killer cell-mediated cytotoxicity and that the therapy-resistant profiles were characterized by decreased expression of N-glycan biosynthesis and by increased expression of bitter-taste transduction receptors, whose stimulation induces bronchodilation. A follow-up study in the same cohort showed that therapy-resistant profiles were also characterized by decreased glucocorticoid receptor signaling and increased activity of the Mitogen Activated Protein Kinase (MAPK) and Jun kinase cascades [40]. Importantly, these profiles could also distinguish adults with and without asthma. A study of mRNA and noncoding micro-RNA (miRNA) expression profiles in circulating CD4+ and CD8+ T-cells from adults showed that the latter class was exclusively activated during severe asthma [41]. Another study of the CD4+ T-cell transcriptome found a male-specific expression of a single interleukin receptor gene (IL17RB) driving a substantial portion of total serum levels of immunoglobulin E in young adults with mild-to-moderate asthma [42]. Finally, a data-clustering approach identified four distinct asthma endotypes in children based on both clinical biomarker and gene-expression data from blood [43], pointing to varying levels of involvement of the adaptive and/or innate immune systems and drawing a connection to features of metabolic syndrome. Work thus far in gene-expression profiling of asthma in blood implicates highly relevant biology related to systemic inflammation.

Lung and other airway tissues are much more directly involved in the local causal mechanisms of asthma than blood is, and so corresponding gene-expression profiles from these tissues would likely be more informative of the underlying pathobiology [44]. However, to retrieve such airway tissues often requires the use of much more expensive and invasive procedures than a simple blood draw (e.g., bronchoscopy). One of the most well-known asthma-specific gene expression signatures has been a set of three bronchial-airway epithelial genes representing key drivers of T-helper type 2 (Th2) inflammation [45], which was subsequently found by the same group to distinguish two ("Th2-high" and "Th2-low") subtypes of mild-to-moderate asthma based on airway epithelial brushings (one eosinophilic and the other not) [46]. In another study of severe and moderate asthmatic adults, airway epithelial cell-expression profiles were used to identify a subset of genes associated with exhaled nitric oxide, a biomarker for atopy, and eosinophilia, which were then clustered to identify five clinical subtypes [47]. A gene-expression profile of ASM tissue identified cell signaling, development, and proliferation pathways enriched among asthmatic adults, as well as four genes associated with AHR [48]. Profiles in whole endobronchial biopsy samples showed pathways related to cell movement, death, and morphology (including MAPK and Nuclear Factor kappa-light-chain-enhancer of activated B cells (NF-κB) pathways) were enriched during mild, atopic asthma [49]. In addition, nasal transcriptomic profiles of childhood asthma [50] and induced sputum profiles from asthmatic adults [51] were both shown to recapitulate the "Th2-high" profile identified in deeper airway tissue profiles. A six-gene signature identified in induced sputum differentiated asthmatic adults by type of underlying inflammation (e.g., eosinophilic versus noneosinophilic) and predicted response to inhaled corticosteroids [52]. Induced sputum profiles and clinical phenotypes in adults were used to identify three subtypes of asthma severity, which also were recapitulated in matched blood samples for a subset of subjects [53].

There is also strong evidence to suggest that asthma susceptibility has some of its origins during fetal development of the pulmonary system, which is highly susceptible to perturbations by environmental exposures (e.g., in-utero smoke, vitamin D, and folate) [54]. Expression profiles in fetal lungs showed that glucocorticoid genes were important for airway development and were then associated with postnatal risk of asthma and treatment response [55]. These results suggest that expression profiles of asthma from a variety of tissue types are indeed modifiable by disease-relevant exposures and highlight particular pathways and genes for further study.

PHARMACOGENETICS

The two main types of asthma drugs are the so-called "reliever" drugs that target the acute bronchoconstriction and the so-called "controller" drugs that are used to reduce the severity of airway inflammation and the severity and frequency of obstruction [56]. The main reliever drugs are short-acting β2-agonists (e.g., albuterol, metaproterenol, pirbuterol, and levalbuterol), which relax the bronchial smooth muscle by activating β2-adrenergic receptors. This is the treatment of choice for very mild asthma. For moderate and severe asthma, the reliever treatment is usually combined with controller treatment. The two commonly used classes of controller agents are the inhaled glucocorticoids and the LT inhibitors. Inhaled glucocorticoids (e.g., budesonide, beclomethasone, flunisolide, and fluticasone) and LT inhibitors (e.g., montelukast and zafirlukast) modify the inflammatory microenvironment of the airway to reduce airway obstruction and hyperresponsiveness.

There is large interindividual variation in the treatment response to asthma medications [5,57]. For example, in a study by Malmstrom et al. comparing the efficacy of the inhaled steroid beclomethasone (200 μg bid) with the LT antagonist montelukast (10 mg/qd), it is clear that both drugs are effective over a 12-week course of treatment with a mean increase in FEV 1 of 13.1% for the beclomethasone group and 7.4% for the montelukast group (Fig. 13.3A) [58]. However, when these same data are viewed from a different perspective, focusing on the number of individuals as a function of percent change in FEV 1 from baseline, it is clear that many patients had little response (Fig. 13.3B). In fact, 22% of patients appear to have had an adverse response to treatment with a

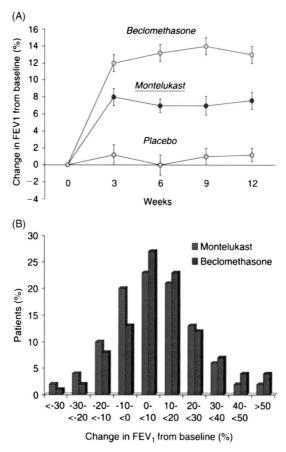

FIGURE 13.3 Panel A depicts the mean change in FEV_1, with two different asthma treatments compared to placebo. Panel B depicts the same patients in the same clinical trial as a percent change from baseline (Adapted from reference 5).

decline in FEV1 at 12 weeks compared with baseline. A unique subset of up to 25% of the nonresponders can be classified as having glucocorticoid-resistant asthma [59]. These patients are nonresponsive to even high doses of oral steroids. Despite these studies, there is no universally accepted definition of a steroid (or β2-agonist) nonresponder, although recent work has sought to combine various metrics of asthma stability into a steroid-response endophenotype [60]. These types of data illustrating variable drug efficacy are not limited to asthma drug trials but can be found for almost all classes of drugs.

The LTs are a family of lipoxygenated eicosatetraenoic acids derived from arachidonic acid and produced in the airways of asthmatics that are potent bronchoconstrictors. Many enzymes involved in the formation of the LTs have been investigated, along with other genes in the LT pathway, for pharmacogenetic effects on the efficacy of LT inhibitors. LT inhibitors target two proteins: 5-lipooxygenase (encoded by ALOX5) and cysteinyl LT receptor (encoded by CYSLTR1). A variant in ALOX5 was the first to be associated with response to these LT inhibitors in asthmatics [61]. Other genes on the pathway between ALOX5 and CYSLTR1 have been investigated. One such gene is LTC4S, and among asthmatic subjects treated with zafirlukast (which targets CYSLTR1), those homozygous for the A allele of a SNP in LTC4S ($n = 10$) had a lower FEV 1 response than those with the C/C or C/A ($n = 13$) genotype [62]. LTC4S was additionally associated with response to zileuton, which targets ALOX5 [63]. CHRM2 was associated with improved FEV1 during montelukast treatment [64], as was MRP1 [65]. Subsequent studies have identified MLLT3 in a GWAS of montelukast response [66]. The membrane protein SLCO2B1 was associated with circulating levels of montelukast, leading to reduced bioavailablility and reduced effectiveness in asthma therapy [67].

Studies evaluating the β2-agonist pathway have originally focused on the β2-adrenergic receptor gene (ADRB2). Numerous clinical [68,69] and cellular [70] studies, including one prospective, genotype-stratified, clinical trial [71], support the association of variation in this gene with response to inhaled β2-agonist therapy. Further trials were unable to support this association [72,73], including a prospective, genotype-stratified clinical trial [74]. Continuing work on this gene has identified many other polymorphisms that are sometimes associated with response to long- or short-acting β2-agonists, sometimes restricted to particular ethnic groups [75]. Additional work has looked at other genes in the β2-adrenergic pathway, including ARG1 [76,77]. Also, CRHR2 was associated with bronchodilator response [78]. Other genes have been associated with bronchodilator response through GWAS, including SPATS2L in European ancestry cohorts [79] and SLC22A15 in a Mexican cohort [80]. Further genes have been associated with response to β2-agonists through GWAS including ASB3. [81].

Genetic associations have also been reported for response to corticosteroids in asthma. In children, the FBXL7 gene was associated with corticosteroid response [82]. The glucocorticoid gene (GLCCI1) [83] was associated with corticosteroid response. Further studies of the corticosteroid molecular pathway identified CRHR1 as associated with corticosteroid response [84], and another investigated STIP1, which was shown to improve lung function under corticosteroid therapy [85]. Variants in TBCD have been associated with bone mineral accretion in response to steroid therapy in asthmatics [86]. CHRM2 was associated with improved symptom score in steroid treated asthmatics, and HSPA8 was associated improved lung function (FEV1) [64]. CYP3A4*22 was associated with improved response to corticosteroids [87]. GWAS have found SNPs in intergenic regions (rs6924808) and in regions with many genes (rs1353649, near nicotine dependence genes) associated with steroid response [88]. Similarly, the T gene was identified as a mediator of corticosteroid response from GWAS [89].

Some variants have been found that impact the effect of β2-agonists in conjunction with corticosteroids, including DUSP1 [90], ARG1 and ARG2 [91], ADCY9 [92], TBX21 [93], and ZNF432 [94].

Emerging work has drawn a link between vitamin D and asthma, and some variants have been associated with an interaction with vitamin D levels and incidence of asthma [95].

GENOMIC PREDICTION IN ASTHMA

While many variants and genes have been associated with asthma incidence or related phenotypes, the effect sizes of these associations are typically small, leading even a combination of many relevant SNPs to have little discriminative ability. For example, using clinical and demographic attributes plus 215 SNPs in candidate genes associated with asthma-related phenotypes, Prosperi et al. were able to predict asthma incidence at an Area Under the Curve (AUC) of 84% (where 100% is perfect prediction and 50% is random guessing) [96]. However, the SNPs themselves contributed almost nothing to the prediction, achieving only 54% AUC by themselves [96]. Experiments using possibly very large numbers of SNPs were similarly unpredictive (AUC 54%) [97]. However, similar methodologies have achieved impressive prediction in other diseases, including celiac disease (AUC ~90%) [98] and oral mucositis (82% accuracy) [99].

CONCLUSIONS

Asthma genomics has the potential to have an even greater clinical impact in the next 5–10 years through the development of genomic predictors of disease occurrence, response to asthma medication, and asthma severity (exacerbations). Future insights into asthma pathogenesis will ultimately come from the large-scale statistical integration of multiple broad "omic" data along the underlying causal pathways [38], namely, elucidating the network of networks connecting the genome, epigenome, transcriptome, proteome, metabolome, and/or microbiome to downstream disease. By more comprehensively modeling, these causal interdependencies, we will obtain a systems genomics view of asthma and allergic disease with the potential to identify personal health profiles with which to guide treatment development and application [100].

ACKNOWLEDGMENTS

The authors wish to acknowledge their colleagues in the Channing Division of Network Medicine.

REFERENCES

[1] Palmer LJ, Cookson WO. Genomic approaches to understanding asthma. Genome Res 2000;10:1280–7.
[2] Weiss KB, Gergen PJ, Wagener DK. Breathing better or wheezing worse? The changing epidemiology of asthma morbidity and mortality. Annu Rev Public Health 1993;14:491–513.

[3] Centers for Disease Control and Prevention (CDC) Vital signs: asthma prevalence, disease characteristics, and self-management education: United States, 2001–2009. MMWR Morb Mortal Wkly Rep 2011;60:547–52.

[4] Barnett SB, Nurmagambetov TA. Costs of asthma in the United States: 2002–2007. J Allergy Clin Immunol 2011;127:145–52.

[5] Drazen JM, Silverman EK, Lee TH. Heterogeneity of therapeutic responses in asthma. Br Med Bull 2000;56:1054–70.

[6] Liggett SB. Pharmacogenetic applications of the Human Genome Project. Nat Med 2001;7:281–3.

[7] Silverman ES, Hjoberg J, Palmer LJ, Tantisira K, Weiss ST, Drazen J. Application of pharmacogenetics to the therapeutics of asthma Huston J, editor. Therapeutic targets of airway inflammation. New York: Marcel Dekker; 2002.

[8] Elias JA, Lee CG, Zheng T, Ma B, Homer RJ, Zhu Z. New insights into the pathogenesis of asthma. J Clin Invest 2003;111:291–7.

[9] Reddel HK, Taylor DR, Bateman ED, et al. An official American Thoracic Society/European Respiratory Society statement: asthma control and exacerbations: standardizing endpoints for clinical asthma trials and clinical practice. Am J Respir Crit Care Med 2009;180:59–99.

[10] Duffy DL, Martin NG, Battistutta D, Hopper JL, Mathews JD. Genetics of asthma and hay fever in Australian twins. Am Rev Respir Dis 1990;142:1351–8.

[11] Edfors-Lubs ML. Allergy in 7000 twin pairs. Acta Allergol 1971;26:249–85.

[12] Hopp RJ, Bewtra AK, Nair NM, Townley RG. Specificity and sensitivity of methacholine inhalation challenge in normal and asthmatic children. J Allergy Clin Immunol 1984;74:154–8.

[13] Nieminen MM, Kaprio J, Koskenvuo M. A population-based study of bronchial asthma in adult twin pairs. Chest 1991;100:70–5.

[14] Sibbald B, Horn ME, Brain EA, Gregg I. Genetic factors in childhood asthma. Thorax 1980;35:671–4.

[15] Sandford A, Weir T, Pare P. The genetics of asthma. Am J Respir Crit Care Med 1996;153:1749–65.

[16] McGeachie MJ, Stahl EA, Himes BE, et al. Polygenic heritability estimates in pharmacogenetics: focus on asthma and related phenotypes. Pharmacogenet Genomics 2013;23:324–8.

[17] Ober C, Hoffjan S. Asthma genetics 2006: the long and winding road to gene discovery. Genes Immun 2006;7:95–100.

[18] Moffatt MF, Gut IG, Demenais F, et al. A large-scale, consortium-based genomewide association study of asthma. N Engl J Med 2010;363:1211–21.

[19] Torgerson DG, Ampleford EJ, Chiu GY, et al. Meta-analysis of genome-wide association studies of asthma in ethnically diverse North American populations. Nat Genet 2011;43:887–92.

[20] Yatagai Y, Sakamoto T, Yamada H, et al. Genomewide association study identifies HAS2 as a novel susceptibility gene for adult asthma in a Japanese population. Clin Exp Allergy 2014;44:1327–34.

[21] Galanter JM, Gignoux CR, Torgerson DG, et al. Genome-wide association study and admixture mapping identify different asthma-associated loci in Latinos: the Genes-environments & Admixture in Latino Americans study. J Allergy Clin Immunol 2014;134:295–305.

[22] Himes BE, Hunninghake GM, Baurley JW, et al. Genome-wide association analysis identifies PDE4D as an asthma-susceptibility gene. Am J Hum Genet 2009;84:581–93.

[23] Hirota T, Takahashi A, Kubo M, et al. Genome-wide association study identifies three new susceptibility loci for adult asthma in the Japanese population. Nat Genet 2011;43:893–6.

[24] Sleiman PM, Flory J, Imielinski M, et al. Variants of DENND1B associated with asthma in children. N Engl J Med 2010;362:36–44.

[25] Noguchi E, Sakamoto H, Hirota T, et al. Genome-wide association study identifies HLA-DP as a susceptibility gene for pediatric asthma in Asian populations. PLoS Genet 2011;7:e1002170.

[26] Ramasamy A, Kuokkanen M, Vedantam S, et al. Genome-wide association studies of asthma in population-based cohorts confirm known and suggested loci and identify an additional association near HLA. PLoS ONE 2012;7:e44008.

[27] Bonnelykke K, Sleiman P, Nielsen K, et al. A genome-wide association study identifies CDHR3 as a susceptibility locus for early childhood asthma with severe exacerbations. Nat Genet 2014;46:51–5.

[28] Ferreira MA, Matheson MC, Duffy DL, et al. Identification of IL6R and chromosome 11q13.5 as risk loci for asthma. Lancet 2011;378:1006–14.

[29] Ferreira MA, Matheson MC, Tang CS, et al. Genome-wide association analysis identifies 11 risk variants associated with the asthma with hay fever phenotype. J Allergy Clin Immunol 2014;133:1564–71.

[30] Hinds DA, McMahon G, Kiefer AK, et al. A genome-wide association meta-analysis of self-reported allergy identifies shared and allergy-specific susceptibility loci. Nat Genet 2013;45:907–11.

[31] Anantharaman R, Andiappan AK, Nilkanth PP, Suri BK, Wang de Y, Chew FT. Genome-wide association study identifies PERLD1 as asthma candidate gene. BMC Med Genet 2011;12:170.

[32] Wan YI, Shrine NR, Soler Artigas M, et al. Genome-wide association study to identify genetic determinants of severe asthma. Thorax 2012;67:762–8.

[33] Schauberger EM, Ewart SL, Arshad SH, et al. Identification of ATPAF1 as a novel candidate gene for asthma in children. J Allergy Clin Immunol 2011;128:753–60. e11.

[34] Forno E, Lasky-Su J, Himes B, et al. Genome-wide association study of the age of onset of childhood asthma. J Allergy Clin Immunol 2012;130:83–90. e4.

[35] Yucesoy B, Kaufman KM, Lummus ZL, et al. Genome-wide association study identifies novel loci associated with diisocyanate-induced occupational asthma. Toxicol Sci 2015;146:192–201.

[36] McGeachie MJ, Wu AC, Tse SM, et al. CTNNA3 and SEMA3D: promising loci for asthma exacerbation identified through multiple genome-wide association studies. J Allergy Clin Immunol 2015;136:1503–10.

[37] Campbell CD, Mohajeri K, Malig M, et al. Whole-genome sequencing of individuals from a founder population identifies candidate genes for asthma. PLoS ONE 2014;9:e104396.

[38] Ritchie MD, Holzinger ER, Li R, Pendergrass SA, Kim D. Methods of integrating data to uncover genotype-phenotype interactions. Nat Rev Genet 2015;16:85–97.

[39] Orsmark-Pietras C, James A, Konradsen JR, et al. Transcriptome analysis reveals upregulation of bitter taste receptors in severe asthmatics. Eur Respir J 2013;42:65–78.

[40] Persson H, Kwon AT, Ramilowski JA, et al. Transcriptome analysis of controlled and therapy-resistant childhood asthma reveals distinct gene expression profiles. J Allergy Clin Immunol 2015;136:638–48.

[41] Tsitsiou E, Williams AE, Moschos SA, et al. Transcriptome analysis shows activation of circulating CD8+ T cells in patients with severe asthma. J Allergy Clin Immunol 2012;129:95–103.

[42] Hunninghake GM, Chu JH, Sharma SS, et al. The CD4+ T-cell transcriptome and serum IgE in asthma: IL17RB and the role of sex. BMC Pulm Med 2011;11:17.

[43] George BJ, Reif DM, Gallagher JE, et al. Data-driven asthma endotypes defined from blood biomarker and gene expression data. PLoS ONE 2015;10:e0117445.

[44] Wesolowska-Andersen A, Seibold MA. Airway molecular endotypes of asthma: dissecting the heterogeneity. Curr Opin Allergy Clin Immunol 2015;15:163–8.

[45] Woodruff PG, Boushey HA, Dolganov GM, et al. Genome-wide profiling identifies epithelial cell genes associated with asthma and with treatment response to corticosteroids. Proc Natl Acad Sci USA 2007;104:15858–63.

[46] Woodruff PG, Modrek B, Choy DF, et al. T-helper type 2-driven inflammation defines major subphenotypes of asthma. Am J Respir Crit Care Med 2009;180:388–95.

[47] Modena BD, Tedrow JR, Milosevic J, et al. Gene expression in relation to exhaled nitric oxide identifies novel asthma phenotypes with unique biomolecular pathways. Am J Respir Crit Care Med 2014;190:1363–72.

[48] Yick CY, Zwinderman AH, Kunst PW, et al. Gene expression profiling of laser microdissected airway smooth muscle tissue in asthma and atopy. Allergy 2014;69:1233–40.

[49] Yick CY, Zwinderman AH, Kunst PW, et al. Transcriptome sequencing (RNA-Seq) of human endobronchial biopsies: asthma versus controls. Eur Respir J 2013;42:662–70.

[50] Poole A, Urbanek C, Eng C, et al. Dissecting childhood asthma with nasal transcriptomics distinguishes subphenotypes of disease. J Allergy Clin Immunol 2014;133:670–8. e12.

[51] Peters MC, Mekonnen ZK, Yuan S, Bhakta NR, Woodruff PG, Fahy JV. Measures of gene expression in sputum cells can identify TH2-high and TH2-low subtypes of asthma. J Allergy Clin Immunol 2014;133:388–94.

[52] Baines KJ, Simpson JL, Wood LG, et al. Sputum gene expression signature of 6 biomarkers discriminates asthma inflammatory phenotypes. J Allergy Clin Immunol 2014;133:997–1007.

[53] Yan X, Chu JH, Gomez J, et al. Noninvasive analysis of the sputum transcriptome discriminates clinical phenotypes of asthma. Am J Respir Crit Care Med 2015;191:1116–25.

[54] Sharma S, Chhabra D, Kho AT, Hayden LP, Tantisira KG, Weiss ST. The genomic origins of asthma. Thorax 2014;69:481–7.

[55] Sharma S, Kho AT, Chhabra D, et al. Glucocorticoid genes and the developmental origins of asthma susceptibility and treatment response. Am J Respir Cell Mol Biol 2015;52:543–53.

[56] A genome-wide search for asthma susceptibility loci in ethnically diverse populations. The Collaborative Study on the Genetics of Asthma (CSGA). Nat Genet 1997;15:389–92.

[57] Szefler SJ, Martin RJ, King TS, et al. Significant variability in response to inhaled corticosteroids for persistent asthma. J Allergy Clin Immunol 2002;109:410–8.

[58] Malmstrom K, Rodriguez-Gomez G, Guerra J, et al. Oral montelukast, inhaled beclomethasone, and placebo for chronic asthma. A randomized, controlled trial. Montelukast/ Beclomethasone Study Group. Ann Intern Med 1999;130:487–95.

[59] Chan MT, Leung DY, Szefler SJ, Spahn JD. Difficult-to-control asthma: clinical characteristics of steroid-insensitive asthma. J Allergy Clin Immunol 1998;101:594–601.

[60] Clemmer GL, Wu AC, Rosner B, et al. Measuring the corticosteroid responsiveness endophenotype in asthmatic patients. J Allergy Clin Immunol 2015.

[61] Drazen JM, Yandava CN, Dube L, et al. Pharmacogenetic association between ALOX5 promoter genotype and the response to anti-asthma treatment. Nat Genet 1999;22:168–70.

[62] Sampson AP, Siddiqui S, Buchanan D, et al. Variant LTC(4) synthase allele modifies cysteinyl leukotriene synthesis in eosinophils and predicts clinical response to zafirlukast. Thorax 2000;55(Suppl 2):S28–31.

[63] Tantisira KG, Lima J, Sylvia J, Klanderman B, Weiss ST. 5-Lipoxygenase pharmacogenetics in asthma: overlap with Cys-leukotriene receptor antagonist loci. Pharmacogenet Genomics 2009;19:244–7.

[64] Mougey EB, Chen C, Tantisira KG, et al. Pharmacogenetics of asthma controller treatment. Pharmacogenomics J 2013;13:242–50.

[65] Lima JJ, Zhang S, Grant A, et al. Influence of leukotriene pathway polymorphisms on response to montelukast in asthma. Am J Respir Crit Care Med 2006;173:379–85.

[66] Dahlin A, Litonjua A, Lima JJ, et al. Genome-wide association study identifies novel pharmacogenomic loci for therapeutic response to montelukast in asthma. PLoS ONE 2015;10:e0129385.

[67] Mougey EB, Feng H, Castro M, Irvin CG, Lima JJ. Absorption of montelukast is transporter mediated: a common variant of OATP2B1 is associated with reduced plasma concentrations and poor response. Pharmacogenet Genomics 2009;19:129–38.

[68] Israel E, Drazen JM, Liggett SB, et al. Effect of polymorphism of the beta(2)-adrenergic receptor on response to regular use of albuterol in asthma. Int Arch Allergy Immunol 2001;124:183–6.

[69] Reihsaus E, Innis M, MacIntyre N, Liggett SB. Mutations in the gene encoding for the beta 2-adrenergic receptor in normal and asthmatic subjects. Am J Respir Cell Mol Biol 1993;8:334–9.

[70] Liggett SB. Pharmacogenetics of beta-1- and beta-2-adrenergic receptors. Pharmacology 2000;61:167–73.

[71] Israel E, Chinchilli VM, Ford JG, et al. Use of regularly scheduled albuterol treatment in asthma: genotype-stratified, randomised, placebo-controlled cross-over trial. Lancet 2004;364:1505–12.

[72] Bleecker ER, Postma DS, Lawrance RM, Meyers DA, Ambrose HJ, Goldman M. Effect of ADRB2 polymorphisms on response to longacting beta2-agonist therapy: a pharmacogenetic analysis of two randomised studies. Lancet 2007;370:2118–25.

[73] Bleecker ER, Meyers DA, Bailey WC, et al. ADRB2 polymorphisms and budesonide/formoterol responses in COPD. Chest 2012;142:320–8.

[74] Wechsler ME, Kunselman SJ, Chinchilli VM, et al. Effect of beta2-adrenergic receptor polymorphism on response to longacting beta2 agonist in asthma (LARGE trial): a genotype-stratified, randomised, placebo-controlled, crossover trial. Lancet 2009;374:1754–64.

[75] Ortega VE, Hawkins GA, Moore WC, et al. Effect of rare variants in ADRB2 on risk of severe exacerbations and symptom control during longacting beta agonist treatment in a multiethnic asthma population: a genetic study. Lancet Respir Med 2014;2:204–13.

[76] Litonjua AA, Lasky-Su J, Schneiter K, et al. ARG1 is a novel bronchodilator response gene: screening and replication in four asthma cohorts. Am J Respir Crit Care Med 2008;178:688–94.

[77] Duan QL, Gaume BR, Hawkins GA, et al. Regulatory haplotypes in ARG1 are associated with altered bronchodilator response. Am J Respir Crit Care Med 2011;183:449–54.

[78] Poon AH, Tantisira KG, Litonjua AA, et al. Association of corticotropin-releasing hormone receptor-2 genetic variants with acute bronchodilator response in asthma. Pharmacogenet Genomics 2008;18:373–82.

[79] Himes BE, Jiang X, Hu R, et al. Genome-wide association analysis in asthma subjects identifies SPATS2L as a novel bronchodilator response gene. PLoS Genet 2012;8:e1002824.

[80] Drake KA, Torgerson DG, Gignoux CR, et al. A genome-wide association study of bronchodilator response in Latinos implicates rare variants. J Allergy Clin Immunol 2014;133:370–8.

[81] Israel E, Lasky-Su J, Markezich A, et al. Genome-wide association study of short-acting beta2-agonists. A novel genome-wide significant locus on chromosome 2 near ASB3. Am J Respir Crit Care Med 2015;191:530–7.

[82] Park HW, Dahlin A, Tse S, et al. Genetic predictors associated with improvement of asthma symptoms in response to inhaled corticosteroids. J Allergy Clin Immunol 2014;133:664–9. e5.

[83] Tantisira KG, Lasky-Su J, Harada M, et al. Genomewide association between GLCCI1 and response to glucocorticoid therapy in asthma. N Engl J Med 2011;365:1173–83.

[84] Tantisira KG, Lake S, Silverman ES, et al. Corticosteroid pharmacogenetics: association of sequence variants in CRHR1 with improved lung function in asthmatics treated with inhaled corticosteroids. Hum Mol Genet 2004;13:1353–9.

[85] Hawkins GA, Lazarus R, Smith RS, et al. The glucocorticoid receptor heterocomplex gene STIP1 is associated with improved lung function in asthmatic subjects treated with inhaled corticosteroids. J Allergy Clin Immunol 2009;123:1376–83. e7.

[86] Park HW, Ge B, Tse S, et al. Genetic risk factors for decreased bone mineral accretion in children with asthma receiving multiple oral corticosteroid bursts. J Allergy Clin Immunol 2015;136:1240–6.

[87] Stockmann C, Fassl B, Gaedigk R, et al. Fluticasone propionate pharmacogenetics: CYP3A4*22 polymorphism and pediatric asthma control. J Pediatr 2013;162:1222–7. 7 e1–2.

[88] Wang Y, Tong C, Wang Z, et al. Pharmacodynamic genome-wide association study identifies new responsive loci for glucocorticoid intervention in asthma. Pharmacogenomics J 2015;15:422–9.

[89] Tantisira KG, Damask A, Szefler SJ, et al. Genome-wide association identifies the T gene as a novel asthma pharmacogenetic locus. Am J Respir Crit Care Med 2012;185:1286–91.

[90] Jin Y, Hu D, Peterson EL, et al. Dual-specificity phosphatase 1 as a pharmacogenetic modifier of inhaled steroid response among asthmatic patients. J Allergy Clin Immunol 2010;126:618–25. e1–2.

[91] Vonk JM, Postma DS, Maarsingh H, Bruinenberg M, Koppelman GH, Meurs H. Arginase 1 and arginase 2 variations associate with asthma, asthma severity and beta2 agonist and steroid response. Pharmacogenet Genomics 2010;20:179–86.

[92] Tantisira KG, Small KM, Litonjua AA, Weiss ST, Liggett SB. Molecular properties and pharmacogenetics of a polymorphism of adenylyl cyclase type 9 in asthma: interaction between beta-agonist and corticosteroid pathways. Hum Mol Genet 2005;14:1671–7.

[93] Tantisira KG, Hwang ES, Raby BA, et al. TBX21: a functional variant predicts improvement in asthma with the use of inhaled corticosteroids. Proc Natl Acad Sci USA 2004;101: 18099–104.

[94] Wu AC, Himes BE, Lasky-Su J, et al. Inhaled corticosteroid treatment modulates ZNF432 gene variant's effect on bronchodilator response in asthmatics. J Allergy Clin Immunol 2014;133:723–8. e3.

[95] Du R, Litonjua AA, Tantisira KG, et al. Genome-wide association study reveals class I MHC-restricted T cell-associated molecule gene (CRTAM) variants interact with vitamin D levels to affect asthma exacerbations. J Allergy Clin Immunol 2012;129:368–73. 73 e1–5.

[96] Prosperi MC, Marinho S, Simpson A, Custovic A, Buchan IE. Predicting phenotypes of asthma and eczema with machine learning. BMC Med Genomics 2014;7(Suppl 1):S7.

[97] Spycher BD, Henderson J, Granell R, et al. Genome-wide prediction of childhood asthma and related phenotypes in a longitudinal birth cohort. J Allergy Clin Immunol 2012;130:503–9. e7.

[98] Abraham G, Tye-Din JA, Bhalala OG, Kowalczyk A, Zobel J, Inouye M. Accurate and robust genomic prediction of celiac disease using statistical learning. PLoS Genet 2014;10:e1004137.

[99] Sonis S, Antin J, Tedaldi M, Alterovitz G. SNP-based Bayesian networks can predict oral mucositis risk in autologous stem cell transplant recipients. Oral Dis 2013;19:721–7.

[100] Bunyavanich S, Schadt EE. Systems biology of asthma and allergic diseases: a multiscale approach. J Allergy Clin Immunol 2015;135:31–42.

Chapter 14

Diabetes

Miriam S. Udler and Jose C. Florez

Massachusetts General Hospital, Boston, MA, United States

Chapter Outline

INTRODUCTION

Diabetes mellitus is defined by the occurrence of hyperglycemia in the basal and/or postprandial states. Hyperglycemia results when the pancreatic β cell is unable to secrete enough insulin to maintain normal glucose levels given the metabolic context of the organism. Therefore, diabetes can be caused by a variety of mechanisms, but the final common pathway results from relative insulin insufficiency. Clinical complications shared by all forms of diabetes are thought to result from long-standing hyperglycemia but are compounded to a greater or lesser extent by comorbid conditions.

The best characterized forms of diabetes include autoimmune diabetes (type 1 diabetes and latent autoimmune diabetes in the adult), which result from autoimmune destruction of pancreatic β cells and rarer monogenic forms, such as

Genomic and Precision Medicine. DOI: http://dx.doi.org/10.1016/B978-0-12-800685-6.00014-X
245

maturity onset diabetes of the young (MODY) or neonatal diabetes, which arise from specific single-gene mutations. However, the most common form (~90% of all diabetes cases worldwide) is type 2 diabetes (T2D), caused by a protean conglomerate of pathophysiological processes which lead to insulin resistance and β-cell failure in varying degrees, and is often accompanied and/or influenced by obesity, hypertension, and dyslipidemia. Therefore, for the purposes of this chapter and its placement within the thematic flow of this work, we will focus solely on T2D.

Readers interested in the fascinating and largely parallel exploration of genetic determinants in type 1 diabetes are referred to excellent reviews on this subject [1–6]. Similarly, monogenic or syndromic forms of diabetes are reviewed elsewhere [7–11].

EPIDEMIOLOGY AND GENETICS

Epidemiology of Type 2 Diabetes and its Complications

T2D is a leading cause of poor health worldwide [12]. In the United States, diabetes prevalence continues to increase [13] and is the leading cause among adults of new blindness, end-stage renal disease, and nontraumatic lower limb amputation [14,15]. As a sedentary lifestyle and excess caloric intake spreads worldwide, diabetes is reaching pandemic proportions in developing nations as well [16–18]. The human toll, opportunity costs, and financial burdens imposed by growing numbers of people with diabetes threaten to severely damage the US and world economy [19–21].

Heritability of Type 2 Diabetes

The quick rise of T2D prevalence over the past century (during which the human genome has remained largely unchanged) suggests that development of the disease is powerfully influenced by the environment and related behaviors, including access to food, dietary intake, and physical activity. However, several lines of evidence suggest that individuals vary in their susceptibility to these exposures, at least in part, as a function of genetics [22–24]. First, the risk of diabetes varies depending on the populations studied, even when these share a similar environment. For example, Mexican- or Native Americans have a higher risk of diabetes than their white compatriots [25]. Second, family history is a key independent diabetes risk factor: in contrast to the population risk of 5–10%, a first-degree relative of someone with T2D has a 40% lifetime risk of developing diabetes, and this risk nearly doubles if he or she has two first-degree diabetic relatives [26,27]. Third, the effect of genes can be distinguished from those of a shared environment early in life through the use of twin studies. Rates of concordance are much higher for monozygotic twins (who share identical copies of the genome sequence) when compared to dizygotic twins, whose rates of concordance are similar to those of other first-degree relatives (parent–offspring or sibling pairs) [28,29]. Finally, as mentioned above, mutations in specific genes transmitted in a Mendelian fashion also have been found to confer a strong risk for diabetes [8,10]. All of the above provides compelling evidence that many forms of diabetes have a strong genetic component.

THE SEARCH FOR GENETIC DETERMINANTS OF TYPE 2 DIABETES

Early Studies

Linkage and Association

Throughout the last 30 years, linkage mapping and positional cloning have allowed for scanning of the human genome prior to the availability of its full sequence and enabled the discovery of hundreds of genetic mutations responsible for relatively rare human diseases. As described earlier in this volume, linkage analysis is predicated on the principle of identity by descent, by which genomic segments containing the causal mutation are inherited from the same ancestor by affected members of a pedigree, while unaffected relatives inherit the same segment from someone else who lacks the mutation. It is particularly powerful in the setting of rare variants (i.e., more likely to segregate via a single uniparental line) that have a strong effect on phenotype, where there is close to a 1:1 relationship between mutation and disease.

In complex diseases, however, the phenotype arises through the interplay of many genetic variants that interact with the environment. Each of those variants has a modest effect on risk, and many of them may be quite common (because their modest effect size, the late age of disease onset, and/or the kind of trait acquired in a different environmental context have not subjected them to strong purifying selection). Linkage analysis met with quick success when extended to type 1 diabetes, where as much as 50% of the genetic contribution to the phenotype can be explained by the human leukocyte antigen (HLA) region [30]. This early finding, now viewed as exceptional, falsely reassured investigators that linkage could be deployed to other complex diseases with relative ease. The large number of inconclusive studies that followed, coupled with statistical simulations, showed that sample sizes were inadequate to generate linkage signals of sufficient magnitude to merit their subsequent exploration [31,32].

Association analysis entails a fundamentally different approach. It simply asks whether a genetic variant is overrepresented in disease versus health. It asks about presence rather than inheritance, so can therefore be carried out in family or unrelated datasets; while it also requires large sample sizes to detect modest effects, it is quite robust to genotyping error. The otherwise unexplained overrepresentation of a genetic variant in cases when compared to controls indicates that the DNA segment tagged by such a variant must be causal in disease, since (as opposed to other epidemiological associations) it precedes disease and is unaffected by it.

The First T2D Genes

Unlike linkage, where anonymous genetic markers anchoring the entire genome facilitated genome-wide analyses, association studies initially required previous knowledge of the genetic variant to be queried. Thus, prior to the completion of the human genome sequence only candidate gene association studies were possible. Investigators chose a particular gene based on its implication in T2D pathophysiology, sequenced it to discover common polymorphisms, and genotyped these in appropriate samples to establish whether a specific allele deviated from the expected frequency distribution under the null hypothesis of no association. From the many candidate gene studies performed in T2D, two associations have stood the test of time: the P12A polymorphism in the peroxisome proliferator-activated receptor γ2 (*PPARG*) [33], and the highly correlated E23K and A1369S variants in the two subunits that comprise the islet ATP-dependent potassium channel, encoded by the adjacent genes *KCNJ11* and *ABCC8*, respectively [34,35]. Interestingly, both of these loci involve missense changes that alter protein function, and both encode targets of antidiabetic medications (thiazolidinediones activate PPARγ, and sulfonylureas bind to the sulfonylurea receptor/potassium channel complex).

Immediately before the advent of genome-wide association studies (GWAS), a combination of approaches led to the identification of the common genetic variant that up until very recently had the strongest effect on T2D risk. A linkage study reported a number of suggestive peaks [36], and saturation fine-mapping of the various peaks uncovered a strong association in chromosome 10, which was replicated in two independent populations and did not explain the original linkage signal [37]. The association mapped to the transcription factor 7-like 2 (*TCF7L2*) gene and seemed to stem from an intronic single nucleotide polymorphism (SNP) (rs7903146). It has since been replicated in most ethnic groups [38,39]; the risk allele is present at a ~30% frequency in populations of European or African origin and confers a ~40% excess risk per allele carried.

Genome-Wide Association Studies

The First Generation

GWAS became possible through several major advances. The complete sequencing of the human genome [40,41], the ensuing cataloging of common human genetic variation [42,43] and the empiric realization that the existing correlation among genetic variants, leads to reduced haplotypic diversity [44–46]—all gave rise to the International HapMap project [47], which by 2005 had produced a comprehensive view of the statistical relationships between common SNPs in four major world populations [48]. Technology development kept apace, with several companies using this information to manufacture arrays that allowed for the generation of accurate genotype calls at hundreds of thousands of genomic sites. And the rising consciousness among scientists that genetic variants of modest effects could only be detected through very large sample sizes (particularly given the high number of tests performed) facilitated international collaborations and the sharing of data, analytical tools and methodologies. By the midpoint of the first decade in the millennium, the field was ripe for genetic association testing to be deployed genome-wide across many human traits.

T2D was one of the first diseases to undergo high-density genome-wide association scanning. In a much anticipated landmark paper, in early 2007, a French and French–Canadian collaboration reported that genetic associations had been detected at *TCF7L2*, the *SLC30A8* gene encoding the zinc transporter ZnT8, a region around genes encoding a pancreatic transcription factor (*HHEX*) and the insulin degrading enzyme (*IDE*), and two other loci that later proved to be irreproducible [49]. The convincing replication of the *TCF7L2* signal at the top of the list validated both the specific genetic association as well as the genome-wide approach.

Shortly thereafter, several other GWAS came to light. A collaboration between the Diabetes Genetics Initiative (DGI) [50], the Wellcome-Trust Case Control Consortium (WTCCC) [51], and the FUSION group [52] jointly confirmed the *TCF7L2*, *SLC30A8*, and *HHEX* associations and identified *CDKAL1*, *CDKN2A/B*, and *IGF2BP2* as additional T2D susceptibility loci. Independently, deCODE investigators also reported the *CDKAL1* association [53]. In the span of a few months, the number of *bona fide* T2D loci had grown from three to eight, an approach had been validated, and a fast-track path had been opened for investigators to begin populating the lists of genetic variants that increase T2D risk.

Meta-analyses

While well-powered candidate gene studies continued to yield associated loci, such as common SNPs in the Wolfram syndrome gene *WFS1* [54,55] or the MODY gene *HNF1B* [56,57], the emphasis had shifted to the integration of genome-wide datasets via meta-analysis. This strategy makes sense, in that many true associations remain hidden below the stringent statistical threshold used to declare genome-wide significance, empirically established at $p < 5 \times 10^{-8}$ for common variants [58]. Larger sample sizes should raise statistical confidence that these false negative results represent real signals. In this manner, the DGI, WTCCC, and FUSION cohorts were combined into the Diabetes Genetics Replication and Meta-analysis (DIAGRAM) consortium, leading to identification of six new T2D loci (*JAZF1*, *CDC123-CAMK1D*, *TSPAN8-LGR5*, *THADA*, *ADAMTS9*, and *NOTCH2-ADAM30*) [59]. In parallel, follow-up of the French/French–Canadian GWAS identified *IRS1* as a biologically plausible T2D gene, through both genetic and functional means [60]. A more comprehensive meta-analysis of GWAS datasets informative for T2D, termed DIAGRAMv2, subsequently reported 12 additional loci (*BCL11A*, *ZBED3*, *KLF14*, *TP53INP1*, *TLE4*, *KCNQ1*, *ARAP1*, *HMGA2*, *HNF1A*, *ZFAND6*, *PRC1*, and *DUSP9*) [61]. Another GWAS that year identified *RBMS1* [62].

Based on initial T2D GWAS findings and those of other metabolic and cardiovascular traits, two different custom arrays, one termed the "ITMAT-Broad-CARe (IBC)" array [63] and the other termed the "Metabochip" [64], were developed to facilitate meta-analyses of thousands of variants. The IBC array was used to genotype participants in 39 studies totaling 17,418 cases and 70,298 controls, including individuals of European, African–American, Hispanic, and Asian ancestry. Through this meta-analysis, *GATAD2A/CILP2/PBX4* was identified as a new locus [63]. The Metabochip was used to genotype individuals in several studies as part of a meta-analyses termed DIAGRAMv3, which included the DIAGRAMv2 GWAS plus 4 additional studies, for a total of 34,840 cases and 114,981 controls, the majority of whom were of European descent. This meta-analysis yielded 10 new diabetes loci: *ZMIZ1, ANK1, KLHDC5, TLE1, ANKRD55, CILP2, MC4R, BCAR1, CCND2,* and *GIPR* [65].

Interestingly, it was recognized that the effect sizes at *CCND2* and *GIPR* differed between men and women, with *CCND2* more strongly associated in men, and *GIPR* more so in women [65]. Four other known T2D loci were also found by Morris et al. to exhibit this sex differentiation: *KCNQ1, DGKB,* and *BCL11A* were more strongly associated with T2D in men, and *GRB14* was more so in women [65]. An Icelandic study showed that individuals had increased risk of diabetes, if they inherited a variant in *DUSP9* from their father, but not if the same variant was inherited from their mother, most likely due to imprinting [66].

By factoring body mass index (BMI) into GWAS meta-analyses, an additional T2D locus, *LAMA1*, was identified by Perry et al. [67]. The authors hypothesized that genetic effects might differ in people with T2D who were "lean" versus those who were "obese," and thus stratified their analysis by BMI. The *LAMA1* variant (rs8090011) had a stronger association in lean [$p = 8.5 \times 10^{-9}$, odds ratio (OR) = 1.13 (95% confidence interval (CI) 1.09–1.18)] than in obese cases [$p = 0.04$, OR = 1.03 (95% CI 1.00–1.06)].

GWAS in Non-European Populations

While the initial GWAS were conducted predominantly in populations of European descent, study of other ethnic groups has also facilitated detection of novel signals. For example, two Japanese groups independently found that variants in *KCNQ1*, which are much more frequent in Asians than in Europeans, also increase T2D risk [68,69]. Additional GWAS and meta-GWAS in East Asian populations have identified 20 further loci: *PTPRD, SRR* [70]; *SPRY2* [71]; *UBE2E2, C2CD4A/4B* [72]; *GRK5, RASGRP1* [73]; *GLIS3, PEPD, FITM2/R3HDML/HNF4A, KCNK16, MAEA, GCC1/PAX4, PSMD6, ZFAND3* [74]; CCDC63, C12orf51 [75]; and *MIR129-LEP, GPSM1,* and *SLC16A11/A13* [76].

The *SLC16A11/A13* locus identified by Hara et al. [76] in a Japanese population was also independently identified by researchers studying Mexican and Latin American individuals through the Slim Initiative in Genomic Medicine for the Americas (SIGMA) T2D consortium [77]. A haplotype containing four amino acid substitutions in *SLC16A11* achieved genome-wide significance and is present in 50% of individuals with Mexican ancestry, 10% of East Asians (consistent with the Japanese GWAS findings), but rare in European and African samples. Population genetics analysis of this haplotype suggested that it arose in modern humans via admixture with Neanderthals [77].

GWAS in South Asian populations have identified 8 new loci: *GRB14*, *ST6GAL1*, *VPS26A*, *HMG20A*, *AP3S2*, *HNF4A* [78]; *TMEM163* [79]; and *SGCG* [80]. A GWAS in American Indians identified a variant in the gene *DNER* as a novel T2D locus, just shy of reaching genome-wide significance ($p = 6.6 \times 10^{-8}$) [81]. African–American GWAS have identified *RND3-RBM43* [82], *HLA-B*, and *INS-IGF2* [83]. The variants at the *HLA-B* (rs2244020) and *INS-IGF2* (rs3842770) loci are not included in Table 14.1 due to their location being within 500 kilobases of the variants in *POU5F1/TCF19* and *DUSP8*, respectively.

A transethnic aggregation of published GWAS of European, East Asian, South Asian, Mexican, and Mexican American ancestry identified seven additional T2D loci (*TMEM154*, *SSRI-RREBI*, *FAF1*, *POU511-TCF19*, *LPP*, *ARL15*, and *MPHOSPH9*) and again highlighted that many loci replicate across the different ethnic groups [84].

T2D loci reported to date with p values $<5 \times 10^{-8}$ are listed in Table 14.1, and include variants identified initially as associated with T2D-related quantitative traits, as described below. For each locus, one representative common variant is reported. The year of discovery with approximate effect size are displayed in Fig. 14.1.

GWAS for T2D-Related Quantitative Traits

A complementary strategy in the quest for T2D genes involves the study of T2D-related quantitative traits. As T2D is defined by hyperglycemia and often results from insulin resistance, a search for genetic determinants of serum glucose or insulin levels may identify other loci implicated in T2D pathogenesis. This approach has the advantage of greater numbers afforded by population cohorts when compared to case/control collections, because nearly all individuals will be informative. Participants with T2D must be excluded, however, because diabetes treatment and the disease process itself may affect the very parameters one intends to measure. Restricting the study of these traits to the subdiabetic range allows for the additional distinction between genetic variants that cause progressive pathology from those that simply regulate homeostatic glycemia.

Early candidate gene studies had suggested that the gene encoding glucokinase (*GCK*), whose role as the rate-limiting enzyme regulating glucose-induced insulin secretion makes it function as the glucose sensor in the pancreatic β cell, not only harbors loss-of-function mutations that cause MODY but also polymorphisms of subtler effects which raise fasting glucose and over time

TABLE 14.1 Genetic Loci Associated with Type 2 Diabetes at Genome-Wide Levels of Statistical Significance ($p < 5 \times 10^{-8}$)

Marker	Chr	Nearest gene(s)	Discovery cohort(s)	Marker	Chr	Nearest gene(s)	Discovery cohort(s)
rs17706184	1	FAF1	[84]	rs17584499	9	PTPRD	[70]
rs2296172	1	MACF1	[85]	rs2796441	9	TLE1	[65]
rs10923931	1	NOTCH2	[59]	rs13292136	9	TLE4 (formerly CHCHD9)	[61]
rs340874	1	PROX1[a]	[86]	rs12779790	10	CDC123/CAMK1D	[59]
rs243021	2	BCL11A	[61]	rs10886471	10	GRK5	[73]
rs3923113	2	COBLL1/GRB14	[78]	rs1111875	10	HHEX	[49]
rs780094	2	GCKR[a]	[86]	rs7903146	10	TCF7L2	[37]
rs2943641	2	IRS1	[60]	rs1802295	10	VPS26A	[78]
rs7560163	2	RBM43/RND3	[82]	rs12571751	10	ZMIZ1	[65]
rs7593730	2	RBMS1	[62]	rs1552224	11	ARAP1 (formerly CENTD2)	[61]
rs7578597	2	THADA	[59]	rs2334499	11	DUSP8	[66]
rs6723108	2	THEM163	[79]	rs5219/rs757110	11	KCNJ11/ABCC8	[34]
rs4607103	3	ADAMTS9/ PSMD6	[59]	rs2237892	11	KCNQ1	[68,69]
rs11708067	3	ADCY5[a]	[86]	rs10830963	11	MTNR1B[a]	[87]
rs4402960	3	IGF2BP2	[50–52]	rs2074356	12	C12orf51	[75]
rs6808574	3	LPP	[84]	rs11065756	12	CCDC63	[75]

rs831571	3	PSMD6	[74]	rs11063069	12	CCND2	[65]
rs1801282	3	PPARG	[33]	rs1531343	12	HMGA2	[61]
rs16861329	3	ST6GAL1	[78]	rs7957197	12	HNF1A	[61]
rs7612463	3	UBE2E2	[72]	rs10842994	12	KLHDC5	[65]
rs6815464	4	MAEA	[74]	rs1727313	12	MPHOSPH9	[84]
rs6813195	4	TMEM154	[84]	rs7961581	12	TSPAN8/LGR5	[59]
rs10010131	4	WFS1	[54]	rs9552911	13	SGCG	[80]
rs459193	5	ANKRD55	[65]	rs1359790	13	SPRY2	[71]
rs702634	5	ARL15	[84]	rs61736969	13	TBC1D4ᵃ	[88]
rs35658696	5	PAM/PPIP5K2	[89]	rs2007084	15	AP3S2	[78]
rs4457053	5	ZBED3	[61]	rs4502156	15	VPS13C/C2CD4A/B	[72]
rs7754840	6	CDKAL1	[50–53]	rs7178572	15	HMG20A	[78]
rs1535500	6	KCNK16	[74]	rs8042680	15	PRC1	[61]
rs3132524	6	POU5F1/TCF19	[84]	rs7403531	15	RASGRP1	[73]
rs9502570	6	SSR1/RREB1	[84]	rs11634397	15	ZFAND6	[61]
rs9470794	6	ZFAND3	[74]	rs7202877	16	BCAR1	[65]
rs2191349	7	DGKB/TMEM195ᵃ	[86]	rs9939609	16	FTO	[90]
rs6467136	7	GCC1/PAX4	[74]	rs4430796	17	HNF1B	[56,57]
rs4607517	7	GCKᵃ	[86]	rs312457	17	SLC16A11/13	[76,77]

(Continued)

TABLE 14.1 Genetic Loci Associated with Type 2 Diabetes at Genome-Wide Levels of Statistical Significance ($p < 5 \times 10^{-8}$) (Continued)

Marker	Chr	Nearest gene(s)	Discovery cohort(s)	Marker	Chr	Nearest gene(s)	Discovery cohort(s)
rs864745	7	JAZF1	[59]	rs391300	17	SRR	[70]
rs972283	7	KLF14	[61]	rs12454712	18	BCL2	[63]
rs791595	7	LEP	[76]	rs8090011	18	LAMA1	[67]
rs516946	8	ANK1	[65]	rs12970134	18	MC4R	[65]
rs13266634	8	SLC30A8	[49]	rs10401969	19	CILP2	[65]
rs896854	8	TP53INP1	[61]	rs8108269	19	GIPR	[65]
rs1081161	9	CDKN2A/B	[50–52]	rs3786897	19	PEPD	[74]
rs7041847	9	GLIS3	[74]	rs4812829	20	HNF4A	[74,78]
rs11787792	9	GPSM1	[76]	rs5945326	X	DUSP9	[61]

Source: Modified from Mohlke KL, Boehnke M. Recent advances in understanding the genetic architecture of type 2 diabetes. Hum Mol Genet 2015;24(R1):R85–R92.
Loci are arranged alphabetically by chromosome number. One representative variant and one or two genes are provided for each locus. Loci are defined as association signals located within 500 kb of each other regardless of linkage disequilibrium.
[a]Discovery of T2D association followed detection in GWAS for quantitative glycemic traits (see Table 14.2).

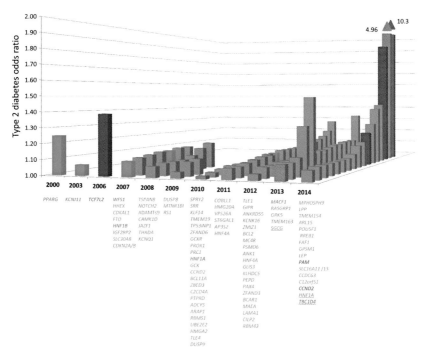

FIGURE 14.1   Chronological listing of type 2 diabetes-associated genes, plotted by year of definitive publication and approximate effect size. Candidate genes are shown in green, genes discovered via agnostic genome-wide association approaches are shown in blue, genes identified by exome sequencing are in orange, and by whole-genome sequencing in red. *TCF7L2* (shown in green) was discovered by dense fine-mapping under a linkage signal. *TBC1D4* (show in pink) was identified by exome sequencing of a locus found to be associated with a diabetes-related quantitative trait. Approximate allelic effect sizes with only risk effects shown were derived from the DIAGRAM meta-GWAS (65) and Asian meta-GWAS (74) when possible. Gene names that are underlined denote identification in population isolates.

can contribute to T2D [92]. In the same pathway, the gene encoding glucose-6-phosphate catalytic subunit (*G6PC2*) was also found to regulate fasting glucose [93,94].

In a complementary approach to the GWAS, several cohorts with glycemic trait data began collaborating to form meta-GWAS, including the Meta-Analysis of Glucose and Insulin-related traits Consortium (MAGIC). An initial finding corroborated by two independent studies was that SNPs in the gene that encodes the melatonin receptor 2 (*MTNR1B*) influenced fasting glucose and increased T2D risk [87,95]. A more formal meta-analysis from the MAGIC investigators led to the discovery and/or confirmation of multiple loci associated with fasting glucose (*ADCY5, MADD, ADRA2A, CRY2, FADS1, GLIS3, SLC2A2, PROX1*, and *C2CD4B*), fasting insulin (*GCKR* and *IGF1*) [86], and 2-h glucose (*ADCY5, GIPR*, and *VPS13C* besides *TCF7L2* and *GCKR*) [96].

Of these, *GCK, MTNR1B, DGKB, GCKR, PROX1*, and *ADCY5* were also associated with T2D [86]. A GWAS for glycated hemoglobin (HbA1c) showed that the known glycemic variants at *GCK, G6PC2*, and *MTNR1B* were also associated with HbA1c, whereas seven others modified HbA1c largely via nonglycemic mechanisms, such as erythrocyte turnover [97]. Subsequently, variants in eight loci (*LARP6, SGSM2, MADD, PCSK1, TCF7L2, SLC30A8, VPS13C/C2CD4A/B*, and *ARAP1*) were found to be associated with proinsulin levels, which is a precursor of mature insulin and C-peptide [98]; variants in five of these loci were previously associated with T2D risk (*TCF7L2, SLC30A8, VPS13C/C2CD4A/B*, and *ARAP1*).

By aggregating MAGIC meta-GWAS data with Metabochip data in up to 133,010 individuals of European ancestry, Scott et al. identified 30 new loci for glycemic traits, raising the total number to 53 [99]. Of these 53 loci, 33 were also associated with T2D ($q < 0.05$). Also using the MAGIC meta-GWAS data, Manning et al. applied a joint meta-analysis approach to simultaneously test both a main genetic effect (adjusted for BMI) and a potential interaction between each genetic variant and BMI, which led to independent identification of several loci from Scott et al., as well as *DPYSL5* associated with fasting glucose [100]. The MAGIC meta-GWAS and Metabochip were further employed by Prokopenko et al. to study measures of insulin response in subjects who underwent an oral glucose tolerance test [101]. Variants in eight loci were significantly associated with insulin response, including rs933360 in *GRB10*, which interestingly had a parent-of-origin affect, whereby, individuals had elevated fasting glucose when the A allele was paternally inherited but had lower fasting glucose and enhanced insulin sensitivity when maternally inherited. The authors went on to show that *GRB10* is imprinted in a parent-of-origin fashion in pancreatic islet cells, and *GRB10* knock-down in these cells results in reduced insulin and glucagon secretion [101].

In another recent meta-GWAS, the European Network for Genetic and Genomic Epidemiology consortium aggregated 22 GWAS of European populations. Through this approach, two additional loci (*RMST* and *EMID2*) were found to be associated with fasting glucose; the variant in *EMID2* is uncommon [minor allele frequency (MAF) 1.7%], and its effect is only seen in female carriers [102]. As the majority of the T2D-associated loci to date seems to influence β-cell function, there has been interest to identify loci related to insulin sensitivity. The GENEtics of Insulin Sensitivity (GENESIS) consortium conducted a GWAS for direct measures of insulin sensitivity and recently identified the *NAT2* locus as associated with insulin sensitivity [103].

A particularly interesting finding came through the work of Moltke et al., who analyzed the Metabochip for association with glycemic traits in the Greenlandic population and identified a variant in *TBC1D4* associated with glucose and insulin values measured 2 h after oral glucose challenge [88]. The authors went on to sequence the exomes of *TBC1D4* and found a nonsense p.Arg684Ter variant with allele frequency of 17% in the Greenlandic population, associated with 2-h glucose and insulin levels. Stop-codon homozygotes had higher concentrations of fasting glucose and insulin, and most notably, a 10-fold increased risk of T2D compared to wild-type allele carriers (OR 10.3, $p = 1.6 \times 10^{-24}$). This variant derives from Inuit ancestry. Skeletal muscle biopsies showed that with increasing numbers of p.Arg684Ter alleles, protein levels of TBC1D4 and glucose transporter GLUT4 decreased. Taken together, these findings suggest that disruption of TBC1D4 in skeletal muscle appears to cause decreased insulin-stimulated glucose uptake, leading to postprandial hyperglycemia, impaired glucose tolerance, and T2D [88].

Studies of additional populations have also helped identify new glycemic trait loci. In African populations, *SC4MOL* and *TCERG1L* were associated with fasting insulin and homeostasis model assessment-insulin resistance (HOMA-IR) [104]. In East-Asian populations, variants in *SIX2-SIX3* [105], *PDK1*, and *IGFR1* [106] were found to be associated with fasting glucose levels. *MYL2*, *C12orf51*, and *OAS1* were found to be significantly associated with 1 h plasma glucose following a 50-g oral glucose challenge [107]. In the Mexican Americans, SNPs in T2D locus *MTNR1B* were found to be significantly associated with acute insulin response as well as disposition index, which is the product of measures of insulin sensitivity and first-phase insulin secretion [108].

Loci discovered by association with quantitative glycemic traits and reaching significance level of 5×10^{-8} are listed in Table 14.2. As in Table 14.1, a representative variant for the locus is listed.

Exome Array and Next-Generation Sequencing

Available Technology

The identification of many loci associated with T2D and related traits was facilitated by the 1000 Genomes Project (http://www.1000genomes.org) [114], which is a publicly available resource of sequence data for over 1000 genomes of individuals of different ancestries intended to provide a comprehensive resource on human genetic variants above 1% frequency. Genetic studies have leveraged the sequenced-based reference panels from the 1000 Genomes Project to predict or "impute" variants beyond those included in their GWAS SNP panels.

As costs of sequencing technology have dropped, large-scale whole-genome sequencing of study subjects has also become feasible, enabling the detection of less common genetic variants (MAF < 5%) and rare variants (MAF < 1%) (Fig. 14.3). However, identification of robust genetic association remains a challenge, as the potential modest gain in effect size is counteracted by the lower frequency and resulting lower sample sizes. Whole-exome sequencing, which involves selectively sequencing the 1–2% of the genome that encodes proteins, has been employed as a focused, cost-effective strategy, particularly in Mendelian diseases [115]. There are two advantages of targeting exonic variants: first, in

TABLE 14.2 Genetic Variants Associated with Quantitative Glycemic Traits at Genome-Wide Levels of Statistical Significance ($p < 5 \times 10^{-8}$)

Marker	Chr	Nearest gene(s)	Trait	Discovery cohort(s)	Marker	Chr	Nearest gene(s)	Trait	Discovery cohort(s)
rs2820436	1	LYPLAL1	I	[99,100]	rs306549	9	DDX31	FP	[98]
rs340874	1	PROX1a	B	[86]	rs3829109	9	DNLZ/GPSM1a	B	[99]
rs9727115	1	SNX7	FP	[98]	rs7034200	9	GLIS3a	B	[86]
rs2779116	1	SPTA1	H	[97]					
rs6684514	1	TMEM79	H	[109]	rs16913693	9	IKBKAP	B	[99]
rs10195252	2	COBLL1/GRB14a	I	[99,100]	rs3824420	9	KANK1	FP	[106,110]
rs1371614	2	DPYSL5	B	[100]	rs10885122	10	ADRA2A	B	[86]
rs560887	2	G6PC2	B	[86,97]	rs7923866	10	HHEXa	B	[101]
rs1402837			H						
rs780094	2	GCKRa	B	[86,96]	rs7072268	10	HKI	H	[97]
rs1260326			I						
rs2972143	2	IRS1a	I	[99,100]	rs10829854	10	TCERG1L	I	[104]

SNP	Chr	Gene	Association	Reference
rs733331	2	PDK1/RAPGEF4	B	[106]
rs895636	2	SIX2/SIX3	B	[105]
rs1530559	2	YSK4	I	[99]
rs11708067	3	ADCY5[a]	B	[86,96]
rs2877716			I	
rs11715915	3	AMT	B	[99]
rs7651090	3	IGF2BP2[a]	B, I	[99]
rs17036328	3	PPARG[a]	I	[99]
rs11920090	3	SLC2A2	B	[86]
rs3822072	4	FAM13A	I	[99]
rs4691380	4	PDGFC	I	[99,100]
rs17046216	4	SC4MOL	I	[109]
rs9884482	4	TET2	I	[99]
rs459193	5	ANKRD55[a]	I	[99]
rs4865796	5	ARL15[a]	I	[99]
rs1019503	5	ERAP2	I	[99]
rs35658696	5	PAM/PPIP5K2[a]	I	[110]
rs7903146	10	TCF7L2[a]	B, FP, I	[86,96,98,99]
rs11603334	11	ARAP1[a]	B, FP	[98–100]
rs11605924	11	CRY2	B	[86]
rs174550	11	FADS1	B	[86]
rs174570	11	FADS2	H	[109]
rs7944584	11	MADD	B, FP	[86,98]
rs10830963	11	MTNR1B[a]	B, H	[87,95,97,101,108]
rs1483121	11	OR4S1	B	[100]
rs2074356	12	C12orf51[a]	B	[107]
rs2657879	12	GLS2	B	[99]
rs2650000	12	HNF1A[a]	B	[110]
rs35767	12	IGF1	I	[86]
rs122229654	12	MYL2	I	[107]
rs11066453	12	OAS1	I	[107]
rs10747083	12	P2RX2	B	[99]
rs17331697	12	RMST	B	[102]

(Continued)

TABLE 14.2 Genetic Variants Associated with Quantitative Glycemic Traits at Genome-Wide Levels of Statistical Significance ($p < 5 \times 10^{-8}$) (Continued)

Marker	Chr	Nearest gene(s)	Trait	Discovery cohort(s)
rs4869272	5	PCSK1	B	[99,100]
rs6235			FP	
rs7708285	5	ZBED3[a]	B	[99]
rs9368222	6	CDKAL1[a]	B, H, I	[99,101,111]
rs7747752				
rs10305492	6	GLP1R	G	[112]
rs9399137	6	HBS1L/MYB	H	[109]
rs1800562	6	HFE	H	[97]
rs2745353	6	RSPO3	I	[99]
rs17762454	6	SSR1/RREB1[a]	B	[99]
rs6912327	6	UHRF1BP1	I	[99,100]
rs2191349	7	DGKB/TMEM195[a]	B	[86]

Marker	Chr	Nearest gene(s)	Trait	Discovery cohort(s)
rs150781447	12	TBCD130	FP	[110]
rs7998202	13	ATP11A	H	[97]
rs576674	13	KL	B	[99]
rs11619319	13	PDX1	B	[89,99,100]
rs61736969	13	TBC1D4[a]	B, I	[88]
rs3783347	14	WARS	B	[99]
rs11071657	15	VPS13C/C2CD4A/B[a]	B, FP, I	[86,96,98,101]
rs2018860	15	IGF1R	B	[106]
rs1549318	15	LARP6	FP	[98]
rs9933309	16	CYBA	H	[109]

SNP	Chr	Gene	Trait	Reference
rs6947345	7	EMID2	B	[102]
rs4607517, rs730497	7	GCK^a	B, H, I	[86,97,99,101]
rs6943153	7	GRB10	B	[99,101]
rs1167800	7	HIP1	I	[99]
rs6474359	8	ANK1^a	B, H	[97,101]
rs983309	8	PPP1R3B	B	[99,100]
rs11782386	8		I	
rs11558471	8	SLC30A8^a	B	[86,98]
rs13266634			FP	
rs651007	9	ABO	B	[110,113]
rs10811661	9	CDKN2A/B^a	B	[99]
rs1421085	16	FTO^a	I	[99]
rs1046896	17	FN3K	H	[97]
rs4790333	17	SGSM2	FP	[98]
rs10423928	19	GIPR^a	B, I	[96,99,101]
rs11667918	19	MYO9B	H	[109]
rs731839	19	PEPD^a	I	[99]
rs6113722	20	FOXA2	B	[99,100]
rs6072275	20	TOP1	B	[99]
rs855791	22	TMPRSS6	H	[97]

Source: Modified from Mohlke KL, Boehnke M. Recent advances in understanding the genetic architecture of type 2 diabetes. Hum Mol Genet. 2015;24(R1):R85–R92. *Loci are arranged alphabetically by chromosome number. One representative variant and one or two genes are provided for each locus for each glycemic trait. Loci are defined as association signals located within 500kb of each other regardless of linkage disequilibrium.B = beta-cell (fasting glucose, HOMA-B, corrected insulin response, disposition index, insulinogenic index, or these traits adjusted for BMI).I = insulin resistance (fasting insulin, 1-h glucose, 2-h glucose, HOMA-IR, or these traits adjusted for BMI).FP = fasting proinsulin.H = hemoglobin A1c.HOMA-B and HOMA-IR = β-cell function and insulin resistance by homeostasis model assessment, respectively.*
^aLocus also associated with type 2 diabetes.

contrast to noncoding variants whose mechanism of action is often challenging to determine, these variants encode amino acids, and the related proteins can be more readily studied to facilitate understanding of gene function; and second, it is presumed that many coding changes may have stronger effects on phenotype, thus increasing statistical power. Thus, these findings might also be more readily transferrable to new drug discoveries [116].

Results of whole-exome sequencing experiments across populations can be combined into more efficient exome arrays, which are a cost-effective alternative to whole-exome sequencing, however do not capture the full diversity of coding variants, particularly in non-European populations.

New Loci Identified

Recently whole-exome arrays have been deployed to identify variants associated with both T2D risk and glycemic traits. Huyghe et al. studied ~8000 nondiabetic Finnish individuals and identified associations between fasting proinsulin concentrations and variants in *SGSM2* (MAF = 1.4%) and *MADD* (MAF = 3.7%), as well as three new genes with low-frequency variants associated with fasting proinsulin or insulinogenic index: *TBC1D30*, *KANK1*, and *PAM* [110]. By identifying a coding variant in *SGSM2* separate from the original proinsulin GWAS signal, this approach supported *SGSM2* as the causal gene out of several possible candidates at this locus. Wessel et al. analyzed exome array data in 60,564 nondiabetic individuals and 16,491 T2D cases, and 81,877 controls without T2D to identify a low-frequency (MAF 1.4%) variant in *GLP1R* (p.Ala316Thr; rs10305492) associated with multiple glycemic traits and T2D risk, as well as a common noncoding variant in ABO rs651007 (MAF 20%) associated with fasting glucose [113]. Mahajan et al. independently and simultaneously utilized exome array data in up to 33,231 nondiabetic individuals of European ancestry to identify the *GLP1R* p.Ala316Thr variant associated with fasting glucose levels, while also identifying a new variant in *URB2* (p.Glu594Val, MAF = 0.1%) as associated with fasting insulin levels [112].

Whole-exome sequencing was performed by Albrechtsen et al. in 2000 Danish participants with and without T2D [85]. The top significant variants were then genotyped in nearly 16,000 Danish individuals and then nearly 64,000 European individuals, resulting in identification of two T2D loci: p.Asn939Asp in *COBLL1* (MAF 12.5%) and p.Met2290Val in *MACF1* (MAF 23.4%) [85]. Of note, a common variant at the *COBLL1/GRB14* locus was previously identified by Kooner et al. [78]. The SIGMA consortium performed whole-exome sequencing in 3756 Mexican and US-Latino T2D cases and controls, leading to the identification of a rare missense variant (p.Glu508Lys) in *HNF1A* associated with T2D (OR = 4.96, $p = 2.4 \times 10^{-9}$) [117]. The variant was observed in 0.36% of participants without T2D and 2.1% of participants with it. Mutations in *HNF1A* cause MODY3, although interestingly, diabetic carriers and noncarriers of the *HNF1A* variant in this study showed no differences in clinical characteristics such as age at diagnosis or BMI [117].

Whole-genome sequencing was employed by Steinthorsdottir et al. in 2630 Icelanders, and then imputation was used to study 11,114 Icelandic cases and 267,140 controls as well as additional Danish and Iranian samples [89]. This effort identified four T2D susceptibility variants in *CCND2* and *PAM*, all with MAF less than 5%; the *PAM* variant, p.Asp563Gly, was the same as that previously associated with Insulinogenic index by Huyghe et al. [110]. In addition, a low-frequency (1.47%) intronic allele of *CCND2* rs76895963 reduces risk of T2D by half (OR $= 0.53, p = 5.0 \times 10^{-21}$) and enhances insulin secretion [89,118].

Rare Variants at Known Loci

Several genes associated with T2D or related glycemic traits have recently been found to harbor rarer variants associated with disease, beyond the initial identifying common GWAS variant. For example, as previously mentioned, multiple GWAS independently identified SNPs in the gene that encodes the melatonin receptor 2 (*MTNR1B*) as influencing fasting glucose, insulin secretion, and T2D risk [87,95,108]. Subsequent resequencing of *MTNR1B* and functional work demonstrated that rare coding variants causing loss of melatonin binding and signaling activity collectively caused a substantial increase in T2D risk (OR 5.67, $p = 4.1 \times 10^{-4}$) [119].

Some confusion had surrounded another GWAS locus, *G6PC2/ABCB11*, containing the lead variant rs560887 associated with fasting glucose levels [93,94]. Interestingly, the glucose-raising allele of rs560887 is also associated with elevated hemoglobin A1c but does not contribute to T2D susceptibility, and in fact is associated with improved glucose-stimulated insulin secretion [86,120]. In vitro functional work by Bouatia-Naji et al. showed that rs560887 is highly correlated with an SNP in the promoter of *G6PC2*, which is also strongly associated with fasting glucose and alters promoter activity. Recently, analysis of exome chip data identified three novel missense variants in *G6PC2* (p.His177Tyr, p.Tyr207Ser, and p.Val219Leu), each independently associated with fasting glucose levels and also resulting in reduced protein abundance via proteasomal degradation [112]. These findings, along with the lack of any fasting glucose association signals among coding variants in *ABCB11*, suggest that *G6PC2* is the causal gene at this locus, and they also confirm directionality to the fasting glucose association with lower protein abundance conferring hypoglycemia, consistent with *G6pc2* knockout mice having a ~15% decrease in fasting glucose levels [121]. In addition, the fact that glucose-lowering Leu219 allele was carried exclusively in cis with the glucose-raising allele of GWAS SNP rs560887 may also explain why the GWAS variant's directionality has been puzzling; further work is needed to fully elucidate this complex locus.

INSIGHTS GAINED FROM GENETIC STUDIES IN TYPE 2 DIABETES

From Genetic Association to Function

Despite robust signals of association at multiple loci, in most cases, the specific DNA sequences that cause the molecular phenotype have not been identified. Indeed, the SNPs identified thus far merely signal genomic regions—at times far away from known genes—where an association has been found but do not necessarily represent the causal variants: fine-mapping and functional studies must be carried out before the true contribution of these loci to T2D can be accurately assessed. The presumed mechanisms for some of the better characterized loci are shown in Fig. 14.2.

KCNJ11/ABCC8

The E23K polymorphism in *KCNJ11* (MAF ~30% in Europeans) was one of the first T2D associations proven beyond reasonable doubt [34]. Its presence in a gene clearly involved in the transduction of glucose signaling in the β cell and the nonconservative nature of the mutation—both made this variant a likely causal candidate. Initial functional studies seemed to support this notion [122]. However, this SNP is highly correlated with another missense variant in a nearby and functionally related gene, *ABCC8*; each of them encodes a separate subunit of the islet ATP-sensitive potassium channel, and because both genes are next to each other, and little recombination has occurred between the two SNPs, carriers of the risk K allele at *KCNJ11* E23K almost always carry the risk A allele at *ABCC8* S1369A in most ethnic groups [35]. Thus, the association cannot be distinguished on genetic grounds alone, and functional experiments require the design of constructs where the paired relationship is experimentally truncated. This has been accomplished recently: at least with regard to response to the sulfonylurea gliclazide, the causal mutation appears to be *ABCC8* S1369A [123]. Interestingly, risk allele carriers at this locus seem to respond better to gliclazide therapy than their wild-type counterparts [124].

GCKR

GCKR encodes the glucokinase regulatory protein (GKRP), which modulates the activity of hepatic hexokinase and thereby gates the entry of glucose into the glycolytic and glycogen synthesis pathways. It is not present in β cells. Interestingly, the index SNP was initially associated with elevated circulating triglycerides [50]; follow-up of this association showed that the triglyceride-raising allele was associated with lower fasting glucose and a lower level of hepatic insulin resistance, as indicated by the HOMA-IR [125,126]. Its association with lower HOMA-IR was confirmed at the genome-wide significant level by MAGIC; the alternate (glucose-raising but triglyceride-lowering) allele

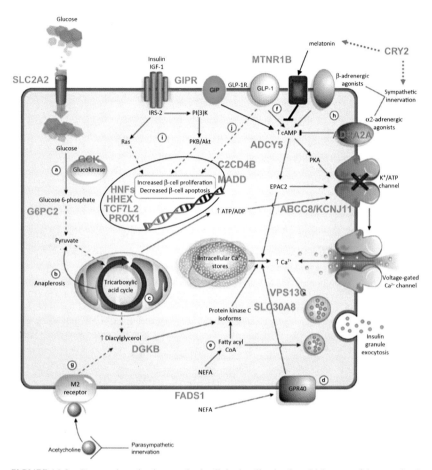

FIGURE 14.2   Proposed mechanisms and subcellular localization by which some of the examined genetic loci could influence glycemic regulation in the β cell, and thereby increase risk for type 2 diabetes. *Modified from Ingelsson E, Langenberg C, Hivert MF, Prokopenko I, Lyssenko V, Dupuis J, et al. Detailed physiologic characterization reveals diverse mechanisms for novel genetic loci regulating glucose and insulin metabolism in humans. Diabetes. 2010;59(5):1266–75.*

was associated with T2D [86]. Fine-mapping in people of African descent implicated a nearby missense polymorphism (P446L) as the likely causal variant [125]. At first sight, the association of one allele with higher triglyceride levels but lower glucose and diminished T2D risk appears counterintuitive, as hypertriglyceridemia and hyperglycemia tend to cluster in metabolic dysregulation. Elegant functional experiments have demonstrated that the presence of the 446L allele makes GKRP less responsive to binding fructose-6-phosphate, a known inducer of GKRP/GCK binding; diminished interaction between the

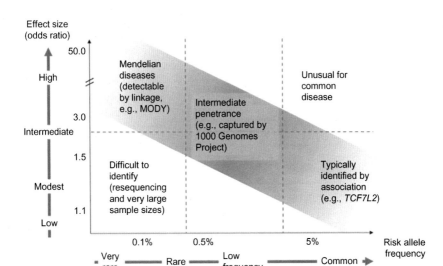

FIGURE 14.3 A portion of the heritability that is missing from present-day GWAS in type 2 diabetes may be explained by low-frequency variants with intermediate penetrance. Current GWAS arrays may not have captured alleles with low frequencies. Next-generation high-throughput sequencing technology and novel genotyping arrays may help detect these low-frequency variants; larger sample sizes may be needed to identify significant signals. *Modified from Billings LK, Florez JC. The genetics of type 2 diabetes: what have we learned from GWAS? Ann N Y Acad Sci 2010;1212:59–77.*

two proteins leads to higher GCK activity and promotes the flux of glucose into glycolysis, thereby augmenting the availability of malonyl-CoA for de novo lipogenesis and providing a unifying mechanism for the apparent epidemiological discrepancy [127].

Discovery of Novel Pathways

The unbiased search for genetic determinants of T2D and the implementation of rigorous statistical criteria before declaring a true association have enabled the identification of loci previously unsuspected to play a role in T2D pathophysiology. While any one of a number of genes under an association signal could give rise to the phenotype, in a few instances fine-mapping and functional work has implicated a specific gene (and even a variant within it) as causal. This type of follow-up validation is more advanced for loci discovered in the early phases of this effort; two such examples are offered here.

TCF7L2

As mentioned above, *TCF7L2* harbors the common genetic variant that confers the highest risk of T2D reported to date [61]. Early fine-mapping efforts, aided by the different haplotype structures in people of African origin, concluded that the intronic SNP rs7903146 was the source of the association signal [37,128]. In vivo phenotyping in humans soon documented that risk allele carriers had impaired β-cell function [129,130]. This is accompanied by higher proinsulin levels relative to fasting insulin [131–134], as well as a diminished incretin response [130,135–137]. Functional work suggests that *TCF7L2* may cause these effects by interfering with incretin receptor expression [138], β-cell proliferation during development [139,140], and/or insulin vesicle transport and fusion [141]. Elegant fine-mapping work has shown that the diabetes risk T allele at rs7903146 affects TCF7L2 expression, pinpointing the exact nucleotide that gives rise to the functional defect [142]. The precise nature of the genes targeted by this transcription factor is currently under active investigation [143]; a recent study suggests that WNT/β-catenin signaling via TCF7L2 regulates GLP-1 effects in β-cells by transcriptionally controlling the GLP-1 receptor [144].

SLC30A8

SLC30A8 encodes an islet ZnT8 zinc transporter that is responsible for zinc transport into β-cell insulin-secretory granules, which is necessary for insulin packaging and secretion [145,146]. Although this gene had been implicated in β-cell biology, it was not known that alterations in this pathway could lead to T2D until GWAS associated the missense variant R325W with T2D [49]. Subsequent animal investigation showed that mice null for this gene exhibit subtle abnormalities in insulin secretion when challenged by glucose [147,148] or a high-fat diet [149]. These phenotypes are obvious even in β-cell specific knockout mice [150], and recent studies suggested that *SLC30A8* regulates hepatic insulin clearance [151]. However, review of the protein's crystal structure indicated that the implicated amino acid substitution may not affect function substantially [152], and the directional relationship between altered ZnT8 function and direction of effect of diabetes risk remained uncertain. A defining study by Flannick et al. identified protein truncating variants in *SLC30A8* and showed that loss-of-function mutations in humans collectively conferred a 65% reduction in T2D risk ($p = 1.7 \times 10^{-6}$), providing strong evidence that *SLC30A8* haploinsufficiency protects against T2D risk [153].

Focus on β-Cell Function

The majority of loci associated with T2D seem to impair β-cell function [154]. While this observation could be partially explained by the ascertainment of lean T2D cases in two early GWAS [49,50] (thus minimizing the relative contribution of obesity-related insulin resistance), the phenomenon still held for more comprehensive recent meta-analyses [61,99]. When an unbiased genomic search for homeostasis estimates of beta-cell function (HOMA-B) or HOMA-IR is conducted in a diabetes-free population of equal size for both traits, the preponderance of the genome-wide significant signals are associated with HOMA-B rather than HOMA-IR [86]. Examination of the *p*-value distributions for each trait shows that many more signals are to be found for HOMA-B than for HOMA-IR, despite the same sample size, the same set of SNPs, and the same biochemical measurements of fasting glucose and insulin. The dearth of "insulin resistance" signals is evident even when fasting insulin is chosen as the surrogate measure, as a way to obviate the need for a mathematical formula to derive the trait. This suggests that there exists a different genetic architecture underlying both traits and is consistent with the higher heritability estimates typically ascribed to insulin secretion when compared with insulin resistance [86].

Put another way, the contribution of the environment seems to be stronger for insulin resistance than for β-cell function. This is consistent with the notion that the recent diabetes pandemic is the result of an environmentally-driven increase in insulin resistance occurring on a background of genetic predisposition to β-cell failure. To discover genetic determinants of insulin resistance, therefore, integration of environmental variables (e.g., those causing obesity) into the statistical analyses may bring out genetic effects more likely to manifest themselves in an obesogenic context. This strategy was taken by Manning et al., as described above, by searching for genetic associations with fasting insulin and HOMA-IR while introducing adjustment for BMI as well as a term for interaction with BMI. This approach led to the discovery of six novel loci associated with insulin resistance [100].

More refined measures of insulin sensitivity, such as those derived from euglycemic clamps and insulin suppression tests, were utilized by the GENESIS consortium GWAS that identified the *NAT2* locus as associated with insulin sensitivity [103]. Aggregation of additional studies would provide greater power to detect new associations (although these measures are available in a much lower number of individuals). Insulin resistance signals that have emerged from GWAS are listed in Table 14.2.

Hyperglycemia and T2D are not Equivalent

Another insight gained from these early genetic studies concerns the lack of a perfect overlap between variants that increase T2D risk and those that raise fasting glucose. While genetic causes of T2D must contribute to hyperglycemia (by definition), the converse is not necessarily true: there may be genetic variants that simply raise the homeostatic point of glycemia slightly, but do so in a stable manner and without reaching thresholds that are injurious to the β cell. On the other hand, other variants that increase glycemia do so by directly impacting β-cell mass or function, eventually leading to β-cell deterioration and ensuing T2D.

This point is illustrated by *G6PC2*, which has a substantial impact on fasting glycemia, and yet displays little-to-no association with T2D (if present, it is of a magnitude much smaller than that exerted on fasting glucose). In contrast, *TCF7L2*, which robustly increases T2D risk, has a much lower effect on fasting glucose. Similarly, *MADD* and *ADCY5* have comparable allele frequencies and the same effect on fasting glucose, while the former has no evidence of T2D association, and the latter does so at genome-wide significance levels [86]. Therefore, as summed up by the MAGIC investigators, "variation in fasting glucose in healthy individuals is not necessarily an endophenotype for T2D, which posits the hypothesis that the mechanism by which glucose is raised, rather than a mere elevation in fasting glucose levels, is a key contributor to disease progression" [86].

Support for Prior Epidemiological Observations

Epidemiologic studies have long supported the clustering of metabolically deleterious factors around the so-called "metabolic syndrome" [155,156]. The suggestion has been made that a "common soil" gave rise to conditions such as insulin resistance, obesity, hypertension, and dyslipidemia. Nevertheless, there has been little molecular evidence linking these various processes. As GWAS have been conducted around several of these traits, there was hope that some genetic markers would have pleiotropic effects across these various phenotypes. Such has been the case for *FADS1* and *HNF1A*, which influence both circulating lipid and glucose levels [61,86,157].

A link between derangements in circadian rhythmicity and glycemic dysregulation is also emerging. Observational [158,159] and interventional [160,161] studies have suggested that alterations in sleep and/or circadian oscillations can increase T2D risk. Human data have been supported by phenotyping studies in mice that lack certain circadian genes and also exhibit metabolic abnormalities, including defects in pancreatic insulin secretion [162,163]. The recent identification of two classical circadian genes, *MTNR1B* (encoding the melatonin receptor expressed in beta cells) and *CRY2* (encoding a cryptochrome isoform) as harboring variants that elevate fasting glucose, with the former also increasing T2D risk, is another genetic building block that connects these two areas of physiology.

As a third illustration, it has long been known that low birth weight predisposes to insulin resistance and T2D in adulthood [164]. The relative contribution of inheritance and the uterine environment to such a phenotype is not completely elucidated, but gestational traits are undergoing the same intense and fruitful scrutiny afforded by modern genetic techniques. The fetal insulin hypothesis maintains that a relative impairment in insulin secretion or action during early development (when insulin also functions as a growth factor) leads to diminished fetal growth, and if permanent also to the organism's inability to compensate for insulin resistance later in life. Whether the phenotype is caused by the maternal or fetal risk allele can be distinguished in studies that have access to both DNA samples and incorporate mother/fetus pairs that are discordant for the risk genotype [92,165]. In a targeted study that examined established T2D loci, the T2D-causing alleles at *CDKAL1* and *HHEX* also caused decreased birth weight [165]. It is particularly intriguing that an initial well-powered meta-analysis of GWAS for birth weight has independently uncovered the fasting glucose and T2D locus *ADCY5* (encoding adenylate cyclase type 5) as also influencing birth weight [166]. These findings have been confirmed in separate populations [167]. In summary, the coalescing of cohorts that have undergone extensive phenotyping, together with dense genotyping platforms and widespread collaboration, have begun to provide potential genetic and molecular bases for heretofore unexplained epidemiological relationships.

Genetic Prediction is not Yet Practical

The presence of genetic information at birth and its essential immutability during the lifetime of an individual make genetic markers potentially attractive variables for use in prediction models. However, our ability to distinguish modest genetic effects in very large population samples does not imply that such effects will be of use at the individual level. When distributions of a quantitative trait (e.g., fasting glucose or risk allele count) overlap substantially between two groups, a significant difference between the population means does not necessarily translate into the unequivocal placement of an individual in one of the two groups. Furthermore, clinical variables obtained in the course of routine medical care already perform quite well at predicting outcomes [168], and therefore, genetic data may not add much new information.

The efficacy of genetic prediction has been tested in multiple populations through the use of the receiver–operator characteristics curves, whose area under the curve (the c-statistic) essentially estimates the proportion of times such predictions will be correct. The most recent of such studies was conducted in two cohorts, one of only white participants with mean baseline age of 35.9 years and a second of white and black Americans selected to be younger with baseline ages 18–30 years [169]. The authors tested a genotype score composed of 62 T2D-associated variants and found that the genetic risk score added little to the clinical prediction in both populations, although it did significantly improve the C statistic in the cohort of only white Americans (C statistic 0.698 and 0.726 for models without and with GRSt, respectively; $p < 0.001$) [169]. These results were consistent with two similar earlier studies in individuals of European descent, which used ~18 T2D-associated variants for the genetic risk scores [170,171].

The limited utility of the genetic risk score beyond clinical prediction may be due to known T2D loci, and therefore, genotype scores explaining only a small proportion of T2D heritability and/or because clinical variables such as fasting glucose or BMI already contain a fair amount of the information furnished by genetic data. On the other hand, genetic markers perform better when tested in younger people or in those in whom diabetes developed much later, highlighting that clinical variables are less informative in those settings, and suggesting that genetic results might be used earlier in life as a way to deploy appropriate prevention strategies.

Pharmacogenetics is in Its Infancy

A general observation in the field of complex diseases is that most common genetic variants only confer a modest effect on risk. This stands to reason, in that strongly deleterious mutations would have been subject to purifying selection and kept at very low frequencies in the population. However, while selection pressure may have been acting for a long time (in evolutionary terms) on certain phenotypes, modern pharmacology is relatively recent. Thus, there may not have been enough time for genetic determinants of drug response to undergo purifying selection, and therefore, genetic effects on drug response may be stronger than those seen for disease pathogenesis [172–174].

In T2D, pharmacogenetic investigation remains at a very early stage. In retrospective studies conducted in the GoDARTS cohort, for which clinical outcomes are electronically available and DNA samples have been collected, carriers of the risk variant at *TCF7L2* showed diminished response to sulfonylurea therapy [175] and were more likely to require insulin therapy [176]. This finding suggests that the impairment on insulin secretion conferred by *TCF7L2* cannot be overcome with an insulin secretagogue, and stands in contrast to that seen for the β-cell sulfonylurea receptor/potassium channel, where carriers of the risk alleles at *KCNJ11/ABCC8* displayed improved response to the specific sulfonylurea drug (gliclazide) that has shown an in vitro allelic effect [123,124].

An early report that a missense polymorphism in the OCT1 metformin transporter (which is responsible for the absorption of the drug into hepatocytes) alters response to metformin [177] has not been substantiated in GoDARTS [178]. On the other hand, a polymorphism in a different metformin transporter (MATE1), which catalyzes the disposition of drug into bile and urine, seemed to affect metformin response in a different small retrospective study [179]; this has been confirmed in the Diabetes Prevention Program, where ~1000 participants at high risk of diabetes who received metformin for approximately 3 years for diabetes prevention showed little benefit from metformin treatment only if they were homozygous for the risk allele [180]. A recent systematic review of the literature by Maruthur et al. has highlighted that additional high-quality controlled studies are needed to identify robust pharmacogenetic results. Nevertheless, the following drugs and loci were noted worthy of additional follow-up: (1) metformin with *SLC22A1*, *SLC22A2*, *SLC47A1*, *PRKAB2*, *PRKAA2*, *PRKAA1*, and *STK11*; (2) sulfonylureas with *CYP2C9* and *TCF7L2*; (3) repaglinide with *KCNJ11*, *SLC30A8*, *NEUROD1/BETA2*, *UCP2*, and *PAX4*; (4) pioglitazone with *PPARG2* and *PTPRD*; (5) rosiglitazone with *KCNQ1* and *RBP4*; and (5) acarbose with *PPARA, HNF4A, LIPC*, and *PPARGC1A* [181]. It is hoped that the integration of clinical datasets where pharmacotherapeutic information is available will allow for the deployment of the same genome-wide methods that have enabled the discovery of disease genes, but this time in a search for genetic determinants of drug response. Such an undertaking would move the field from the current "trial and error" or "one size fits all" mentality in T2D therapeutics to a setting where drug choices will be predicated on individual characteristics.

CONCLUSIONS AND FUTURE DIRECTIONS

In conclusion, contemporary genomic approaches have yielded a rich trove of novel findings that illuminate new pathways of T2D pathogenesis. The laborious fine-mapping of these association signals, in a search for the actual causal variants that produce the observed phenotypic changes at the molecular level, is currently ongoing. Now deeper sequencing and identification of rarer variants is possible on a larger scale, already providing some exciting findings, particularly in population isolates. Once the precise mechanism is identified, cellular and animal models can be created for in-depth characterization of the defect involved and assessment of potential therapeutic strategies to overcome it. Ultimately, whether this newfound genetic knowledge will lead to better patient outcomes will need to be tested in appropriately designed clinical trials.

REFERENCES

[1] Bluestone JA, Herold K, Eisenbarth G. Genetics, pathogenesis and clinical interventions in type 1 diabetes. Nature 2010;464(7293):1293–300.
[2] Pociot F, Akolkar B, Concannon P, Erlich HA, Julier C, Morahan G, et al. Genetics of type 1 diabetes: what's next? Diabetes 2010;59(7):1561–71.
[3] Todd JA. Etiology of type 1 diabetes. Immunity 2010;32(4):457–67.
[4] Naik RG, Brooks-Worrell BM, Palmer JP. Latent autoimmune diabetes in adults. J Clin Endocrinol Metab 2009;94(12):4635–44.
[5] Concannon P, Rich SS, Nepom GT. Genetics of type 1A diabetes. N Engl J Med 2009;360(16):1646–54.
[6] Grant SF, Hakonarson H. Genome-wide association studies in type 1 diabetes. Curr Diab Rep 2009;9(2):157–63.
[7] Ellard S, Bellanne-Chantelot C, Hattersley AT. Best practice guidelines for the molecular genetic diagnosis of maturity-onset diabetes of the young. Diabetologia 2008;51(4):546–53.
[8] Greeley SA, Tucker SE, Worrell HI, Skowron KB, Bell GI, Philipson LH. Update in neonatal diabetes. Curr Opin Endocrinol Diabetes Obes 2010;17(1):13–19.
[9] Hattersley A, Bruining J, Shield J, Njolstad P, Donaghue KC. The diagnosis and management of monogenic diabetes in children and adolescents. Pediatr Diabetes 2009;10(Suppl. 12):33–42.
[10] Vaxillaire M DP, Bonnefond A, Froguel P. Breakthroughs in monogenic diabetes genetics: from pediatric forms to young adulthood diabetes. Pediatr Endocrinol Rev 2009;6(3):405–17.

[11] De Franco E, Ellard S. Genome, exome, and targeted next-generation sequencing in neonatal diabetes. Pediatr Clin North Am 2015;62(4):1037–53.

[12] Bonow RO, Gheorghiade M. The diabetes epidemic: a national and global crisis. Am J Med 2004;116(Suppl 5A):2S–10S.

[13] Menke A, Casagrande S, Geiss L, Cowie CC. Prevalence of and trends in diabetes among adults in the United States, 1988–2012. JAMA 2015;314(10):1021–9.

[14] Engelgau MM, Geiss LS, Saaddine JB, Boyle JP, Benjamin SM, Gregg EW, et al. The evolving diabetes burden in the United States. Ann Intern Med 2004;140(11):945–50.

[15] NIDDK. National Diabetes Statistics fact sheet: general information and national estimates on diabetes in the United States, 2005. Bethesda, MD: U.S. Department of Health and Human Services, National Institute of Health, 2005.

[16] Yoon KH, Lee JH, Kim JW, Cho JH, Choi YH, Ko SH, et al. Epidemic obesity and type 2 diabetes in Asia. Lancet 2006;368(9548):1681–8.

[17] Chan JC, Malik V, Jia W, Kadowaki T, Yajnik CS, Yoon KH, et al. Diabetes in Asia: epidemiology, risk factors, and pathophysiology. JAMA 2009;301(20):2129–40.

[18] Mbanya JC, Motala AA, Sobngwi E, Assah FK, Enoru ST. Diabetes in sub-Saharan Africa. Lancet 2010;375(9733):2254–66.

[19] Ritz E, Rychlik I, Locatelli F, Halimi S. End-stage renal failure in type 2 diabetes: a medical catastrophe of worldwide dimensions. Am J Kidney Dis 1999;34(5):795–808.

[20] American Diabetes Association. Economic costs of diabetes in the U.S. in 2007. Diabetes Care 2008;31(3):596–615.

[21] Hamer RA, El Nahas AM. The burden of chronic kidney disease is rising rapidly worldwide. BMJ 2006;332:563–4.

[22] Groop LC. The molecular genetics of non-insulin-dependent diabetes mellitus. J Intern Med 1997;241(2):95–101.

[23] Elbein SC. The search for genes for type 2 diabetes in the post-genomic era. Endocrinology 2002;143(6):2012–8.

[24] Florez JC, Hirschhorn JN, Altshuler D. The inherited basis of diabetes mellitus: implications for the genetic analysis of complex traits. Annu Rev Genomics Hum Genet 2003;4:257–91.

[25] Carter JS, Pugh JA, Monterrosa A. Non-insulin-dependent diabetes mellitus in minorities in the United States. Ann Intern Med 1996;125(3):221–32.

[26] Meigs JB, Cupples LA, Wilson PWF. Parental transmission of type 2 diabetes mellitus: the Framingham Offspring Study. Diabetes 2000;49:2201–7. 2201–7.

[27] Weijnen CF, Rich SS, Meigs JB, Krolewski AS, Warram JH. Risk of diabetes in siblings of index cases with Type 2 diabetes: implications for genetic studies. Diabet Med 2002;19:41–50.

[28] Barnett AH, Eff C, Leslie RD, Pyke DA. Diabetes in identical twins. A study of 200 pairs. Diabetologia 1981;20(2):87–93.

[29] Poulsen P, Kyvik KO, Vaag A, Beck-Nielsen H. Heritability of type II (non-insulin-dependent) diabetes mellitus and abnormal glucose tolerance—a population-based twin study. Diabetologia 1999;42(2):139–45.

[30] Davies JL, Kawaguchi Y, Bennett ST, Copeman JB, Cordell HJ, Pritchard LE, et al. A genome-wide search for human type 1 diabetes susceptibility genes. Nature 1994;371:130–6.

[31] Risch NJ. Searching for genetic determinants in the new millennium. Nature 2000;405(6788):847–56.

[32] Risch N, Merikangas K. The future of genetic studies of complex human diseases. Science 1996;273(5281):1516–7.

[33] Altshuler D, Hirschhorn JN, Klannemark M, Lindgren CM, Vohl MC, Nemesh J, et al. The common PPARγ Pro12Ala polymorphism is associated with decreased risk of type 2 diabetes. Nat Genet 2000;26(1):76–80.

[34] Gloyn AL, Weedon MN, Owen KR, Turner MJ, Knight BA, Hitman G, et al. Large-scale association studies of variants in genes encoding the pancreatic β-cell KATP channel subunits Kir6.2 (*KCNJ11*) and SUR1 (*ABCC8*) confirm that the *KCNJ11* E23K variant is associated with type 2 diabetes. Diabetes 2003;52(2):568–72.

[35] Florez JC, Burtt N, de Bakker PIW, Almgren P, Tuomi T, Holmkvist J, et al. Haplotype structure and genotype-phenotype correlations of the sulfonylurea receptor and the islet ATP-sensitive potassium channel gene region. Diabetes 2004;53(5):1360–8.

[36] Reynisdottir I, Thorleifsson G, Benediktsson R, Sigurdsson G, Emilsson V, Einarsdottir AS, et al. Localization of a susceptibility gene for type 2 diabetes to chromosome 5q34-q35.2. Am J Hum Genet 2003;73(2):323–35.

[37] Grant SFA, Thorleifsson G, Reynisdottir I, Benediktsson R, Manolescu A, Sainz J, et al. Variant of transcription factor 7-like 2 (*TCF7L2*) gene confers risk of type 2 diabetes. Nat Genet 2006;38(3):320–3.

[38] Cauchi S, El Achhab Y, Choquet H, Dina C, Krempler F, Weitgasser R, et al. *TCF7L2* is reproducibly associated with type 2 diabetes in various ethnic groups: a global meta-analysis. J Mol Med 2007;85(7):777–82.

[39] Florez JC. The new type 2 diabetes gene *TCF7L2*. Curr Opin Clin Nutr Metab Care 2007;10(4):391–6.

[40] Lander ES, Linton LM, Birren B, Nusbaum C, Zody MC, Baldwin J, et al. Initial sequencing and analysis of the human genome. Nature 2001;409(6822):860–921.

[41] Venter JC, Adams MD, Myers EW, Li PW, Mural RJ, Sutton GG, et al. The sequence of the human genome. Science 2001;291(5507):1304–51.

[42] Sachidanandam R, Weissman D, Schmidt SC, Kakol JM, Stein LD, Marth G, et al. A map of human genome sequence variation containing 1.42 million single nucleotide polymorphisms. Nature 2001;409(6822):928–33.

[43] Reich DE, Gabriel SB, Altshuler D. Quality and completeness of SNP databases. Nat Genet 2003;33(4):457–8.

[44] Jorde LB. Linkage disequilibrium and the search for complex disease genes. Genome Res 2000;10(10):1435–44.

[45] Daly MJ, Rioux JD, Schaffner SF, Hudson TJ, Lander ES. High-resolution haplotype structure in the human genome. Nat Genet 2001;29:229–32.

[46] Gabriel SB, Schaffner SF, Nguyen H, Moore JM, Roy J, Blumenstiel B, et al. The structure of haplotype blocks in the human genome. Science 2002;296(5576):2225–9.

[47] The International HapMap C. The International HapMap Project. Nature 2003;426(6968): 789–96.

[48] The International HapMap C. A haplotype map of the human genome. Nature 2005;437: 1299–320.

[49] Sladek R, Rocheleau G, Rung J, Dina C, Shen L, Serre D, et al. A genome-wide association study identifies novel risk loci for type 2 diabetes. Nature 2007;445(7130):828–30.

[50] Diabetes Genetics Initiative of Broad Institute of Harvard and MIT Lund University and Novartis Institutes for BioMedical R Saxena R, Voight BF, Lyssenko V, Burtt NP, et al. Genome-wide association analysis identifies loci for type 2 diabetes and triglyceride levels. Science 2007;316:1331–6.

[51] Zeggini E, Weedon MN, Lindgren CM, Frayling TM, Elliott KS, Lango H, et al. Replication of genome-wide association signals in U.K. samples reveals risk loci for type 2 diabetes. Science 2007;316:1336–41.

[52] Scott LJ, Mohlke KL, Bonnycastle LL, Willer CJ, Li Y, Duren WL, et al. A genome-wide association study of type 2 diabetes in Finns detects multiple susceptibility variants. Science 2007;316:1341–5.

[53] Steinthorsdottir V, Thorleifsson G, Reynisdottir I, Benediktsson R, Jonsdottir T, Walters GB, et al. A variant in *CDKAL1* influences insulin response and risk of type 2 diabetes. Nat Genet 2007;39(6):770–5.

[54] Sandhu MS, Weedon MN, Fawcett KA, Wasson J, Debenham SL, Daly A, et al. Common variants in *WFS1* confer risk of type 2 diabetes. Nat Genet 2007;39(8):951–3.

[55] Franks PW, Rolandsson O, Debenham SL, Fawcett KA, Payne F, Dina C, et al. Replication of the association between variants in *WFS1* and risk of type 2 diabetes in European populations. Diabetologia 2008;51(3):458–63.

[56] Winckler W, Weedon MN, Graham RR, McCarroll SA, Purcell S, Almgren P, et al. Evaluation of common variants in the six known Maturity-Onset Diabetes of the Young (MODY) genes for association with type 2 diabetes. Diabetes 2007;56(3):685–93.

[57] Gudmundsson J, Sulem P, Steinthorsdottir V, Bergthorsson JT, Thorleifsson G, Manolescu A, et al. Two variants on chromosome 17 confer prostate cancer risk, and the one in *TCF2* protects against type 2 diabetes. Nat Genet 2007;39(8):977–83.

[58] Pe'er I, Yelensky R, Altshuler D, Daly MJ. Estimation of the multiple testing burden for genomewide association studies of nearly all common variants. Genet Epidemiol 2008;32(4):381–5.

[59] Zeggini E, Scott LJ, Saxena R, Voight BF, Marchini JL, Hu T, et al. Meta-analysis of genome-wide association data and large-scale replication identifies additional susceptibility loci for type 2 diabetes. Nat Genet 2008;40(5):638–45.

[60] Rung J, Cauchi S, Albrechtsen A, Shen L, Rocheleau G, Cavalcanti-Proenca C, et al. Genetic variant near *IRS1* is associated with type 2 diabetes, insulin resistance and hyperinsulinemia. Nat Genet 2009;41(10):1110–5.

[61] Voight BF, Scott LJ, Steinthorsdottir V, Morris AP, Dina C, Welch RP, et al. Twelve type 2 diabetes susceptibility loci identified through large-scale association analysis. Nat Genet 2010;42(7):579–89.

[62] Qi L, Cornelis MC, Kraft P, Stanya KJ, Linda Kao WH, Pankow JS, et al. Genetic variants at 2q24 are associated with susceptibility to type 2 diabetes. Hum Mol Genet 2010;19(13):2706–15.

[63] Saxena R, Elbers CC, Guo Y, Peter I, Gaunt TR, Mega JL, et al. Large-scale gene-centric meta-analysis across 39 studies identifies type 2 diabetes loci. Am J Hum Genet 2012;90(3):410–25.

[64] Voight BF, Kang HM, Ding J, Palmer CD, Sidore C, Chines PS, et al. The metabochip, a custom genotyping array for genetic studies of metabolic, cardiovascular, and anthropometric traits. PLoS Genet 2012;8(8):e1002793.

[65] Morris AP, Voight BF, Teslovich TM, Ferreira T, Segre AV, Steinthorsdottir V, et al. Large-scale association analysis provides insights into the genetic architecture and pathophysiology of type 2 diabetes. Nat Genet 2012;44(9):981–90.

[66] Kong A, Steinthorsdottir V, Masson G, Thorleifsson G, Sulem P, Besenbacher S, et al. Parental origin of sequence variants associated with complex diseases. Nature 2009;462(7275):868–74.

[67] Perry JR, Voight BF, Yengo L, Amin N, Dupuis J, Ganser M, et al. Stratifying type 2 diabetes cases by BMI identifies genetic risk variants in LAMA1 and enrichment for risk variants in lean compared to obese cases. PLoS Genet 2012;8(5):e1002741.

[68] Yasuda K, Miyake K, Horikawa Y, Hara K, Osawa H, Furuta H, et al. Variants in *KCNQ1* are associated with susceptibility to type 2 diabetes mellitus. Nat Genet 2008;40(9):1092–7.

[69] Unoki H, Takahashi A, Kawaguchi T, Hara K, Horikoshi M, Andersen G, et al. SNPs in *KCNQ1* are associated with susceptibility to type 2 diabetes in East Asian and European populations. Nat Genet 2008;40(9):1098–102.

[70] Tsai FJ, Yang CF, Chen CC, Chuang LM, Lu CH, Chang CT, et al. A genome-wide association study identifies susceptibility variants for type 2 diabetes in Han Chinese. PLoS Genet 2010;6(2):e1000847.

[71] Shu XO, Long J, Cai Q, Qi L, Xiang YB, Cho YS, et al. Identification of new genetic risk variants for type 2 diabetes. PLoS Genet 2010;6(9):e1001127.

[72] Yamauchi T, Hara K, Maeda S, Yasuda K, Takahashi A, Horikoshi M, et al. A genome-wide association study in the Japanese population identifies susceptibility loci for type 2 diabetes at UBE2E2 and C2CD4A-C2CD4B. Nat Genet 2010;42(10):864–8.

[73] Li H, Gan W, Lu L, Dong X, Han X, Hu C, et al. A genome-wide association study identifies GRK5 and RASGRP1 as type 2 diabetes loci in Chinese Hans. Diabetes 2013;62(1):291–8.

[74] Cho YS, Chen CH, Hu C, Long J, Ong RT, Sim X, et al. Meta-analysis of genome-wide association studies identifies eight new loci for type 2 diabetes in East Asians. Nat Genet 2012;44(1):67–72.

[75] Go MJ, Hwang JY, Park TJ, Kim YJ, Oh JH, Kim YJ, et al. Genome-wide association study identifies two novel Loci with sex-specific effects for type 2 diabetes mellitus and glycemic traits in a Korean population. Diabetes Metab J 2014;38(5):375–87.

[76] Hara K, Fujita H, Johnson TA, Yamauchi T, Yasuda K, Horikoshi M, et al. Genome-wide association study identifies three novel loci for type 2 diabetes. Hum Mol Genet 2014;23(1):239–46.

[77] Consortium STD, Williams AL, Jacobs SB, Moreno-Macias H, Huerta-Chagoya A, Churchhouse C, et al. Sequence variants in SLC16A11 are a common risk factor for type 2 diabetes in Mexico. Nature 2014;506(7486):97–101.

[78] Kooner JS, Saleheen D, Sim X, Sehmi J, Zhang W, Frossard P, et al. Genome-wide association study in individuals of South Asian ancestry identifies six new type 2 diabetes susceptibility loci. Nat Genet 2011;43(10):984–9.

[79] Tabassum R, Chauhan G, Dwivedi OP, Mahajan A, Jaiswal A, Kaur I, et al. Genome-wide association study for type 2 diabetes in Indians identifies a new susceptibility locus at 2q21. Diabetes 2013;62(3):977–86.

[80] Saxena R, Saleheen D, Been LF, Garavito ML, Braun T, Bjonnes A, et al. Genome-wide association study identifies a novel locus contributing to type 2 diabetes susceptibility in Sikhs of Punjabi origin from India. Diabetes 2013;62(5):1746–55.

[81] Hanson RL, Muller YL, Kobes S, Guo T, Bian L, Ossowski V, et al. A genome-wide association study in American Indians implicates DNER as a susceptibility locus for type 2 diabetes. Diabetes 2014;63(1):369–76.

[82] Palmer ND, McDonough CW, Hicks PJ, Roh BH, Wing MR, An SS, et al. A genome-wide association search for type 2 diabetes genes in African Americans. PLoS One 2012;7(1):e29202.

[83] Ng MC, Shriner D, Chen BH, Li J, Chen WM, Guo X, et al. Meta-analysis of genome-wide association studies in African Americans provides insights into the genetic architecture of type 2 diabetes. PLoS Genet 2014;10(8):e1004517.

[84] DIAbetes Genetics Replication And Meta-analysis (DIAGRAM) Consortium, Asian Genetic Epidemiology Network Type 2 Diabetes (AGEN-T2D) Consortium, South Asian Type 2 Diabetes (SAT2D) Consortium, Mexican American Type 2 Diabetes (MAT2D) Consortium, Type 2 Diabetes Genetic Exploration by Next-generation sequencing in multi-Ethnic Samples (T2D-GENES) Consortium. Genome-wide trans-ancestry meta-analysis provides insight into the genetic architecture of type 2 diabetes susceptibility. Nat Genet 2014;46(3):234–44.

[85] Albrechtsen A, Grarup N, Li Y, Sparso T, Tian G, Cao H, et al. Exome sequencing-driven discovery of coding polymorphisms associated with common metabolic phenotypes. Diabetologia 2013;56(2):298–310.

[86] Dupuis J, Langenberg C, Prokopenko I, Saxena R, Soranzo N, Jackson AU, et al. New genetic loci implicated in fasting glucose homeostasis and their impact on type 2 diabetes risk. Nat Genet 2010;42(2):105–16.

[87] Prokopenko I, Langenberg C, Florez JC, Saxena R, Soranzo N, Thorleifsson G, et al. Variants in *MTNR1B* influence fasting glucose levels. Nat Genet 2009;41(1):77–81.

[88] Moltke I, Grarup N, Jorgensen ME, Bjerregaard P, Treebak JT, Fumagalli M, et al. A common Greenlandic TBC1D4 variant confers muscle insulin resistance and type 2 diabetes. Nature 2014;512(7513):190–3.

[89] Steinthorsdottir V, Thorleifsson G, Sulem P, Helgason H, Grarup N, Sigurdsson A, et al. Identification of low-frequency and rare sequence variants associated with elevated or reduced risk of type 2 diabetes. Nat Genet 2014;46(3):294–8.

[90] Frayling TM, Timpson NJ, Weedon MN, Zeggini E, Freathy RM, Lindgren CM, et al. A common variant in the *FTO* gene is associated with body mass index and predisposes to childhood and adult obesity. Science 2007;316(5826):889–94.

[91] Mohlke KL, Boehnke M. Recent advances in understanding the genetic architecture of type 2 diabetes. Hum Mol Genet 2015;24(R1):R85–92.

[92] Weedon MN, Clark VJ, Qian Y, Ben-Shlomo Y, Timpson N, Ebrahim S, et al. A common haplotype of the glucokinase gene alters fasting glucose and birth weight: association in six studies and population-genetics analyses. Am J Hum Genet 2006;79(6):991–1001.

[93] Bouatia-Naji N, Rocheleau G, Van Lommel L, Lemaire K, Schuit F, Cavalcanti-Proenca C, et al. A polymorphism within the *G6PC2* gene is associated with fasting plasma glucose levels. Science 2008;320(5879):1085–8.

[94] Chen W-M, Erdos MR, Jackson AU, Saxena R, Sanna S, Silver KD, et al. Association studies in Caucasians identify variants in the *G6PC2/ABCB11* region regulating fasting glucose levels. J Clin Invest 2008;118(7):2620–8.

[95] Bouatia-Naji N, Bonnefond A, Cavalcanti-Proenca C, Sparso T, Holmkvist J, Marchand M, et al. A variant near *MTNR1B* is associated with increased fasting plasma glucose levels and type 2 diabetes risk. Nat Genet 2009;41(1):89–94.

[96] Saxena R, Hivert MF, Langenberg C, Tanaka T, Pankow JS, Vollenweider P, et al. Genetic variation in *GIPR* influences the glucose and insulin responses to an oral glucose challenge. Nat Genet 2010;42(2):142–8.

[97] Soranzo N, Sanna S, Wheeler E, Gieger C, Radke D, Dupuis J, et al. Common variants at 10 genomic loci influence hemoglobin A(1)(C) levels via glycemic and nonglycemic pathways. Diabetes 2010;59(12):3229–39.

[98] Strawbridge RJ, Dupuis J, Prokopenko I, Barker A, Ahlqvist E, Rybin D, et al. Genome-wide association identifies nine common variants associated with fasting proinsulin levels and provides new insights into the pathophysiology of type 2 diabetes. Diabetes 2011;60(10):2624–34.

[99] Scott RA, Lagou V, Welch RP, Wheeler E, Montasser ME, Luan J, et al. Large-scale association analyses identify new loci influencing glycemic traits and provide insight into the underlying biological pathways. Nat Genet 2012;44(9):991–1005.

[100] Manning AK, Hivert MF, Scott RA, Grimsby JL, Bouatia-Naji N, Chen H, et al. A genome-wide approach accounting for body mass index identifies genetic variants influencing fasting glycemic traits and insulin resistance. Nat Genet 2012;44(6):659–69.

[101] Prokopenko I, Poon W, Magi R, Prasad BR, Salehi SA, Almgren P, et al. A central role for GRB10 in regulation of islet function in man. PLoS Genet 2014;10(4):e1004235.

[102] Horikoshi M, Mgi R, van de Bunt M, Surakka I, Sarin AP, Mahajan A, et al. Discovery and fine-mapping of glycaemic and obesity-related trait loci using high-density imputation. PLoS Genet 2015;11(7):e1005230.

[103] Knowles JW, Xie W, Zhang Z, Chennemsetty I, Assimes TL, Paananen J, et al. Identification and validation of *N*-acetyltransferase 2 as an insulin sensitivity gene. J Clin Invest 2015;125(4):1739–51.

[104] Chen G, Bentley A, Adeyemo A, Shriner D, Zhou J, Doumatey A, et al. Genome-wide association study identifies novel loci association with fasting insulin and insulin resistance in African Americans. Hum Mol Genet 2012;21(20):4530–6.

[105] Kim YJ, Go MJ, Hu C, Hong CB, Kim YK, Lee JY, et al. Large-scale genome-wide association studies in East Asians identify new genetic loci influencing metabolic traits. Nat Genet 2011;43(10):990–5.

[106] Hwang JY, Sim X, Wu Y, Liang J, Tabara Y, Hu C, et al. Genome-wide association meta-analysis identifies novel variants associated with fasting plasma glucose in East Asians. Diabetes 2015;64(1):291–8.

[107] Go MJ, Hwang JY, Kim YJ, Hee Oh J, Kim YJ, Heon Kwak S, et al. New susceptibility loci in MYL2, C12orf51 and OAS1 associated with 1-h plasma glucose as predisposing risk factors for type 2 diabetes in the Korean population. J Hum Genet 2013;58(6):362–5.

[108] Palmer ND, Goodarzi MO, Langefeld CD, Wang N, Guo X, Taylor KD, et al. Genetic variants associated with quantitative glucose homeostasis traits translate to type 2 diabetes in Mexican Americans: the GUARDIAN (genetics underlying diabetes in Hispanics) consortium. Diabetes 2015;64(5):1853–66.

[109] Chen P, Takeuchi F, Lee JY, Li H, Wu JY, Liang J, et al. Multiple nonglycemic genomic loci are newly associated with blood level of glycated hemoglobin in East Asians. Diabetes 2014;63(7):2551–62.

[110] Huyghe JR, Jackson AU, Fogarty MP, Buchkovich ML, Stancakova A, Stringham HM, et al. Exome array analysis identifies new loci and low-frequency variants influencing insulin processing and secretion. Nat Genet 2013;45(2):197–201.

[111] Ryu J, Lee C. Association of glycosylated hemoglobin with the gene encoding CDKAL1 in the Korean Association Resource (KARE) study. Hum Mutat 2012;33(4):655–9.

[112] Mahajan A, Sim X, Ng HJ, Manning A, Rivas MA, Highland HM, et al. Identification and functional characterization of G6PC2 coding variants influencing glycemic traits define an effector transcript at the G6PC2-ABCB11 locus. PLoS Genet 2015;11(1):e1004876.

[113] Wessel J, Chu AY, Willems SM, Wang S, Yaghootkar H, Brody JA, et al. Low-frequency and rare exome chip variants associate with fasting glucose and type 2 diabetes susceptibility. Nat Commun 2015;6:5897.

[114] Genomes Project C, Auton A, Brooks LD, Durbin RM, Garrison EP, Kang HM, et al. A global reference for human genetic variation. Nature 2015;526(7571):68–74.

[115] Bamshad MJ, Ng SB, Bigham AW, Tabor HK, Emond MJ, Nickerson DA, et al. Exome sequencing as a tool for Mendelian disease gene discovery. Nat Rev Genet 2011;12(11):745–55.

[116] Plenge RM, Scolnick EM, Altshuler D. Validating therapeutic targets through human genetics. Nat Rev Drug Discov 2013;12(8):581–94.

[117] Consortium STD, Estrada K, Aukrust I, Bjorkhaug L, Burtt NP, Mercader JM, et al. Association of a low-frequency variant in HNF1A with type 2 diabetes in a Latino population. JAMA 2014;311(22):2305–14.

[118] Yaghootkar H, Stancakova A, Freathy RM, Vangipurapu J, Weedon MN, Xie W, et al. Association analysis of 29,956 individuals confirms that a low-frequency variant at

CCND2 halves the risk of type 2 diabetes by enhancing insulin secretion. Diabetes 2015;64(6):2279–85.

[119] Bonnefond A, Clement N, Fawcett K, Yengo L, Vaillant E, Guillaume JL, et al. Rare MTNR1B variants impairing melatonin receptor 1B function contribute to type 2 diabetes. Nat Genet 2012;44(3):297–301.

[120] Ingelsson E, Langenberg C, Hivert MF, Prokopenko I, Lyssenko V, Dupuis J, et al. Detailed physiologic characterization reveals diverse mechanisms for novel genetic loci regulating glucose and insulin metabolism in humans. Diabetes 2010;59(5):1266–75.

[121] Pound LD, Oeser JK, O'Brien TP, Wang Y, Faulman CJ, Dadi PK, et al. G6PC2: a negative regulator of basal glucose-stimulated insulin secretion. Diabetes 2013;62(5): 1547–56.

[122] Schwanstecher C, Meyer U, Schwanstecher M. KIR6.2 polymorphism predisposes to type 2 diabetes by inducing overactivity of pancreatic β-cell ATP-sensitive K+ channels. Diabetes 2002;51(3):875–9.

[123] Hamming KS, Soliman D, Matemisz LC, Niazi O, Lang Y, Gloyn AL, et al. Coexpression of the type 2 diabetes susceptibility gene variants KCNJ11 E23K and ABCC8 S1369A alter the ATP and sulfonylurea sensitivities of the ATP-sensitive K(+) channel. Diabetes 2009;58(10):2419–24.

[124] Feng Y, Mao G, Ren X, Xing H, Tang G, Li Q, et al. Ser1369Ala variant in sulfonylurea receptor gene *ABCC8* is associated with antidiabetic efficacy of gliclazide in Chinese type 2 diabetic patients. Diabetes Care 2008;31(10):1939–44.

[125] Orho-Melander M, Melander O, Guiducci C, Perez-Martinez P, Corella D, Roos C, et al. Common missense variant in the glucokinase regulatory protein gene is associated with increased plasma triglyceride and C-reactive protein but lower fasting glucose concentrations. Diabetes 2008;57(11):3112–21.

[126] Sparso T, Andersen G, Nielsen T, Burgdorf KS, Gjesing AP, Nielsen AL, et al. The *GCKR* rs780094 polymorphism is associated with elevated fasting serum triacylglycerol, reduced fasting and OGTT-related insulinaemia, and reduced risk of type 2 diabetes. Diabetologia 2008;51(1):70–5.

[127] Beer NL, Tribble ND, McCulloch LJ, Roos C, Johnson PR, Orho-Melander M, et al. The P446L variant in *GCKR* associated with fasting plasma glucose and triglyceride levels exerts its effect through increased glucokinase activity in liver. Hum Mol Genet 2009;18(21):4081–8.

[128] Helgason A, Palsson S, Thorleifsson G, Grant SF, Emilsson V, Gunnarsdottir S, et al. Refining the impact of *TCF7L2* gene variants on type 2 diabetes and adaptive evolution. Nat Genet 2007;39(2):218–25.

[129] Florez JC, Jablonski KA, Bayley N, Pollin TI, de Bakker PIW, Shuldiner AR, et al. *TCF7L2* polymorphisms and progression to diabetes in the Diabetes Prevention Program. N Engl J Med 2006;355(3):241–50.

[130] Lyssenko V, Lupi R, Marchetti P, del Guerra S, Orho-Melander M, Almgren P, et al. Mechanisms by which common variants in the *TCF7L2* gene increase risk of type 2 diabetes. J Clin Invest 2007;117:2155–63.

[131] Loos RJF, Franks PW, Francis RW, Barroso I, Gribble FM, Savage DB, et al. *TCF7L2* polymorphisms modulate proinsulin levels and β-cell function in a British Europid population. Diabetes 2007;56:1943–7.

[132] Kirchhoff K, Machicao F, Haupt A, Schafer SA, Tschritter O, Staiger H, et al. Polymorphisms in the *TCF7L2*, *CDKAL1* and *SLC30A8* genes are associated with impaired proinsulin conversion. Diabetologia 2008;51(4):597–601.

[133] Gonzalez-Sanchez JL, Martinez-Larrad MT, Zabena C, Perez-Barba M, Serrano-Rios M. Association of variants of the *TCF7L2* gene with increases in the risk of type 2 diabetes and the proinsulin:insulin ratio in the Spanish population. Diabetologia 2008;51(11): 1993–7.

[134] Stolerman ES, Manning AK, McAteer JB, Fox CS, Dupuis J, Meigs JB, et al. *TCF7L2* variants are associated with increased proinsulin/insulin ratios but not obesity traits in the Framingham Heart Study. Diabetologia 2009;52(4):614–20.

[135] Schafer SA, Tschritter O, Machicao F, Thamer C, Stefan N, Gallwitz B, et al. Impaired glucagon-like peptide-1-induced insulin secretion in carriers of transcription factor 7-like 2 (*TCF7L2*) gene polymorphisms. Diabetologia 2007;50(12):2443–50.

[136] Pilgaard K, Jensen CB, Schou JH, Lyssenko V, Wegner L, Brons C, et al. The T allele of rs7903146 *TCF7L2* is associated with impaired insulinotropic action of incretin hormones, reduced 24 h profiles of plasma insulin and glucagon, and increased hepatic glucose production in young healthy men. Diabetologia 2009;52(7):1298–307.

[137] Villareal DT, Robertson H, Bell GI, Patterson BW, Tran H, Wice B, et al. *TCF7L2* variant rs7903146 affects the risk of type 2 diabetes by modulating incretin action. Diabetes 2010;59(2):479–85.

[138] Shu L, Matveyenko AV, Kerr-Conte J, Cho JH, McIntosh CH, Maedler K. Decreased TCF7L2 protein levels in type 2 diabetes mellitus correlate with downregulation of GIP- and GLP-1 receptors and impaired beta-cell function. Hum Mol Genet 2009;18(13):2388–99.

[139] Liu Z, Habener JF. Glucagon-like peptide-1 activation of TCF7L2-dependent Wnt signaling enhances pancreatic beta cell proliferation. J Biol Chem 2008;283(13):8723–35.

[140] Shu L, Sauter NS, Schulthess FT, Matveyenko AV, Oberholzer J, Maedler K. Transcription factor 7-like 2 regulates β-cell survival and function in human pancreatic islets. Diabetes 2008;57(3):645–53.

[141] da Silva Xavier G, Loder MK, McDonald A, Tarasov AI, Carzaniga R, Kronenberger K, et al. TCF7L2 regulates late events in insulin secretion from pancreatic islet beta-cells. Diabetes 2009;58(4):894–905.

[142] Gaulton KJ, Nammo T, Pasquali L, Simon JM, Giresi PG, Fogarty MP, et al. A map of open chromatin in human pancreatic islets. Nat Genet 2010;42(3):255–9.

[143] Zhou Y, Park SY, Su J, Bailey K, Ottosson-Laakso E, Shcherbina L, et al. TCF7L2 is a master regulator of insulin production and processing. Hum Mol Genet 2014;23(24):6419–31.

[144] Shao W, Xiong X, Ip W, Xu F, Song Z, Zeng K, et al. The expression of dominant negative TCF7L2 in pancreatic beta cells during the embryonic stage causes impaired glucose homeostasis. Mol Metab 2015;4(4):344–52.

[145] Chimienti F, Devergnas S, Favier A, Seve M. Identification and cloning of a β-cell-specific Zinc transporter, ZnT-8, localized into insulin secretory granules. Diabetes 2004;53(9):2330–7.

[146] Chimienti F, Devergnas S, Pattou F, Schuit F, Garcia-Cuenca R, Vandewalle B, et al. In vivo expression and functional characterization of the zinc transporter ZnT8 in glucose-induced insulin secretion. J Cell Sci 2006;119(20):4199–206.

[147] Nicolson TJ, Bellomo EA, Wijesekara N, Loder MK, Baldwin JM, Gyulkhandanyan AV, et al. Insulin storage and glucose homeostasis in mice null for the granule zinc transporter ZnT8 and studies of the type 2 diabetes-associated variants. Diabetes 2009;58(9):2070–83.

[148] Pound LD, Sarkar SA, Benninger RK, Wang Y, Suwanichkul A, Shadoan MK, et al. Deletion of the mouse Slc30a8 gene encoding zinc transporter-8 results in impaired insulin secretion. Biochem J 2009;421(3):371–6.

[149] Lemaire K, Ravier MA, Schraenen A, Creemers JW, Van de Plas R, Granvik M, et al. Insulin crystallization depends on zinc transporter ZnT8 expression, but is not required for normal glucose homeostasis in mice. Proc Natl Acad Sci USA 2009;106(35):14872–7.

[150] Wijesekara N, Dai FF, Hardy AB, Giglou PR, Bhattacharjee A, Koshkin V, et al. Beta cell-specific Znt8 deletion in mice causes marked defects in insulin processing, crystallisation and secretion. Diabetologia 2010;53(8):1656–68.

[151] Tamaki M, Fujitani Y, Hara A, Uchida T, Tamura Y, Takeno K, et al. The diabetes-susceptible gene SLC30A8/ZnT8 regulates hepatic insulin clearance. J Clin Invest 2013;123(10):4513–24.

[152] Weijers RN. Three-dimensional structure of beta-cell-specific zinc transporter, ZnT-8, predicted from the type 2 diabetes-associated gene variant SLC30A8 R325W. Diabetol Metab Syndr 2010;2(1):33.

[153] Flannick J, Thorleifsson G, Beer NL, Jacobs SB, Grarup N, Burtt NP, et al. Loss-of-function mutations in SLC30A8 protect against type 2 diabetes. Nat Genet 2014;46(4):357–63.

[154] Florez JC. Newly identified loci highlight beta cell dysfunction as a key cause of type 2 diabetes: Where are the insulin resistance genes? Diabetologia 2008;51:1100–10.

[155] Reaven G. Banting lecture 1988. Role of insulin resistance in human disease. Diabetes 1988;37(12):1595–607.

[156] Meigs JB, D'Agostino Sr. RB, Wilson PW, Cupples LA, Nathan DM, Singer DE. Risk variable clustering in the insulin resistance syndrome. The Framingham Offspring Study. Diabetes 1997;46(10):1594–600.

[157] Kathiresan S, Willer CJ, Peloso GM, Demissie S, Musunuru K, Schadt EE, et al. Common variants at 30 loci contribute to polygenic dyslipidemia. Nat Genet 2009;41(1):56–65.

[158] Nilsson PM, Roost M, Engstrom G, Hedblad B, Berglund G. Incidence of diabetes in middle-aged men is related to sleep disturbances. Diabetes Care 2004;27(10):2464–9.

[159] Knutson KL, Ryden AM, Mander BA, Van Cauter E. Role of sleep duration and quality in the risk and severity of type 2 diabetes mellitus. Arch Intern Med 2006;166(16):1768–74.

[160] Spiegel K, Leproult R, Van Cauter E. Impact of sleep debt on metabolic and endocrine function. Lancet 1999;354(9188):1435–9.

[161] Scheer FA, Hilton MF, Mantzoros CS, Shea SA. Adverse metabolic and cardiovascular consequences of circadian misalignment. Proc Natl Acad Sci USA 2009;106(11):4453–8.

[162] Turek FW, Joshu C, Kohsaka A, Lin E, Ivanova G, McDearmon E, et al. Obesity and metabolic syndrome in circadian Clock mutant mice. Science 2005;308(5724):1043–5.

[163] Marcheva B, Ramsey KM, Buhr ED, Kobayashi Y, Su H, Ko CH, et al. Disruption of the clock components CLOCK and BMAL1 leads to hypoinsulinaemia and diabetes. Nature 2010;466(7306):627–31.

[164] Barker DJ, Hales CN, Fall CH, Osmond C, Phipps K, Clark PM. Type 2 (non-insulin-dependent) diabetes mellitus, hypertension and hyperlipidaemia (syndrome X): relation to reduced fetal growth. Diabetologia 1993;36(1):62–7.

[165] Freathy RM, Bennett AJ, Ring SM, Shields B, Groves CJ, Timpson NJ, et al. Type 2 diabetes risk alleles are associated with reduced size at birth. Diabetes 2009;58(6):1428–33.

[166] Freathy RM, Mook-Kanamori DO, Sovio U, Prokopenko I, Timpson NJ, Berry DJ, et al. Variants in ADCY5 and near CCNL1 are associated with fetal growth and birth weight. Nat Genet 2010;42(5):430–5.

[167] Andersson EA, Pilgaard K, Pisinger C, Harder MN, Grarup N, Faerch K, et al. Type 2 diabetes risk alleles near ADCY5, CDKAL1 and HHEX-IDE are associated with reduced birthweight. Diabetologia 2010;53(9):1908–16.

[168] Wilson PW, Meigs JB, Sullivan L, Fox CS, Nathan DM, D'Agostino Sr. RB. Prediction of incident diabetes mellitus in middle-aged adults: the Framingham Offspring Study. Arch Intern Med 2007;167(10):1068–74.

[169] Vassy JL, Hivert MF, Porneala B, Dauriz M, Florez JC, Dupuis J, et al. Polygenic type 2 diabetes prediction at the limit of common variant detection. Diabetes 2014;63(6):2172–82.

[170] Lyssenko V, Jonsson A, Almgren P, Pulizzi N, Isomaa B, Tuomi T, et al. Clinical risk factors, DNA variants, and the development of type 2 diabetes. N Engl J Med 2008;359(21):2220–32.

[171] Meigs JB, Shrader P, Sullivan LM, McAteer JB, Fox CS, Dupuis J, et al. Genotype score in addition to common risk factors for prediction of type 2 diabetes. N Engl J Med 2008;359(21):2208–19.

[172] Link E, Parish S, Armitage J, Bowman L, Heath S, Matsuda F, et al. *SLCO1B1* variants and statin-induced myopathy—a genomewide study. N Engl J Med 2008;359(8):789–99.

[173] Mega JL, Close SL, Wiviott SD, Shen L, Hockett RD, Brandt JT, et al. Cytochrome P-450 polymorphisms and response to clopidogrel. N Engl J Med 2008 NEJMoa0809171.

[174] Shuldiner AR, O'Connell JR, Bliden KP, Gandhi A, Ryan K, Horenstein RB, et al. Association of cytochrome P450 2C19 genotype with the antiplatelet effect and clinical efficacy of clopidogrel therapy. JAMA 2009;302(8):849–57.

[175] Pearson ER, Donnelly LA, Kimber C, Whitley A, Doney ASF, McCarthy MI, et al. Variation in *TCF7L2* influences therapeutic response to sulfonylureas: a GoDARTs study. Diabetes 2007;56:2178–82.

[176] Kimber CH, Doney AS, Pearson ER, McCarthy MI, Hattersley AT, Leese GP, et al. *TCF7L2* in the Go-DARTS study: evidence for a gene dose effect on both diabetes susceptibility and control of glucose levels. Diabetologia 2007;50(6):1186–91.

[177] Shu Y, Sheardown SA, Brown C, Owen RP, Zhang S, Castro RA, et al. Effect of genetic variation in the organic cation transporter 1 (OCT1) on metformin action. J Clin Invest 2007;117(5):1422–31.

[178] Zhou K, Donnelly LA, Kimber CH, Donnan PT, Doney AS, Leese G, et al. Reduced function *SLC22A1* polymorphisms encoding Organic Cation Transporter 1 (OCT1) and glycaemic response to metformin: a Go-DARTS study. Diabetes 2009;58(6):1434–9.

[179] Becker ML, Visser LE, van Schaik RHN, Hofman A, Uitterlinden AG, Stricker BHC. Genetic variation in the multidrug and toxin extrusion 1 transporter protein influences the glucose-lowering effect of metformin in patients with diabetes: a preliminary study. Diabetes 2009;58(3):745–9.

[180] Jablonski KA, McAteer JB, de Bakker PI, Franks PW, Pollin TI, Hanson RL, et al. Common variants in 40 genes assessed for diabetes incidence and response to metformin and lifestyle interventions in the Diabetes Prevention Program. Diabetes 2010;59(10):2672–81.

[181] Maruthur NM, Gribble MO, Bennett WL, Bolen S, Wilson LM, Balakrishnan P, et al. The pharmacogenetics of type 2 diabetes: a systematic review. Diabetes Care 2014;37(3):876–86.

[182] Billings LK, Florez JC. The genetics of type 2 diabetes: what have we learned from GWAS? Ann N Y Acad Sci 2010;1212:59–77.

Chapter 15

Metabolic Syndrome

Matthew B. Lanktree[1] and Robert A. Hegele[2]

[1]Mcmaster University, Hamilton, ON, Canada, [2]University of Western Ontario, London, ON, Canada

Chapter Outline

INTRODUCTION

The clustering of several metabolic abnormalities, including dyslipidemia (elevated serum triglycerides and depressed high-density lipoprotein (HDL) cholesterol), dysglycemia, hypertension, and central obesity has been termed the metabolic syndrome (MetS). The term syndrome, originating from Greek literally meaning "running together," refers to a set of signs or symptoms occurring together where the underlying pathophysiology leading to the concurrence is unknown. Several lines of evidence suggest that a MetS definition is important both clinically and as a research tool, though some controversy exists regarding the importance of a MetS diagnosis compared to the sum of the risk factors

Genomic and Precision Medicine. DOI: http://dx.doi.org/10.1016/B978-0-12-800685-6.00015-1

TABLE 15.1 Criteria for Metabolic Syndrome Diagnosis

Metabolic Syndrome Component	Threshold for Criteria
Waist circumference	>80 cm in women; >94 cm in men
HDL cholesterol	<1.0 mmol/L
Triglycerides	≥1.7 mmol/L or drug therapy for high triglycerides[a]
Blood pressure	SBP ≥130 mmHg or DBP ≥85 mmHg or drug therapy for hypertension
Fasting glucose	≥5.5 mmol/L or drug therapy for elevated glucose

Abbreviations: HDL, high-density lipoprotein; SBP, systolic blood pressure; DBP, diastolic blood pressure. Data taken from stakeholder consensus on metabolic syndrome definition [1].
[a]*Fibrates and nicotinic acid are the most commonly used triglyceride lowering therapies.*

independently, for prediction of the development of diabetes and atherosclerotic cardiovascular disease. Nonetheless, it is clinically apparent that these risk factors occur together more often than one would expect if they were independent processes, and a fivefold increased risk of diabetes and twofold increased risk for cardiovascular disease is observed with a diagnosis of MetS [1]. In addition, in patients with an extreme perturbation of one of the components of MetS, as observed in lipodystrophy or monogenic obesity, disruption of the other components almost inevitably follows. The common form of MetS is a classic complex genetic trait involving the interaction of a multitude of genetic and environmental factors, and genetic investigations into MetS may yield insights into the responsible mechanisms.

DEFINING METABOLIC SYNDROME

At least six different organizations have published criteria for MetS diagnosis, but a recent consensus statement of stakeholders has unified the definition for clinical use, as well as for epidemiological and basic research studies [1]. The debate over the criteria for MetS has largely revolved around whether elevated abdominal obesity should be a mandatory component, and what the threshold for continuous variables should be. The revised definition includes five criteria, three of which must be met for a diagnosis (Table 15.1). The criteria include elevated waist circumference, triglycerides, blood pressure, fasting glucose, and depressed HDL cholesterol. Differences in baseline waist circumference observed between the sexes and ethnicities have also been a concern, and sex- and ethnicity-specific guidelines have been specified [1].

Regardless of the definition used, MetS is a common diagnosis. Data from the Third National Health and Nutrition Examination Survey, which used the

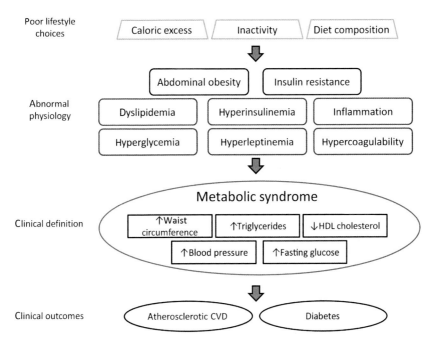

FIGURE 15.1   Metabolic syndrome. Poor lifestyle choices lead to the development of inciting factors for metabolic syndrome (black boxes) and abnormal physiology. Disrupted metabolism is clinically measured by the components of the metabolic syndrome. Metabolic syndrome subsequently increases risk for development of atherosclerotic cardiovascular disease (CVD) and diabetes.

National Cholesterol Education Program Adult Treatment Panel III MetS criteria, found the prevalence of MetS to be 24% in the United States [2]. Using the International Diabetes Foundation MetS definition, in a large study including >26,000 participants from 52 countries, the average MetS prevalence was 16.8% [3]. Due to its high prevalence, an improvement to our understanding of MetS could yield large benefits to public health.

PATHOPHYSIOLOGY OF METABOLIC SYNDROME

The inciting factors in the development of MetS are abdominal obesity and insulin resistance (Fig. 15.1). The accumulation of visceral fat, typically caused by overnutrition and physical inactivity, results in the release of free fatty acids leading to lipotoxicity and insulin resistance [4], and eventually hyperinsulinemia and hyperglycemia [5]. Insulin has numerous molecular effects beyond glucose homeostasis including upregulation of amino acid uptake and protein synthesis, activation of lipoprotein lipase and inhibition of very low-density lipoprotein secretion [6]. Abundance of fatty acids and diacylglycerol within skeletal muscle inhibits insulin signaling and reduces its ability to transport and utilize glucose [4]. Visceral fat accumulation creates a dysregulation of adipokine secretion, specifically hyposecretion of adiponectin and hypersecretion of both leptin (*LEP*) and proinflammatory cytokines (such as tumor necrosis factor alpha and interleukin-6 (*IL6*)), each of which may contribute to the MetS pathophysiology. Genetic analysis supports a causal role for adiponectin, which is secreted by adipocytes, in the pathogenesis of MetS [7]. In response to hyperinsulinemia and hyperglycemia, the liver secretes C-reactive protein and prothrombotic molecules such as fibrinogen and plasminogen activator inhibitor-1 [6]. Clearly, MetS is a complicated phenotype involving a complex network of causative and associated biochemical players, disentangling their relationships to further our understanding and develop novel therapeutics is thus a difficult task.

HERITABILITY OF METABOLIC SYNDROME

Strong evidence exists for the heritability of both MetS and its components arising from studies of twins, families, and more recently using population-based cohorts. In a study of 2508 pairs of American male twins, the MetS concordance between monozygotic pairs was 31.6% compared to 6.3% for dizygotic twins [8]. In 1942 Korean twin pairs and their families, significant heritability of both MetS (51–60%) and all of its components (46–77%), were observed [9]. Among 803 individuals from 89 Caribbean–Hispanic families, the heritability of a MetS diagnosis was 24%, with significant heritability for lipid/glucose/obesity (44%) and hypertension (20%) components [10]. In a study of 1277 Omani Arab individuals in five large consanguineous families, a large degree of heritability was observed for MetS (38%), while the heritability of the MetS components ranged from 38 to 63% [11]. Study of family members in the African American Jackson Heart Study yielded a MetS heritability of 32%, with higher heritability estimates for waist circumference (45%), HDL (43%), and triglycerides (42%), but lower estimates for systolic blood pressure (SBP), diastolic blood pressure, and fasting blood glucose (16, 15, and 14%, respectively) [12]. Finally, studying the Framingham Heart Study and Atherosclerosis Risk in Communities populations, the estimates for MetS component heritability were 34% for body mass index (BMI), 28% for waist-to-hip ratio, 47% for triglycerides, 48% for HDL, and 30% for SBP [13].

Variation in the reported heritability of MetS and its components is likely partially due to differences in ethnicity, variation in environmental exposures between family members, and the statistical techniques employed. Based upon the demonstrated heritability of MetS and its components, investigations to identify responsible genetic variants have been undertaken using both linkage analysis and association mapping strategies, the results of which are discussed in this chapter.

MONOGENIC MODELS OF METABOLIC SYNDROME

Monogenic diseases, also termed Mendelian diseases, are the result of a single genetic mutation, and thus have high penetrance, follow Mendelian inheritance patterns, and include small or nonexistent environmental components [14]. Contrarily, in complex diseases such as MetS, there is no apparent inheritance pattern, many genetic loci are involved, and there are large environmental components. In the last 30 years, the genetic basis of several monogenic diseases that include multiple components of MetS have been discovered, garnering insight into potential mechanisms of MetS, despite being responsible for a very small proportion of MetS in the population.

Since monogenic disorders segregate through families, linkage analysis of severely affected families followed by sequencing of genes within the identified region has discovered the responsible genetic variant for many monogenic models of MetS. Whole exome studies, in which the transcribed portion of the genome is sequenced, have already been successful identifying variants responsible for Mendelian disorders which are monogenic models for MetS. With the increasing availability of next-generation sequencing technologies, including whole exome and whole genome sequencing, our ability to identify rare genetic variants responsible for monogenic models of MetS will continue to improve.

Lipodystrophy

Lipodystrophy is a heterogeneous group of disorders characterized by selective or generalized atrophy of anatomical adipose tissue stores (Table 15.2). Loss of the ability to retain excess lipids in "classical" adipose tissue stores leads to the overdevelopment of ectopic fat stores such as within and around skeletal muscle, heart, liver, pancreas and kidneys, and within the arterial wall presenting as atherosclerosis. In patients with congenital generalized lipodystrophy, an absence of adipose tissue is noted in early infancy. In familial partial lipodystrophy (FPLD), patients have normal fat distribution during childhood, but during or shortly after puberty there is a progressive and gradual loss of subcutaneous adipose tissue of the extremities, a triggering event for the development of the other components of MetS including extreme insulin resistance, dyslipidemia, and hypertension.

A total of 16 genes have been identified to contain mutations causative of at least one form of lipodystrophy, with four genes recently identified with whole exome sequencing approaches: *PIK3R1* [15], *POLD1* [16], *PIK3R1* [17], and *PCYT1A* [18] (Table 15.2). The two most commonly mutated genes causing lipodystrophy are nuclear lamin A/C (*LMNA)* and peroxisome proliferator-activated receptor γ (*PPARG*). Over 200 mutations have been identified in *LMNA*, causing 13 different diseases together termed laminopathies: FPLD, Emery–Dreifuss muscular dystrophy, limb-girdle muscular dystrophy type 1B,

TABLE 15.2 Monogenic Lipodystrophy Observed in Humans

Disease	OMIM#
Generalized Lipodystrophy	
Congenital generalized lipodystrophy (CGL) (Berardinelli–Seip)	
CGL1—*AGPAT2*	608594
CGL2—*BSCL2*	269700
CGL3—*CAV1, PTRF*	612526
FOS	164810
PCYT1A	123695
Partial Lipodystrophy	
Familial partial lipodystrophy (FPLD)	
FPLD2 (Dunnigan)—*LMNA*	151660
FPLD3—*PPARG*	604367
FPLD4—*PLIN1*	170290
FPLD5—*CIDEC*	615238
Partial lipodystrophy with congenital cataracts, neurodegeneration—*CAV1*	606721
Partial lipodystrophy associated with *AKT2* mutations	164731
Acquired partial lipodystrophy (APL) (Barraquer–Simmons)	
APL—some cases associated with *LMNB2* mutations	608709
Syndromes that Include Lipodystrophy as a Component	
Mandibuloacral dysplasia (MAD)	
MADA—*LMNA*	248370
MADB—*ZMPSTE24*	608612
SHORT syndrome—*PIK3R1*	269880
Hutchinson–Gilford progeria syndrome (HGPS)—*LMNA*	176670
Werner syndrome (WRN)—*RECQL2, LMNA*	277700
Autoinflammation, lipodystrophy and dermatosis (ALDD)—*PSMB8*	256040
Mandibular hypoplasia, deafness, progeroid lipodystrophy—*POLD1*	615381

dilated cardiomyopathy type 1A, Charcot–Marie–Tooth, Hutchinson–Gilford progeria syndrome, atypical Werner syndrome, and a range of overlapping syndromes (www.hgmd.cf.ac.uk). *LMNA* encodes an intermediate filament protein vital for the structural integrity of the nuclear envelope, transcriptional regulation, nuclear pore functioning, and heterochromatin organization. However, it remains unknown how the mutations observed in *LMNA* specifically cause the observed phenotypes.

PPARG was selected as a candidate for sequencing in FPLD patients due to its important role as a ligand-inducible transcription factor regulating adipogenesis, and the repartitioning of fat stores observed in patients taking the thiazolidinedione class of drugs, which are agonists for PPARG. Sequencing of *PPARG* resulted in the discovery of causative mutations [19,20]. Through functional studies, different *PPARG* mutations were observed to work through both a dominant negative mechanism, in which the mutant receptor is able to inhibit the action of the wild-type receptor [21], and a haploinsufficiency mechanism, in which the 50% reduction in wild-type expression was sufficient to create the phenotype [20].

Monogenic Diseases of Obesity and Insulin Resistance

Over 20 genes have been implicated in rare monogenic diseases that include extreme obesity and/or insulin resistance that could provide insight into more common forms of MetS (Table 15.3). The most common monogenic cause of extreme obesity is mutations in melanocortin receptor 4 (*MC4R*) (OMIM: 155541), accounting for approximately 4% of extremely obese individuals [22]. *MC4R* is primarily expressed in the brain and is thought to impact obesity through central effects on appetite and satiety [23]. Proopiomelanocortin (POMC) (OMIM: 176830) is a precursor protein for multiple biologically active peptide hormones, including but not limited to adrenocorticotropic hormone, β-liptropin, β-endorphin, and α-, β-, and γ-melanocyte-stimulating hormone, which bind with varying affinity to five homologous melanocortin receptors, including MC4R [24]. *Pomc* −/− mice develop obesity and abnormal pigmentation [25], and complete loss-of-function mutations in the *POMC* were first described in patients with hypocortisolism, red hair, and early-onset extreme obesity [26]. Melanocortin signaling has also been implicated in modulating both blood pressure and lipid metabolism, independent of weight and insulin signaling, indicating a potentially greater role for *MC4R* and *POMC* in MetS [27,28].

Bardet–Biedl syndrome (BBS) is a group of at least 16 molecularly distinct but clinically similar diseases that include obesity as a central component (OMIM: 209900). As well as abdominal obesity, patients with BBS have diabetes, hypertension, mental retardation, dysmorphic extremities, retinal dystrophy or pigmentary retinopathy, hypogonadism, and abnormal kidney structure and/or function [14]. Knockout mice for three of the responsible genes have been developed: *Bbs2* −/−, *Bbs4* −/−, and *Bbs6* −/− [29]. All three mice strains

TABLE 15.3 Monogenic Diseases of Obesity and Insulin Resistance Observed in Humans

Disease	OMIM #
Obesity	
Alstrom syndrome—*ALMS1*	203800
Bardet–Biedl syndrome—*BBS1-16*	209900
Cohen syndrome—*COH1*	216550
Leptin deficiency—*LEP*	164160
Leptin receptor deficiency—*LEPR*	601007
Melanocortin-4 receptor deficiency—*MC4R*	155541
NTRK2 mutation	600456
Prader–Willi syndrome—*SNRPN*	176720
Prohormone convertase-1 deficiency—*PCSK1*	600955
Proopiomelanocortin deficiency—*POMC*	609734
Pseudohypoparathyroidism—*GNAS*	103580
SIM1 deletion	603128
WAGR syndrome—*BDNF*	612469
Insulin Resistance	
Donohue syndrome—*INSR*	246200
Disrupted insulin signaling—*AKT2*	164731

were hyperphagic, had low locomotor activity, and elevated circulating *LEP* concentrations [29]. Further examination of the knockout mice, as well as cellular work, suggests that the proteins mutated in BBS are required for *LEP* receptor (*LEPR*) signaling, and the impaired *LEP* signaling was associated with decreased *Pomc* expression [30]. Rare mutations in the genes encoding both leptin (*LEP*) (OMIM: 164160) and the *LEPR* (OMIM: 6011007) also cause obesity, hyperinsulinemia, and insulin resistance and are knocked out in the commonly studied ob^-/ob^- and db^-/db^- mice, respectively.

Alstrom syndrome is an autosomal recessive disorder characterized by childhood obesity, insulin resistance, hyperglycemia, hyperlipidemia, and neurosensory defects caused by mutations in a gene at chromosomal locus 2p13 subsequently named *ALMS1* (OMIM: 203800). Dilated cardiomyopathy, hepatic dysfunction, and hyperthyroidism are also variably present. Mice with *ALMS1* knocked out recapitulate the findings observed in humans providing evidence that loss of *ALMS1* alone is sufficient to cause the phenotype [31].

Mutations in the insulin receptor (*INSR*) can directly cause insulin resistance, creating several syndromes with variable *INSR* dysfunction: leprechaunism (Donahue syndrome), Rabson–Mendenhall syndrome and insulin resistance (OMIM: 147670). Patients with the most severe syndrome, Donahue syndrome, have marked hyperinsulinemia, fasting hypoglycemia, postprandial hyperglycemia, growth restriction, and premature mortality [32].

Using targeted sequence-enrichment (or capture) strategies, coupled with next-generation sequencing platforms, clinicians, and researchers can now interrogate the sequence of all genes identified to harbor mutations in monogenic forms of obesity [33]. However, interpreting the clinical significance of identified rare variants is challenging, even with the best bioinformatic analysis and the development of large databases of sequenced controls (http://exac.broadinstitute.org/; http://www.1000genomes.org/).

GENETICS OF COMMON METABOLIC SYNDROME

Linkage Analysis

Studies to identify genes contributing to common MetS began by searching for chromosomal regions that were transmitted between affected individuals in large families using linkage analysis. While many regions have been reported to be linked with one or more MetS components, few conclusions can be drawn for three reasons: (1) linkage analysis is poorly powered in the context of a genetic locus that explains only a small percentage of variation in the trait; (2) resolution is very poor and identified regions often include 100 genes; and (3) there has been little concordance between regions identified.

Candidate Gene-Association Studies

Genetic-association studies test if alleles at a single nucleotide polymorphism (SNP) are found more often in cases than in controls, or in individuals with elevated levels of a quantitative trait. To find genetic variations of small effect, association studies have greater resolution and are better powered than linkage analysis. The first genetic association studies were candidate gene studies, with more than 20 studies focused upon investigations into genes with known roles in MetS [5]. Genes that have been replicated in multiple studies include those in: lipid metabolism pathways, such as apolipoprotein A5 (*APOA5*), and apolipoprotein C3 (*APOC3*); inflammation, such as *IL6*; and adipose tissue partitioning, such as *LMNA*, *PPARG*, and *CAV1*. The results of these studies created fodder for discussion but were largely limited by inadequate sample size and were likely subject to publication bias.

Genome-Wide Association Studies

Advancements in high-throughput genotyping technologies and databases of common genetic variation have enabled genome-wide association studies (GWAS) to test over a million SNPs in a single experiment, allowing for an unbiased assessment of the genome. GWAS began by examining the individual components of MetS, identifying robust associations that have been replicated across populations, investigators, and genotyping platforms. In an iterative approach, utilizing larger and larger sample sizes now including >150,000 participants, GWAS of the components of MetS including blood lipid and lipoprotein concentrations [34], blood pressure [35], BMI [36], and fasting blood glucose [37] have identified >200 genes involved in metabolic pathways relevant to MetS. Many of the identified genes have previously known functional roles or are mutated in monogenic disease. In addition, many novel genes with no previously known roles in metabolic pathways have been robustly associated and their biological functions are under study.

Pleiotropy describes the situation where a gene impacts multiple phenotypic traits. In the interacting metabolic pathways involved in MetS pathophysiology, pleiotropy is to be expected. However, no genes have been identified to be associated with all five components of the MetS, and identifying genetic variants with strong associations with more than two components of the MetS has proven difficult. With respect to genes associated with more than one of the other MetS components, glucokinase (hexokinase 4) regulatory protein (*GCKR*) is associated with both fasting triglyceride and glucose concentrations [34,38]. The fat mass- and obesity-associated gene (*FTO*) is associated with both adiposity and measures of insulin sensitivity [39]. Variants in *INSR* substrate 1 (*IRS1*) are associated with type 2 diabetes risk, insulin resistance, HDL cholesterol, and triglycerides [34,38]. As the list of genes associated with each MetS component grows due to growing sample sizes, the list of genes associated with multiple components has also grown. Contrarily, it is interesting that some genes that are associated with one of the MetS components with large effect and are not associated with other components. For example, common variants in endothelial lipase (*LIPG*) are robustly associated with HDL concentration, but have not been identified to be associated with plasma triglycerides [34].

To date, four GWAS and one GWAS meta-analysis of MetS have been published (Table 15.4) [40–44]. As universally observed in GWAS, required sample sizes are much larger than initial power calculations predict, and success identifying SNPs with pleotropic associations across MetS components have come in larger studies. Genes associated with multiple components of MetS were all first identified through GWAS of the individual MetS components, but pleiotropic effects are evident. Among the most reliable MetS associations are key players in lipid metabolism (*GCKR, LPL, APOA1/C3/A4/A5, CETP, TRIB1,* and *LIPC*). Of genes identified in monogenic forms of MetS described above, only *FTO* has been implicated [41,42].

Risk-Score Analysis

In risk-score analysis, the cumulative effect of multiple common small-effect polymorphisms is examined. The polymorphisms selected to be included in the risk score are typically selected from the results of previous GWAS. Each individual is assigned a score, which is the sum of the number of risk alleles at each of the group of polymorphisms. Each SNP is weighted by the observed effect on the trait of interest in the previously published data. Yaghootkar et al. created a risk score composed of 11 SNPs selected for previously reported association with insulin resistance [45]. The generated score was significantly associated with higher triglycerides, lower HDL, hepatic steatosis, reduced adiponectin, lower BMI, and increased risk of type 2 diabetes and coronary artery disease. Yaghootkar et al. suggest that patients with an uncommon combination of risk alleles represent a polygenic "lipodystrophy-like" phenotype [45].

FINDING THE MISSING HERITABILITY

Despite the success of GWAS of MetS components, the genetic variants identified to be associated with MetS components typically explain <10% of variation in the traits across the population, despite heritability measurements of >50% in family studies. Termed the "missing heritability," efforts to identify genetic variants that explain additional trait variation is the focus of a tremendous research effort. Examples of potential sources of missing heritability are rare variants, copy number variation (CNV), microRNA (*miR*), epigenetics and gene–gene, and gene–environment interactions.

CNV involves the duplication and deletion of genomic DNA, typically defined as >1000 base pairs in size. CNVs are common in the population and have been associated with extreme obesity [46]. *miR* is short transcribed RNA segments that bind complementary messenger RNA causing translational repression and transcript degradation. Increased *miR-33b* expression was found to lead to increased fatty acid synthesis and decreased cholesterol efflux to HDL, two of the hallmarks of MetS [47]. DNA methylation, an epigenetic modification, impacts gene expression, and genome-wide methylation studies have reported differences between type 2 diabetes patients and controls [48]. Sequencing of genes identified by GWAS reveals individuals with rare reduced- or loss-of-function mutations, especially in individuals in the extremes of the distribution of the trait [49]. Using next-generation sequencing techniques, exome sequencing, in which only the gene coding portion of the genome is sequenced, or gene panels, in which a curated collection of genes is sequenced, are becoming more accessible due to decreasing cost. NGS techniques allow for genotyping rare mutations, common SNPs and CNVs all in one experiment. Large databases of control sequences are publically available (http://exac.broadinstitute.org/), and large-scale sequencing efforts examining not only monogenic disease, but across the spectrum of complex traits, are currently ongoing.

TABLE 15.4 Genome-Wide Association Studies of Metabolic Syndrome

Reference	Population	Phenotype	Findings
Zabaneh et al. [43]	4560 Asian Indians	MetS (IDF 2005)	No SNPs associated with dichotomous MetS diagnosis
Kraja et al. [40]	22,161 Europeans	Pairs of MetS components	29 SNPs in 16 genes associated with two MetS components (APOA5, BUD13, ZNF259, LPL, CETP, GCKR, C2orf16, ZNF512, CCDC121, ABCB11, TFAP2B, TRIB1, LIPC, LOC100129150, LOC100128354, and LOC100129500). None associated with more than two components
Avery et al. [42]	19,458 European Americans 6287 African Americans	MetS phenotype domains (dyslipidemia, vascular dysfunction, inflammation, dysglycemia, central obesity, and thrombosis)	11 genes associated with at least two trait domains (APOC1, BRAP, PLCG1, GCKR, ABCB11, LPL, ABCA1, ABO, HNF1A, FTO, and SUGP1)
Kristiansson et al. [44]	2637 MetS cases, 7927 controls Finnish Europeans	MetS (IDF 2005)	APOA5 associated with MetS, none with more than 2 MetS components
Kraja et al. [41]	85,500 previously published Europeans	Eight metabolic and nine inflammatory markers, MetS (NCEP)	25 pleotropic genes associated with MetS and inflammation (MACF1, KIAA0754, GCKR, GRB14, COBLL1, LOC646736, BAZ1B, BCL7B, TBL2, MLXIPL, LPL, TRIB1, ZNF664, TOMM40, TFAP2B, HECTD4, PTPN11, FTO, SLC39A8, NELFE, SKIV2L, STK19, ATXN2, PDXDC1, and MC4R)

Since a relatively small percentage of variation in MetS components and MetS risk has been explained through variants identified in genetic studies, and environment undoubtedly plays a role in MetS etiology, research into potential gene–environment interactions is required. In order to ensure valid assessment of gene–environment interactions in studies of MetS, suitable sample sizes and appropriate statistical methodologies must be employed.

THE "THRIFTY-GENE" HYPOTHESIS

The "thrifty-gene" hypothesis has been proposed to explain the high prevalence of obesity, diabetes, and subsequent MetS in modern times. In the thrifty-gene hypothesis, genetic variants that lead to the accumulation of adipose stores for preservation of nutrition until times when it would be required are under positive selection, and thus becoming more common in the population. Thus, in current times of caloric excess, previously beneficial alleles have become deleterious. Contradicting this hypothesis is the fact that in times of starvation, death is primarily caused by infection, not loss of adipose tissue, and that obese individuals may be at higher risk of succumbing to predation. Furthermore, as many common genetic variants have now been identified that contribute to variation in obesity and diabetes, it has become possible to test the hypothesis by examining the obesity-, diabetes-, and metabolic-risk alleles. Work by Southam et al. examined the characteristics of risk alleles as either minor or major alleles, ancestral or derived alleles, as well as the population differentiation statistics (fixation statistic (F_{st})). Ultimately, they were able to uncover little evidence in support of the "thrifty-gene" hypothesis [21]. Additional efforts using SNPs identified in GWAS, the international HapMap project populations, and long-range haplotypes was also unable to find evidence of positive selection bias for SNPs associated with fat accumulation [50]. While providing an attractive accessible hypothesis in lay terms, little empiric evidence to support the thrifty-gene hypothesis has been developed to date.

CLINICAL IMPLICATIONS TO GENETIC FINDINGS IN METABOLIC SYNDROME

Genetic testing of common variants associated with small changes in MetS components, such as lipid concentration or fasting glucose, or associated with risk of MetS complications, such as coronary artery disease or type 2 diabetes, remains of little clinical utility at this point. Identification of large effect mutations may assist with diagnosis in individuals with monogenic forms of MetS, such as lipodystrophy. However, it is premature to currently recommend genetic testing in the context of general MetS diagnosis or treatment.

CONCLUSION

MetS is a complex, heterogeneous diagnosis with numerous causative and associated biochemical and environmental players including a pleotropic, polygenic architecture. Much can be learned about MetS pathophysiology by the identification and characterization of individuals with extreme disturbance of one of the MetS components. GWAS of the MetS components have been very successful, verifying known genes, and uncovering novel genes. Large GWAS meta-analyses and risk-score analyses are examining the pleotropic role of SNPs in MetS pathophysiology. As our understanding of MetS improves, we need to begin to understand how these pathways interact and combine to form MetS and the subsequent risk for atherosclerosis and diabetes.

ACKNOWLEDGMENTS

This work was supported by the Edith Schulich Vinet Research Chair, the Jacob J. Wolfe Distinguished Medical Research Chair, the Martha G. Blackburn Chair in Cardiovascular Research, and operating grants from the Canadian Institutes for Health Research (Foundation Program), the Heart and Stroke Foundation of Ontario (NA-6059, T-000353), and Genome Canada through Genome Quebec (award 4530).

GLOSSARY TERMS

Allelic heterogeneity Different mutations in the same gene cause the same phenotype.
Exome The portion of the genome transcribed into mature RNA.
Heritability The portion of trait variability than can be attributed to genetics.
Pleiotropy The situation in which a single gene affects multiple phenotypes, therefore mutations in the gene could affect multiple phenotypes.
Lipodystrophy A heterogeneous group of disorders characterized by selective or generalized atrophy of anatomical adipose tissue stores.

REFERENCES

[1] Alberti KG, Eckel RH, Grundy SM, Zimmet PZ, Cleeman JI, Donato KA, et al. Harmonizing the metabolic syndrome: a joint interim statement of the International Diabetes Federation Task Force on Epidemiology and Prevention; National Heart, Lung, and Blood Institute; American Heart Association; World Heart Federation; International Atherosclerosis Society; and International Association for the Study of Obesity. Circulation 2009;120(16):1640–5.

[2] Ford ES, Giles WH, Dietz WH. Prevalence of the metabolic syndrome among US adults: findings from the third National Health and Nutrition Examination Survey. JAMA 2002;287(3):356–9.

[3] Mente A, Yusuf S, Islam S, McQueen MJ, Tanomsup S, Onen CL, et al. Metabolic syndrome and risk of acute myocardial infarction a case-control study of 26,903 subjects from 52 countries. J Am Coll Cardiol 2010;55(21):2390–8.

[4] Samuel VT, Petersen KF, Shulman GI. Lipid-induced insulin resistance: unravelling the mechanism. Lancet. 2010;375(9733):2267–77.

[5] Pollex RL, Hegele RA. Genetic determinants of the metabolic syndrome. Nat Clin Pract Cardiovasc Med 2006;3(9):482–9.

[6] Cornier MA, Dabelea D, Hernandez TL, Lindstrom RC, Steig AJ, Stob NR, et al. The metabolic syndrome. Endocr Rev 2008;29(7):777–822.

[7] Mente A, Meyre D, Lanktree MB, Heydarpour M, Davis AD, Miller R, et al. Causal relationship between adiponectin and metabolic traits: a Mendelian randomization study in a multiethnic population. PLoS One 2013;8(6):e66808.

[8] Carmelli D, Cardon LR, Fabsitz R. Clustering of hypertension, diabetes, and obesity in adult male twins: same genes or same environments? Am J Hum Genet 1994;55(3):566–73.

[9] Sung J, Lee K, Song YM. Heritabilities of the metabolic syndrome phenotypes and related factors in Korean twins. J Clin Endocrinol Metab 2009;94(12):4946–52.

[10] Lin HF, Boden-Albala B, Juo SH, Park N, Rundek T, Sacco RL. Heritabilities of the metabolic syndrome and its components in the Northern Manhattan Family Study. Diabetologia 2005;48(10):2006–12.

[11] Bayoumi RA, Al-Yahyaee SA, Albarwani SA, Rizvi SG, Al-Hadabi S, Al-Ubaidi FF, et al. Heritability of determinants of the metabolic syndrome among healthy Arabs of the Oman family study. Obesity (Silver Spring) 2007;15(3):551–6.

[12] Khan RJ, Gebreab SY, Sims M, Riestra P, Xu R, Davis SK. Prevalence, associated factors and heritabilities of metabolic syndrome and its individual components in African Americans: the Jackson Heart Study. BMJ Open 2015;5(10):e008675.

[13] Vattikuti S, Guo J, Chow CC. Heritability and genetic correlations explained by common SNPs for metabolic syndrome traits. PLoS Genet 2012;8(3):e1002637.

[14] Moore SJ, Green JS, Fan Y, Bhogal AK, Dicks E, Fernandez BA, et al. Clinical and genetic epidemiology of Bardet–Biedl syndrome in Newfoundland: a 22-year prospective, population-based, cohort study. Am J Med Genet A 2005;132(4):352–60.

[15] Thauvin-Robinet C, Auclair M, Duplomb L, Caron-Debarle M, Avila M, St-Onge J, et al. PIK3R1 mutations cause syndromic insulin resistance with lipoatrophy. Am J Hum Genet 2013;93(1):141–9.

[16] Weedon MN, Ellard S, Prindle MJ, Caswell R, Lango Allen H, Oram R, et al. An in-frame deletion at the polymerase active site of POLD1 causes a multisystem disorder with lipodystrophy. Nat Genet 2013;45(8):947–50.

[17] Dyment DA, Smith AC, Alcantara D, Schwartzentruber JA, Basel-Vanagaite L, Curry CJ, et al. Mutations in PIK3R1 cause SHORT syndrome. Am J Hum Genet 2013;93(1):158–66.

[18] Payne F, Lim K, Girousse A, Brown RJ, Kory N, Robbins A, et al. Mutations disrupting the Kennedy phosphatidylcholine pathway in humans with congenital lipodystrophy and fatty liver disease. Proc Natl Acad Sci U S A 2014;111(24):8901–6.

[19] Agarwal AK, Garg A. A novel heterozygous mutation in peroxisome proliferator-activated receptor-gamma gene in a patient with familial partial lipodystrophy. J Clin Endocrinol Metab 2002;87(1):408–11.

[20] Hegele RA, Cao H, Frankowski C, Mathews ST, Leff T. PPARG F388L, a transactivation-deficient mutant, in familial partial lipodystrophy. Diabetes 2002;51(12):3586–90.

[21] Southam L, Soranzo N, Montgomery SB, Frayling TM, McCarthy MI, Barroso I, et al. Is the thrifty genotype hypothesis supported by evidence based on confirmed type 2 diabetes- and obesity-susceptibility variants? Diabetologia 2009;52(9):1846–51.

[22] Tan K, Pogozheva ID, Yeo GS, Hadaschik D, Keogh JM, Haskell-Leuvano C, et al. Functional characterization and structural modeling of obesity associated mutations in the melanocortin 4 receptor. Endocrinology. 2009;150(1):114–25.

[23] Farooqi IS, Keogh JM, Yeo GS, Lank EJ, Cheetham T, O'Rahilly S. Clinical spectrum of obesity and mutations in the melanocortin 4 receptor gene. N Engl J Med 2003;348(12):1085–95.

[24] Krude H, Biebermann H, Schnabel D, Tansek MZ, Theunissen P, Mullis PE, et al. Obesity due to proopiomelanocortin deficiency: three new cases and treatment trials with thyroid hormone and ACTH4-10. J Clin Endocrinol Metab 2003;88(10):4633–40.

[25] Yaswen L, Diehl N, Brennan MB, Hochgeschwender U. Obesity in the mouse model of pro-opiomelanocortin deficiency responds to peripheral melanocortin. Nat Med. 1999;5(9):1066–70.

[26] Krude H, Biebermann H, Luck W, Horn R, Brabant G, Gruters A. Severe early-onset obesity, adrenal insufficiency and red hair pigmentation caused by POMC mutations in humans. Nat Genet. 1998;19(2):155–7.

[27] Greenfield JR, Miller JW, Keogh JM, Henning E, Satterwhite JH, Cameron GS, et al. Modulation of blood pressure by central melanocortinergic pathways. N Engl J Med 2009;360(1):44–52.

[28] Nogueiras R, Wiedmer P, Perez-Tilve D, Veyrat-Durebex C, Keogh JM, Sutton GM, et al. The central melanocortin system directly controls peripheral lipid metabolism. J Clin Invest 2007;117(11):3475–88.

[29] Rahmouni K, Fath MA, Seo S, Thedens DR, Berry CJ, Weiss R, et al. Leptin resistance contributes to obesity and hypertension in mouse models of Bardet–Biedl syndrome. J Clin Invest 2008;118(4):1458–67.

[30] Seo S, Guo DF, Bugge K, Morgan DA, Rahmouni K, Sheffield VC. Requirement of Bardet–Biedl syndrome proteins for leptin receptor signaling. Hum Mol Genet 2009;18(7):1323–31.

[31] Collin GB, Cyr E, Bronson R, Marshall JD, Gifford EJ, Hicks W, et al. Alms1-disrupted mice recapitulate human Alstrom syndrome. Hum Mol Genet 2005;14(16):2323–33.

[32] Longo N, Wang Y, Smith SA, Langley SD, DiMeglio LA, Giannella-Neto D. Genotype-phenotype correlation in inherited severe insulin resistance. Hum Mol Genet 2002;11(12):1465–75.

[33] Bonnefond A, Philippe J, Durand E, Muller J, Saeed S, Arslan M, et al. Highly sensitive diagnosis of 43 monogenic forms of diabetes or obesity through one-step PCR-based enrichment in combination with next-generation sequencing. Diabetes Care 2014;37(2):460–7.

[34] Willer CJ, Schmidt EM, Sengupta S, Peloso GM, Gustafsson S, Kanoni S, et al. Discovery and refinement of loci associated with lipid levels. Nat Genet 2013;45(11):1274–83.

[35] Ehret GB, Munroe PB, Rice KM, Bochud M, Johnson AD, Chasman DI, et al. Genetic variants in novel pathways influence blood pressure and cardiovascular disease risk. Nature. 2011;478(7367):103–9.

[36] Locke AE, Kahali B, Berndt SI, Justice AE, Pers TH, Day FR, et al. Genetic studies of body mass index yield new insights for obesity biology. Nature. 2015;518(7538):197–206.

[37] Scott RA, Lagou V, Welch RP, Wheeler E, Montasser ME, Luan J, et al. Large-scale association analyses identify new loci influencing glycemic traits and provide insight into the underlying biological pathways. Nat Genet 2012;44(9):991–1005.

[38] Morris AP, Voight BF, Teslovich TM, Ferreira T, Segrè AV, Steinthorsdottir V, et al. Large-scale association analysis provides insights into the genetic architecture and pathophysiology of type 2 diabetes. Nat Genet 2012;44(9):981–90.

[39] Do R, Bailey SD, Desbiens K, Belisle A, Montpetit A, Bouchard C, et al. Genetic variants of FTO influence adiposity, insulin sensitivity, leptin levels, and resting metabolic rate in the Quebec Family Study. Diabetes 2008;57(4):1147–50.

[40] Kraja AT, Vaidya D, Pankow JS, Goodarzi MO, Assimes TL, Kullo IJ, et al. A bivariate genome-wide approach to metabolic syndrome: STAMPEED consortium. Diabetes 2011;60(4):1329–39.

[41] Kraja AT, Chasman DI, North KE, Reiner AP, Yanek LR, Kilpeläinen TO, et al. Pleiotropic genes for metabolic syndrome and inflammation. Mol Genet Metab 2014;112(4):317–38.

[42] Avery CL, He Q, North KE, Ambite JL, Boerwinkle E, Fornage M, et al. A phenomics-based strategy identifies loci on APOC1, BRAP, and PLCG1 associated with metabolic syndrome phenotype domains. PLoS Genet. 2011;7(10):e1002322.

[43] Zabaneh D, Balding DJ. A genome-wide association study of the metabolic syndrome in Indian Asian men. PLoS One 2010;5(8):e11961.

[44] Kristiansson K, Perola M, Tikkanen E, Kettunen J, Surakka I, Havulinna AS, et al. Genome-wide screen for metabolic syndrome susceptibility Loci reveals strong lipid gene contribution but no evidence for common genetic basis for clustering of metabolic syndrome traits. Circ Cardiovasc Genet 2012;5(2):242–9.

[45] Yaghootkar H, Scott RA, White CC, Zhang W, Speliotes E, Munroe PB, et al. Genetic evidence for a normal-weight "metabolically obese" phenotype linking insulin resistance, hypertension, coronary artery disease, and type 2 diabetes. Diabetes 2014;63(12):4369–77.

[46] Walters RG, Jacquemont S, Valsesia A, de Smith AJ, Martinet D, Andersson J, et al. A new highly penetrant form of obesity due to deletions on chromosome 16p11.2. Nature 2010;463(7281):671–5.

[47] Brown MS, Ye J, Goldstein JL. Medicine. HDL miR-ed down by SREBP introns. Science 2010;328(5985):1495–6.

[48] Dayeh T, Volkov P, Salö S, Hall E, Nilsson E, Olsson AH, et al. Genome-wide DNA methylation analysis of human pancreatic islets from type 2 diabetic and non-diabetic donors identifies candidate genes that influence insulin secretion. PLoS Genet 2014;10(3):e1004160.

[49] Johansen CT, Wang J, Lanktree MB, Cao H, McIntyre AD, Ban MR, et al. Excess of rare variants in genes identified by genome-wide association study of hypertriglyceridemia. Nat Genet 2010;42(8):684–7.

[50] Koh XH, Liu X, Teo YY. Can evidence from genome-wide association studies and positive natural selection surveys be used to evaluate the thrifty gene hypothesis in East Asians? PLoS One 2014;9(10):e110974.

Chapter 16

Autism Spectrum Disorder

Akanksha Saxena and Maria Chahrour
University of Texas Southwestern Medical Center, Dallas, TX, United States

Chapter Outline

INTRODUCTION

In 1943, child psychiatrist Leo Kanner first described the condition of 11 children with extreme withdrawal and inability to form normal relationships with others as early infantile autism. He characterized individuals with this condition as desiring extreme solitude and adhering to a strict routine. They could distract themselves for hours with simple, repetitive tasks and were easily alarmed by the slightest deviation from what they were accustomed to. These children varied in their social abilities, some unable to speak altogether [1]. A year after Kanner's report, Hans Asperger independently documented four of his patients who displayed those same characteristics, but differed markedly in their intellectual abilities by demonstrating advanced aptitudes in science and math. Despite this difference, Asperger described his patients as also having autism [2]. Today, a clinical diagnosis of autism effectively encompasses three main symptoms: deficits in verbal communication, impaired social interaction, and restricted interests, with autism spectrum disorder (ASD) representing a continuum of any one or a combination of these symptoms. The ASD core symptoms rarely occur in isolation and are associated with a host of comorbidities.

Genomic and Precision Medicine. DOI: http://dx.doi.org/10.1016/B978-0-12-800685-6.00016-3

ASD affects ~1% of the population and is four times more prevalent in males than in females [3,4]. The economic burden of ASD in the United States is $175 billion a year [5], exceeding the cost of cancer, stroke, and heart disease combined. ASD is projected to cost the United States economy $461 billion by 2025. Both genetic as well as environmental factors are involved in the etiology of ASD [6–8]. The past decade has witnessed major strides in the genetics of ASD, partly owing to advances in genomics and the study of complex disorders. Although effective and specific treatment options are still unavailable, genetic findings have provided insights into the molecular pathogenesis of ASD and inched us closer to developing effective therapeutics.

CLINICAL OVERVIEW

A key feature of ASD is its phenotypic heterogeneity. Patients present with varying degrees of severity in their core symptoms and manifest different comorbidities that can change over time. The most common cooccurrences with ASD are intellectual disability (ID) (~60%) and seizures (~20%). Other common phenotypes include anxiety, attention deficit hyperactivity disorder (ADHD), motor abnormalities, sleep disturbances, and gastrointestinal problems.

The most recent set of guidelines for classifying mental disorders, the fifth edition of the Diagnostic and Statistical Manual of Mental Disorders [9,10], combines four pervasive neurodevelopmental disorders that are significantly similar in their diagnostic criteria under the umbrella of ASD: autistic disorder, Asperger syndrome, pervasive developmental disorder not otherwise specified, and childhood disintegrative disorder.

THE COMPLEX GENETICS OF ASD

ASD is one of the most heritable of neuropsychiatric disorders based on twin studies that provided clear evidence for the genetic component of ASD. The most recent study conducted on a large-population-based cohort that was systematically screened calculated 77–99% concordance rate in diagnosis for monozygotic twins and 22–65% for dizygotic twins, resulting in heritability estimates of 56–95% [6]. Another population-based epidemiological study of ~2.5 million families assessed familial recurrence risk for ASD and found an increase in the individual's risk with increasing genetic relatedness to a patient with ASD and a heritability estimate of ~50% [11]. In all the studies to date, concordance in phenotype between monozygotic twins is incomplete, indicating that nongenetic factors, such as epigenetic and/or environmental factors, play a role in the etiology of ASD.

ASD is highly genetically heterogeneous, with no single genetic aberration accounting for more than 1% of cases. Genetic causes of ASD include large chromosomal abnormalities, submicroscopic deletions or duplications (copy number variants (CNVs)), and mutations in genes associated with neurodevelopmental syndromes that have one or more of the ASD phenotypes. The search for common variation in ASD has yet to yield any consistently reproducible results, and genome-wide association studies performed to date remain underpowered. Although two regions on chromosome 5 have been implicated [12–14], subsequent analyses did not confirm these loci [15–17]. It is estimated that common variation contributes 40–60% of the risk for ASD, and that there are likely many common variants, each with a very small effect size [15,18]. Given the genetic heterogeneity in ASD, much larger cohorts will be needed to identify common variants, in the order of tens of thousands of subjects. To date, rare variants in over 700 genes, both *de novo* and inherited, have been demonstrated to contribute to ASD risk, highlighting its complex genetic architecture.

Chromosomal Abnormalities and Copy Number Variants

Large chromosomal abnormalities have been reported in 5% of ASD cases [19,20], and the most frequent occur at 15q11-q13, 22q11.2, and 22q13.3. Duplications of 15q11–q13 occur in ~1% of all ASD cases and are maternally inherited [21,22]. Although abnormalities at these regions substantially contribute to ASD risk, they are not exclusive to ASD and are involved in a range of neurodevelopmental phenotypes. In addition to ASD, abnormalities in 15q11–q13 dosage result in Angelman syndrome (Mendelian Inheritance in Man (MIM) 105830), Prader–Willi syndrome (MIM 176270), ID, developmental delay, ataxia, and seizures. The 22q11.2 deletion can also result in DiGeorge syndrome (MIM 188400) or velocardiofacial syndrome (MIM 192430) and is associated with developmental delay and neurobehavioral abnormalities. Duplications of the same region can also give rise to a range of neurodevelopmental and neurobehavioral phenotypes [23–26]. Deletions in the 22q13.3 region result in Phelan–McDermid syndrome (MIM 606232), a developmental disorder with severe speech delays and autistic behaviors among other phenotypes [27,28], while duplications in the same region are associated with cases of ADHD, Asperger syndrome, and hyperkinetic neuropsychiatric phenotypes [29].

Advances in microarray technology allowed for a higher resolution scan of the genome and identification of smaller submicroscopic deletions and duplications (typically >1 kb in length). These CNVs are highly heterogeneous, can be inherited or arise *de novo*, and contribute to a range of disorders including ASD. The latest estimates indicate that rare *de novo* CNVs contribute to phenotype in ~5–10% of individuals with ASD [30,31]. Furthermore, homozygous CNVs have also been reported in ASD [32,33]. Loci across the genome have been confirmed in ASD, including 1q21.1, 7q11.23 (Williams–Beuren syndrome region), 15q11–13, 16p11.2, and 22q11.2 (DiGeorge syndrome region). The majority of CNVs have very low recurrence in ASD, and often times, a specific CNV can be unique to a single patient. Furthermore, although the overall burden of *de novo*

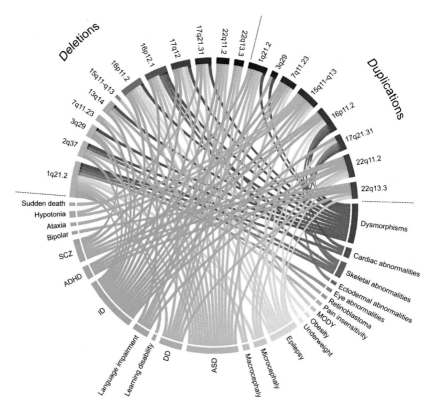

FIGURE 16.1 **ASD-associated copy number variants (CNVs) are highly pleiotropic.** Circos plot depicting the phenotypic diversity resulting from the most common CNVs associated with ASD susceptibility. Deletions and duplications of the same region can give rise to overlapping or distinct phenotypes. Dysmorphisms encompass variable systemic or craniofacial dysmorphisms. Skeletal abnormalities include limb, hand, or foot abnormalities. Ectodermal abnormalities include hirsutism and atopic dermatitis. Several CNVs are associated with specific syndromes that are discussed in the text. Abbreviations: *ADHD*, attention deficit hyperactivity disorder; *ASD*, autism spectrum disorder; *DD*, developmental delay; *ID*, intellectual disability; *MODY*, maturity onset diabetes of the young; *SCZ*, schizophrenia.

CNVs is higher in affected compared to unaffected individuals, these CNVs also occur in the unaffected individuals [34], making it difficult to determine which specific alterations are likely to be disease causing. In addition, ASD-associated CNVs are not specific to the core ASD phenotypes but are rather associated with a wide range of neurodevelopmental phenotypes including ID, epilepsy, and schizophrenia. The heterogeneity of CNV-associated phenotypes can also manifest within the same family. The CNVs most commonly reported in ASD, and the additional phenotypes associated with them, are depicted in Fig. 16.1. It remains to be determined whether this is due to the presence of additional mutations in the genome that act as modifiers. Despite the heterogeneity of CNVs in ASD, clinical chromosomal microarray (CMA) testing remains the single genetic test with the highest diagnostic yield in ASD [35].

One of the most recurring copy number changes in ASD involves the 16p11.2 locus [29]. Both deletions as well as duplications of 16p11.2 are associated with a spectrum of neurodevelopmental disorders. The deletions are frequently *de novo*, but can be inherited in ~50% of cases, from either mildly affected or clinically normal parents [36]. About one-third of deletion patients meet strict criteria for autism, with no additional syndromic phenotypes, another third of patients manifest some ASD phenotypes but not others, and the last third of patients do not have ASD at all. Instead, they may present with other neurodevelopmental conditions such as ID of varying severity or epilepsy, or they can be clinically normal [37]. Duplications of the identical region on 16p11.2 are even more heterogeneous, being associated with brain malformations in some cases (agenesis of the corpus callosum, microcephaly), schizophrenia, or ADHD. Furthermore, copy number alterations at 16p11.2 give rise to mirrored phenotypes, with the deletion being associated with macrocephaly and obesity, and the duplication with microcephaly and being underweight. Studies in zebrafish demonstrated that *KCTD13* within the 16p11.2 locus is responsible for driving the brain size phenotype [38].

Syndromic Genes

Fragile X Syndrome

The fragile X mental retardation 1 (*FMR1*) gene is the most commonly disrupted single gene in ASD, accounting for ~2% of cases [39]. Fragile X syndrome (FXS; MIM 300624) results from the expansion of a Cytosine Guanine Guanine (CGG) trinucleotide repeat at the 5′ end of *FMR1* [40]. FXS is associated with severe ID, characteristic dysmorphisms, and the majority of males with FXS (~90%) demonstrate at least one ASD phenotype.

Rett Syndrome

Rett syndrome (RTT; MIM 312750) is a neurodevelopmental disorder caused by mutations in the X-linked gene *MECP2*, which encodes the transcriptional modulator methyl-CpG binding protein 2 [41]. RTT is characterized by normal development in the first 6–18 months of life, followed by developmental arrest, loss of any acquired speech, social withdrawal, and the replacement of purposeful hand use with stereotypichand wringing movements. Alterations in *MECP2* dosage can result in a range of neurobehavioral abnormalities, including *MECP2* duplication syndrome [42,43], ASD, mild learning disabilities, X-linked ID, and infantile encephalopathy [44,45]. *MECP2* mutations have been reported in ~4% of females diagnosed with autism [34], and males with *MECP2* duplications often present with autism. MeCP2 levels are critical for neurodevelopment, and it is likely that as clinical whole exome (WES) and whole genome sequencing

(WGS) become common practice, additional *MECP2* mutations, including hypomorphic alleles and noncoding regulatory mutations, will be identified that give rise to various neuropsychiatric phenotypes.

Tuberous Sclerosis Complex

Tuberous sclerosis complex (TSC) is an inherited, autosomal dominant disorder that arises from mutations in either the TSC1 or TSC2 tumor suppressor genes. It is characterized by the growth of benign tumors throughout various parts of the body, including the brain. TSC neurological manifestations include epilepsy, ID, behavioral abnormalities, learning difficulties, and autism. Approximately 50% of TSC patients present with ASD, and up to 5% of individuals with ASD have a diagnosis of TSC [46,47]. *TSC1* and *TSC2* encode for the heterodimeric complex that negatively regulates the mammalian target of rapamycin (mTOR) signaling pathway, a key network that regulates various cellular processes, including cell growth, proliferation, and protein synthesis. Mutations in *TSC1* or *TSC2* lead to overactivation of the mTOR signaling pathway. So far, there is no clear understanding of the determinants of ASD phenotypes in patients with TSC, although studies in *Tsc1* and *Tsc2* mouse models suggest that abnormal cerebellar function underlies the ASD features [48].

PTEN-*Related Disorders*

Mutations in the gene encoding the tumor suppressor phosphatase and tensin homolog, *PTEN*, have been identified in ~7% of patients with ASD and macrocephaly greater than three standard deviations above the mean [49]. Genetic disorders caused by *PTEN* germline haploinsufficiency give rise to a group of tumor syndromes, including Cowden syndrome and Bannayan–Riley–Ruvalcaba syndrome, among others. PTEN functions as a negative regulator of the mTOR signaling pathway, and loss of PTEN results in a constitutively active signaling cascade.

Neurofibromatosis

Neurofibromatosis type 1 (NF1; MIM 162200) arises from heterozygous mutations in the neurofibromin gene, *NF1*. Patients present with spots of discoloration on the skin (cafe-au-lait spots), melanocytic hamartomas affecting the iris, fibromatous tumors of the skin, bone deformities, and various benign and malignant tumors of the central nervous system. In addition, NF1 patients develop a range of neurodevelopmental abnormalities, including cognitive and motor deficits [50]. Recent data suggests that up to 40% of NF1 patients have ASD phenotypes, with more pronounced social behavior impairments, fewer repetitive behaviors, and a higher cooccurrence of ADHD [51,52]. NF1 functions as a negative regulator of the proto-oncogene Ras, and mutations in *NF1* result in overactivation of the Ras-MAPK signaling pathway that regulates various key cellular processes including cell growth and proliferation. Abnormalities in this pathway, or RASopathies, are associated with ASD phenotypes [53], suggesting the involvement of this pathway in ASD etiology.

Rare Single Nucleotide Variants, *De Novo* and Inherited

As syndromic genes began emerging and were identified as major contributors to ASD, several groups focused on identifying rare mutations in nonsyndromic, or idiopathic, ASD. Genes encoding several key components of the synapse were initially identified based on small deletions in patients with ASD and in large pedigrees with ASD and ID. It is important to note that mutations in these nonsyndromic ASD genes are not strictly associated with the core phenotypes of ASD and can result in other neurodevelopmental phenotypes. Just as some mutations in syndromic genes can also give rise to nonsyndromic ASD, mutations in "nonsyndromic" genes can be associated with additional phenotypes. Several pathways have emerged as being disrupted in ASD, including synaptic signaling and chromatin remodeling.

The first nonsyndromic ASD gene, *NLGN4X* (neuroligin (NLGN) 4, X-linked), was identified based on *de novo* deletions spanning the gene in patients with ASD and mutations in a family with ASD [54,55]. Subsequently, additional *NLGN4X* mutations were identified, not only in patients with ASD but also with ASD and ID [56]. A second family member, *NLGN3*, is a strong ASD gene candidate. NLGNs comprise a group of postsynaptic cell adhesion molecules that play a significant role in the development and function of synapses. Neurexins, the presynaptic binding partners for NLGNs, are also mutated in ASD. Disruption of the gene encoding the neurexin family member, contactin-associated protein 2 (*CNTNAP2*), was initially identified in a family with Tourette syndrome, obsessive–compulsive disorder (OCD), ID, speech abnormalities, and developmental delay. The children inherited part of an abnormal chromosome 7 that interrupted *CNTNAP2*, arising from a complex rearrangement in the father who also had OCD [57]. Other disruptions of *CNTNAP2* have been identified in ID and ASD, and recessive mutations result in cortical dysplasia, focal epilepsy, and macrocephaly [58]. Recessive mutations in another member of the neurexin family, *NRXN1* (neurexin 1), are associated with Pitt–Hopkins-like syndrome 2 (MIM 614325), characterized by severe ID, epilepsy, and autistic behaviors, while heterozygous deletions are associated with ASD. Of note is the high rate of asymptomatic carriers of *NRXN1* deletions, highlighting the effect of genetic background in modifying the penetrance of ASD phenotypes.

Members of the Shank family are scaffolding proteins localized in the postsynaptic density and that are crucial to synapse development. In addition to regulating dynamics of the cytoskeleton, they gather and anchor synaptic proteins and associate postsynaptic receptors with their downstream signaling components. Mutations in genes encoding members of the Shank family, *SHANK1*, *SHANK2*, and *SHANK3*, have been implicated in ASD [59]. *SHANK3* haploinsufficiency causes Phelan–McDermid syndrome [28], while duplications of the 22q13 region spanning *SHANK3* have been found in individuals with schizophrenia, bipolar, epilepsy, ADHD, and developmental delay [28]. *SHANK3* was the first family member associated with ASD [28] and has the strongest genetic evidence supporting a causal role in ASD [59]. Mutations in *SHANK1* and *SHANK2* have also been associated with ASD [60], highlighting the role of synaptic signaling pathways in ASD etiology.

De Novo Single Nucleotide Variants

Large-scale WES studies indicate that there is a significant excess of *de novo* single nucleotide variants (SNVs) in ASD cases compared to controls. The data suggest that *de novo* mutations are likely to cause disease in at least ~5% of ASD cases, and that between 370 and 1034 genes are affected by these mutations in total [61–64]. More recent studies provided strong evidence for *de novo* loss of function SNVs affecting a subset of these genes (~100 genes). In addition to genes involved in synaptic signaling and chromatin remodeling, other genes that have emerged include transcriptional regulators [65] and genes that function in early embryonic development, particularly cortical development [66]. In Fig. 16.2, we list ASD genes that have strong evidence linking them to disease, and we categorize them based on their associated level of confidence. Genes with *de novo* ASD mutations that are strong candidates for being disease causing include: *ANK2, ARID1B, ASH1L, CHD8, CUL3, DYRK1A, GRIN2B, KATNAL2, POGZ, SUV420H1,* and *TBR1*.

Inherited Single Nucleotide Variants

The contribution of rare recessive mutations to ASD has been highlighted in recent WES studies, with estimates of a ~5% contribution of complete loss of function variants to ASD risk [68]. It is important to note that to date the contribution of recessive missense variants to ASD has not been quantified, as this is difficult since assessing the functional impact of such variants is not straightforward. The impact of recessive mutations on ASD risk will likely be higher than 5%, and it is possible that recessive mutations might contribute to ASD susceptibility and phenotype penetrance as either protective or risk-conferring alleles. Rare recessive ASD mutations have been identified in consanguineous families [69,70], where the parents are closely related and their offspring have higher homozygosity in their genome and are at an increased risk of recessive disorders. Examples of ASD genes with recessive mutations include *SLC9A9* (solute carrier family 9 member A9) [32] and *BCKDK* (branched-chain ketoacid dehydrogenase kinase) [70], both identified in consanguineous families with autism and epilepsy. *SLC9A9* encodes an endosomal sodium/proton exchanger, while the protein encoded by *BCKDK* functions in branched-chain amino acid metabolism. WES studies in nonconsanguineous ("outbred") families also showed a role for recessive mutations in ASD [68,71] and identified candidate ASD

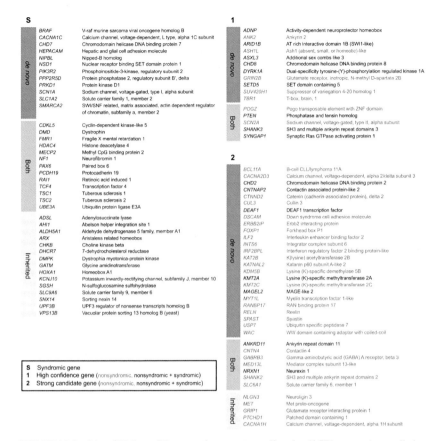

FIGURE 16.2   List of high confidence autism spectrum disorder (ASD) genes. A compilation of ASD genes categorized on a graded scale of confidence according to the association between mutations within those genes and an increased risk of ASD. The scale is based on the Simons Foundation Gene Scoring Module [67]. "S" represents genes identified on the basis of syndromic ASD, "1" represents ASD genes with the highest confidence, and "2" represents strong candidates. Genes are subcategorized by the type of ASD-associated variants reported for them in the current literature: *de novo*, inherited, or both. Some genes in categories 1 and 2 are also associated with syndromes (black), while others have no reported cases of syndromic ASD associated with them to date (blue) (gray in print versions). Information on reported *de novo* and inherited variants, and on gene associations with syndromic and nonsyndromic ASD, was curated from the Human Gene Module of the SFARI Gene database (https://gene.sfari.org/autdb/HG_Home.do) and through literature searches.

genes, including a missense mutation in a gene encoding an E3 ubiquitin ligase, *UBE3B* [71], that was subsequently associated with severe ID, microcephaly, and facial dysmorphisms [72]. Furthermore, some of the recessive ASD mutations are hypomorphic alleles of syndromic genes that lead to severe neurological abnormalities when completely inactivated, but result in ASD when retaining partial activity [69]. Examples of these mutations include missense variants in the genes *AMT* (aminomethyltransferase; the gene responsible for nonketotic hyperglycinemia), *PEX7* (peroxisomal biogenesis factor 7; the gene mutated in rhizomelic chondrodysplasia punctata), and *VPS13B* (vacuolar protein sorting 13 homolog B (yeast); the Cohen syndrome gene) [69]. Similar to other types of ASD mutations, recessive ASD mutations reflect the genetic heterogeneity of ASD and affect genes that function in various pathways. A recently characterized recessive ASD gene, *CC2D1A* (coiled-coil and C2-domain containing 1A), encodes a molecular scaffold that regulates endosomal trafficking and signaling, functions in several key biochemical pathways, and is critical for normal neuronal differentiation and function [73]. Mutations in *CC2D1A* have been identified in families with severe ID, ASD, and seizures, often with phenotypic heterogeneity within the same family [74].

GENOMICS AND DIAGNOSIS IN ASD

The current knowledge in ASD genetics estimates that ~35% of clinical cases will receive a molecular diagnosis [35,75]. Even though the majority of molecular findings in ASD will not result in treatment, the knowledge gained provides families with a definitive diagnosis, allowing access to the appropriate and necessary medical services, and recurrence-risk counseling. An identified molecular basis also allows clinicians to recognize and manage medical risks and conditions associated with syndromic cases, ultimately preventing morbidity in some cases.

The latest guidelines from the American College of Medical Genetics and Genomics (ACMG) recommend a tiered approach in evaluating individuals with ASD [35]. Advances in microarray technology have improved conventional cytogenetic testing and enhanced the diagnostic yield in ASD cases with underlying chromosomal abnormalities and CNVs. Currently, CMA analysis (both array comparative genomic hybridization and single nucleotide polymorphism arrays) is the single test with the highest diagnostic yield in ASD and is the recommended first test for a genetic evaluation. Single-gene testing is indicated for syndromic ASD cases, depending on the patient's phenotype and gender. Although metabolic disorders are rarely associated with an ASD phenotype, recent data suggests the importance of testing for metabolic disorders in ASD. Metabolic disorders are generally recessive and often associated with additional phenotypes (seizures, movement disorders, and regression, among others). However, some individuals with ASD might not present with the clinical manifestation of a metabolic disorder, for example, if they carry hypomorphic alleles in metabolic disorder genes [69]. The ACMG recommends a metabolic evaluation for ASD patients with unidentified etiology. We summarize the patient pathway in Fig. 16.3.

With current technologies in clinical genetic testing, and given the genetic heterogeneity of ASD, family history remains the best predictor for the risk of

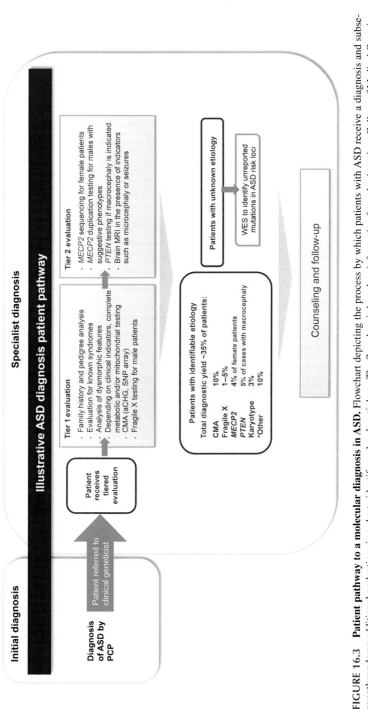

FIGURE 16.3 Patient pathway to a molecular diagnosis in ASD. Flowchart depicting the process by which patients with ASD receive a diagnosis and subsequently undergo additional evaluations in order to identify molecular etiology. The flowchart is based on information from the American College of Medical Genetics and Genomics latest recommendations [35]. Patients initially receive a diagnosis of ASD from their primary care physician (PCP), who discusses the process, expectations, and possible diagnostic outcomes of a clinical genetic evaluation. Using phenotypic criteria and testing, a clinical geneticist then performs a tiered evaluation to screen for causes of syndromic and/or nonsyndromic ASD. Both groups of patients, those with and without an identifiable etiology, receive counseling, management, and follow-up, which is coordinated between clinical geneticists and PCPs. In addition, some patients with an unknown etiology might receive whole exome sequencing to identify unreported mutations in ASD risk loci. Abbreviations: *aCGH*, array comparative genomic hybridization; *ASD*, autism spectrum disorder; *CMA*, chromosomal microarray; *MRI*, magnetic resonance imaging; *SNP*, single nucleotide polymorphism; *WES*, whole exome sequencing.

sibling recurrence. As advances in high throughput next-generation sequencing technology continue to dissect the molecular pathways underlying ASD, clinical WES and eventually WGS applications will no doubt improve the rate of genetic diagnosis. The incomplete penetrance of ASD-associated mutations, variable expressivity, and pleiotropy of ASD genes will eventually be addressed, and we will be able to assign the contribution of specific gene(s) to phenotype(s).

CONCLUSION

ASD genes identified to date encode molecules involved in synaptic development and function, chromatin remodeling and transcriptional control, and protein synthesis and degradation, among other mechanisms. Large-scale sequencing efforts have contributed to our understanding of the complex architecture of ASD genetics. However, the role of rare inherited variation associated with nonsyndromic ASD is still understudied compared to that of rare *de novo* variation, and requires further efforts. The next emerging phase in the field is the development of high-throughput functional validation screens to assess the impact of identified variants on protein function and phenotype development. Functional validation is especially important for missense variants that are currently largely ignored unless they occur in known disease-causing genes. Moreover, studying the normal function of ASD risk genes in model organisms will clearly elucidate the molecular workings of brain function and development and inform us on the molecular basis of social behavior and language. For example, to date, very few genes have been identified that are master regulators of language development (e.g., *FOXP2*) [76]. Advances in ASD genetics are also beginning to define specific pathways and neuronal circuits disrupted in groups of ASD patients sharing common molecular etiologies. As more genes are identified and categorized, the development of precise therapeutics targeted to these subsets of patients will be feasible. On the other hand, with the genetic landscape becoming more defined, individuals with a molecular finding can be reverse-phenotyped and grouped into better-characterized clinical subtypes. Early examples include the clinical characterization of individuals with mutations in *MAGEL2* [77], *CHD8* [78], or *POGZ* [79]. And finally, as the field of translational bioinformatics (TBI) matures, it holds promise for ASD and other complex neurobehavioral disorders. By merging information and data from genomics, statistical genetics, and clinical informatics, including electronic medical records and data collected from clinical wearables, TBI will allow complex genetics to be translated into individualized treatments.

REFERENCES

[1] Kanner L. Autistic disturbances of affective contact. Nerv Child 1943;2:217–50.
[2] Asperger H. Die autistisehen psychopathen im kindesalter. Arch Psych Nervenkrankh 1944;117:76–136.

[3] Elsabbagh M, Divan G, Koh YJ, Kim YS, Kauchali S, Marcin C, et al. Global prevalence of autism and other pervasive developmental disorders. Autism Res 2012;5(3):160–79.

[4] Centers for Disease Control and Prevention Prevalence of autism spectrum disorders—autism and developmental disabilities monitoring network, United States, 2006. MMWR Surveill Summ 2009;58(10):1–20.

[5] Leigh JP, Du J. Brief report: forecasting the economic burden of autism in 2015 and 2025 in the United States. J Autism Dev Disord 2015;45(12):4135–9.

[6] Colvert E, Tick B, McEwen F, Stewart C, Curran SR, Woodhouse E, et al. Heritability of autism spectrum disorder in a UK population-based twin sample. JAMA Psychiatry 2015;72(5):415–23.

[7] Hallmayer J, Cleveland S, Torres A, Phillips J, Cohen B, Torigoe T, et al. Genetic heritability and shared environmental factors among twin pairs with autism. Arch Gen Psychiatry 2011;68(11):1095–102.

[8] Gaugler T, Klei L, Sanders SJ, Bodea CA, Goldberg AP, Lee AB, et al. Most genetic risk for autism resides with common variation. Nat Genet 2014;46(8):881–5.

[9] Regier DA, Kuhl EA, Kupfer DJ. The DSM-5: classification and criteria changes. World Psychiatry 2013;12(2):92–8.

[10] Association AP. Diagnostic and statistical manual of mental disorders, 5 ed. Washington, DC: American Psychiatric Press; 2013.

[11] Sandin S, Lichtenstein P, Kuja-Halkola R, Larsson H, Hultman CM, Reichenberg A. The familial risk of autism. JAMA 2014;311(17):1770–7.

[12] Wang K, Zhang H, Ma D, Bucan M, Glessner JT, Abrahams BS, et al. Common genetic variants on 5p14.1 associate with autism spectrum disorders. Nature 2009;459(7246):528–33.

[13] Weiss LA, Arking DE, Daly MJ, Chakravarti A. A genome-wide linkage and association scan reveals novel loci for autism. Nature 2009;461(7265):802–8.

[14] Ma D, Salyakina D, Jaworski JM, Konidari I, Whitehead PL, Andersen AN, et al. A genome-wide association study of autism reveals a common novel risk locus at 5p14.1. Ann Hum Genet 2009;73(Pt 3):263–73.

[15] Anney R, Klei L, Pinto D, Almeida J, Bacchelli E, Baird G, et al. Individual common variants exert weak effects on the risk for autism spectrum disorders. Hum Mol Genet 2012;21(21):4781–92.

[16] Anney R, Klei L, Pinto D, Regan R, Conroy J, Magalhaes TR, et al. A genome-wide scan for common alleles affecting risk for autism. Hum Mol Genet 2010;19(20):4072–82.

[17] Xia K, Guo H, Hu Z, Xun G, Zuo L, Peng Y, et al. Common genetic variants on 1p13.2 associate with risk of autism. Mol Psychiatry 2014;19(11):1212–9.

[18] Klei L, Sanders SJ, Murtha MT, Hus V, Lowe JK, Willsey AJ, et al. Common genetic variants, acting additively, are a major source of risk for autism. Mol Autism 2012;3(1):9.

[19] Wassink TH, Piven J, Patil SR. Chromosomal abnormalities in a clinic sample of individuals with autistic disorder. Psychiatr Genet 2001;11(2):57–63.

[20] Kielinen M, Rantala H, Timonen E, Linna SL, Moilanen I. Associated medical disorders and disabilities in children with autistic disorder: a population-based study. Autism 2004;8(1):49–60.

[21] Cook Jr. EH, Lindgren V, Leventhal BL, Courchesne R, Lincoln A, Shulman C, et al. Autism or atypical autism in maternally but not paternally derived proximal 15q duplication. Am J Hum Genet 1997;60(4):928–34.

[22] Depienne C, Moreno-De-Luca D, Heron D, Bouteiller D, Gennetier A, Delorme R, et al. Screening for genomic rearrangements and methylation abnormalities of the 15q11–q13 region in autism spectrum disorders. Biol Psychiatry 2009;66(4):349–59.

[23] Vorstman JA, Morcus ME, Duijff SN, Klaassen PW, Heineman-de Boer JA, Beemer FA, et al. The 22q11.2 deletion in children: high rate of autistic disorders and early onset of psychotic symptoms. J Am Acad Child Adolesc Psychiatry 2006;45(9):1104–13.

[24] Fine SE, Weissman A, Gerdes M, Pinto-Martin J, Zackai EH, McDonald-McGinn DM, et al. Autism spectrum disorders and symptoms in children with molecularly confirmed 22q11.2 deletion syndrome. J Autism Dev Disord 2005;35(4):461–70.

[25] Mukaddes NM, Herguner S. Autistic disorder and 22q11.2 duplication. World J Biol Psychiatry 2007;8(2):127–30.

[26] Ou Z, Berg JS, Yonath H, Enciso VB, Miller DT, Picker J, et al. Microduplications of 22q11.2 are frequently inherited and are associated with variable phenotypes. Genet Med 2008;10(4):267–77.

[27] Prasad C, Prasad AN, Chodirker BN, Lee C, Dawson AK, Jocelyn LJ, et al. Genetic evaluation of pervasive developmental disorders: the terminal 22q13 deletion syndrome may represent a recognizable phenotype. Clin Genet 2000;57(2):103–9.

[28] Phelan MC, Rogers RC, Saul RA, Stapleton GA, Sweet K, McDermid H, et al. 22q13 deletion syndrome. Am J Med Genet 2001;101(2):91–9.

[29] Han K, Holder Jr. JL, Schaaf CP, Lu H, Chen H, Kang H, et al. SHANK3 overexpression causes manic-like behaviour with unique pharmacogenetic properties. Nature 2013;503(7474):72–7.

[30] Sanders SJ, He X, Willsey AJ, Ercan-Sencicek AG, Samocha KE, Cicek AE, et al. Insights into autism spectrum disorder genomic architecture and biology from 71 risk loci. Neuron 2015;87(6):1215–33.

[31] Pinto D, Pagnamenta AT, Klei L, Anney R, Merico D, Regan R, et al. Functional impact of global rare copy number variation in autism spectrum disorders. Nature 2010;466(7304):368–72.

[32] Morrow EM, Yoo SY, Flavell SW, Kim TK, Lin Y, Hill RS, et al. Identifying autism loci and genes by tracing recent shared ancestry. Science 2008;321(5886):218–23.

[33] Levy D, Ronemus M, Yamrom B, Lee YH, Leotta A, Kendall J, et al. Rare de novo and transmitted copy-number variation in autistic spectrum disorders. Neuron 2011;70(5):886–97.

[34] Sebat J, Lakshmi B, Troge J, Alexander J, Young J, Lundin P, et al. Large-scale copy number polymorphism in the human genome. Science 2004;305(5683):525–8.

[35] Schaefer GB, Mendelsohn NJ, Professional P, Guidelines C. Clinical genetics evaluation in identifying the etiology of autism spectrum disorders: 2013 guideline revisions. Genet Med 2013;15(5):399–407.

[36] Weiss LA, Shen Y, Korn JM, Arking DE, Miller DT, Fossdal R, et al. Association between microdeletion and microduplication at 16p11.2 and autism. N Engl J Med 2008;358(7):667–75.

[37] Hanson E, Nasir RH, Fong A, Lian A, Hundley R, Shen Y, et al. Cognitive and behavioral characterization of 16p11.2 deletion syndrome. J Dev Behav Pediatr 2010;31(8):649–57.

[38] Golzio C, Willer J, Talkowski ME, Oh EC, Taniguchi Y, Jacquemont S, et al. KCTD13 is a major driver of mirrored neuroanatomical phenotypes of the 16p11.2 copy number variant. Nature 2012;485(7398):363–7.

[39] Zafeiriou DI, Ververi A, Dafoulis V, Kalyva E, Vargiami E. Autism spectrum disorders: the quest for genetic syndromes. Am J Med Genet B Neuropsychiatr Genet 2013;162B(4):327–66.

[40] Pieretti M, Zhang FP, Fu YH, Warren ST, Oostra BA, Caskey CT, et al. Absence of expression of the FMR-1 gene in fragile X syndrome. Cell 1991;66(4):817–22.

[41] Amir RE, Van den Veyver IB, Wan M, Tran CQ, Francke U, Zoghbi HY. Rett syndrome is caused by mutations in X-linked MECP2, encoding methyl-CpG-binding protein 2. Nat Genet 1999;23(2):185–8.

[42] Ramocki MB, Tavyev YJ, Peters SU. The MECP2 duplication syndrome. Am J Med Genet A 2010;152A(5):1079–88.

[43] Ramocki MB, Peters SU, Tavyev YJ, Zhang F, Carvalho CM, Schaaf CP, et al. Autism and other neuropsychiatric symptoms are prevalent in individuals with MeCP2 duplication syndrome. Ann Neurol 2009;66(6):771–82.

[44] Chahrour M, Zoghbi HY. The story of Rett syndrome: from clinic to neurobiology. Neuron 2007;56(3):422–37.

[45] Lugtenberg D, Kleefstra T, Oudakker AR, Nillesen WM, Yntema HG, Tzschach A, et al. Structural variation in Xq28: MECP2 duplications in 1% of patients with unexplained XLMR and in 2% of male patients with severe encephalopathy. Eur J Hum Genet 2009;17(4):444–53.

[46] Wiznitzer M. Autism and tuberous sclerosis. J Child Neurol 2004;19(9):675–9.

[47] Jeste SS, Sahin M, Bolton P, Ploubidis GB, Humphrey A. Characterization of autism in young children with tuberous sclerosis complex. J Child Neurol 2008;23(5):520–5.

[48] Sundberg M, Sahin M. Cerebellar development and autism spectrum disorder in tuberous sclerosis complex. J Child Neurol 2015;30(14):1954–62.

[49] Mcbride KL, Varga EA, Pastore MT, Prior TW, Manickam K, Atkin JF, et al. Confirmation study of PTEN mutations among individuals with autism or developmental delays/mental retardation and macrocephaly. Autism Res 2010;3(3):137–41.

[50] Williams VC, Lucas J, Babcock MA, Gutmann DH, Korf B, Maria BL. Neurofibromatosis type 1 revisited. Pediatrics 2009;123(1):124–33.

[51] Plasschaert E, Descheemaeker MJ, Van Eylen L, Noens I, Steyaert J, Legius E. Prevalence of autism spectrum disorder symptoms in children with neurofibromatosis type 1. Am J Med Genet B Neuropsychiatr Genet 2015;168B(1):72–80.

[52] Constantino JN, Zhang Y, Holzhauer K, Sant S, Long K, Vallorani A, et al. Distribution and within-family specificity of quantitative autistic traits in patients with neurofibromatosis type I. J Pediatr 2015;167(3):621–6. e621.

[53] Adviento B, Corbin IL, Widjaja F, Desachy G, Enrique N, Rosser T, et al. Autism traits in the RASopathies. J Med Genet 2014;51(1):10–20.

[54] Thomas NS, Sharp AJ, Browne CE, Skuse D, Hardie C, Dennis NR. Xp deletions associated with autism in three females. Hum Genet 1999;104(1):43–8.

[55] Jamain S, Quach H, Betancur C, Rastam M, Colineaux C, Gillberg IC, et al. Mutations of the X-linked genes encoding neuroligins NLGN3 and NLGN4 are associated with autism. Nat Genet 2003;34(1):27–9.

[56] Laumonnier F, Bonnet-Brilhault F, Gomot M, Blanc R, David A, Moizard MP, et al. X-linked mental retardation and autism are associated with a mutation in the NLGN4 gene, a member of the neuroligin family. Am J Hum Genet 2004;74(3):552–7.

[57] Verkerk AJ, Mathews CA, Joosse M, Eussen BH, Heutink P, Oostra BA, et al. CNTNAP2 is disrupted in a family with Gilles de la Tourette syndrome and obsessive compulsive disorder. Genomics 2003;82(1):1–9.

[58] Strauss KA, Puffenberger EG, Huentelman MJ, Gottlieb S, Dobrin SE, Parod JM, et al. Recessive symptomatic focal epilepsy and mutant contactin-associated protein-like 2. N Engl J Med 2006;354(13):1370–7.

[59] Guilmatre A, Huguet G, Delorme R, Bourgeron T. The emerging role of SHANK genes in neuropsychiatric disorders. Dev Neurobiol 2014;74(2):113–22.

[60] Leblond CS, Nava C, Polge A, Gauthier J, Huguet G, Lumbroso S, et al. Meta-analysis of SHANK mutations in autism spectrum disorders: a gradient of severity in cognitive impairments. PLoS Genet 2014;10(9):e1004580.

[61] Sanders SJ, Murtha MT, Gupta AR, Murdoch JD, Raubeson MJ, Willsey AJ, et al. De novo mutations revealed by whole-exome sequencing are strongly associated with autism. Nature 2012;485(7397):237–41.

[62] Neale BM, Kou Y, Liu L, Ma'ayan A, Samocha KE, Sabo A, et al. Patterns and rates of exonic de novo mutations in autism spectrum disorders. Nature 2012;485(7397):242–5.

[63] O'Roak BJ, Deriziotis P, Lee C, Vives L, Schwartz JJ, Girirajan S, et al. Exome sequencing in sporadic autism spectrum disorders identifies severe de novo mutations. Nature genetics 2011;43(6):585–9.

[64] Iossifov I, Ronemus M, Levy D, Wang Z, Hakker I, Rosenbaum J, et al. De novo gene disruptions in children on the autistic spectrum. Neuron 2012;74(2):285–99.

[65] De Rubeis S, He X, Goldberg AP, Poultney CS, Samocha K, Cicek AE, et al. Synaptic, transcriptional and chromatin genes disrupted in autism. Nature 2014;515(7526):209–15.

[66] Parikshak NN, Luo R, Zhang A, Won H, Lowe JK, Chandran V, et al. Integrative functional genomic analyses implicate specific molecular pathways and circuits in autism. Cell 2013;155(5):1008–21.

[67] Basu SN, Kollu R, Banerjee-Basu S. AutDB: a gene reference resource for autism research. Nucl Acids Res 2009;37(Database issue):D832–836.

[68] Lim ET, Raychaudhuri S, Sanders SJ, Stevens C, Sabo A, Macarthur DG, et al. Rare complete knockouts in humans: population distribution and significant role in autism spectrum disorders. Neuron 2013;77(2):235–42.

[69] Yu TW, Chahrour MH, Coulter ME, Jiralerspong S, Okamura-Ikeda K, Ataman B, et al. Using whole-exome sequencing to identify inherited causes of autism. Neuron 2013;77(2):259–73.

[70] Novarino G, El-Fishawy P, Kayserili H, Meguid NA, Scott EM, Schroth J, et al. Mutations in BCKD-kinase lead to a potentially treatable form of autism with epilepsy. Science 2012;338(6105):394–7.

[71] Chahrour MH, Yu TW, Lim ET, Ataman B, Coulter ME, Hill RS, et al. Whole-exome sequencing and homozygosity analysis implicate depolarization-regulated neuronal genes in autism. PLoS Genet 2012;8(4):e1002635.

[72] Basel-Vanagaite L, Dallapiccola B, Ramirez-Solis R, Segref A, Thiele H, Edwards A, et al. Deficiency for the ubiquitin ligase UBE3B in a blepharophimosis-ptosis-intellectual-disability syndrome. Am J Hum Genet 2012;91(6):998–1010.

[73] Oaks AW, Zamarbide M, Tambunan DE, Santini E, Di Costanzo S, Pond HL, et al. Cc2d1a loss of function disrupts functional and morphological development in forebrain neurons leading to cognitive and social deficits. Cereb Cortex 2016;pii:bhw009.

[74] Manzini MC, Xiong L, Shaheen R, Tambunan DE, Di Costanzo S, Mitisalis V, et al. CC2D1A regulates human intellectual and social function as well as NF-kappaB signaling homeostasis. Cell Rep 2014;8(3):647–55.

[75] Iossifov I, O'Roak BJ, Sanders SJ, Ronemus M, Krumm N, Levy D, et al. The contribution of de novo coding mutations to autism spectrum disorder. Nature 2014;515(7526):216–21.

[76] Lai CS, Fisher SE, Hurst JA, Vargha-Khadem F, Monaco AP. A forkhead-domain gene is mutated in a severe speech and language disorder. Nature 2001;413(6855):519–23.

[77] Schaaf CP, Gonzalez-Garay ML, Xia F, Potocki L, Gripp KW, Zhang B, et al. Truncating mutations of MAGEL2 cause Prader–Willi phenotypes and autism. Nat Genet 2013;45(11):1405–8.

[78] Bernier R, Golzio C, Xiong B, Stessman HA, Coe BP, Penn O, et al. Disruptive CHD8 mutations define a subtype of autism early in development. Cell 2014;158(2):263–76.

[79] Stessman HA, Willemsen MH, Fenckova M, Penn O, Hoischen A, Xiong B, et al. Disruption of POGZ is associated with intellectual disability and autism spectrum disorders. Am J Hum Genet 2016;98(3):541–52.

Chapter 17

Viral Hepatitis

Thomas Tu[1], Keyur Patel[2,3] and Nicholas A. Shackel[4,5,6]

[1]*Universitätsklinikum Heidelberg, Heidelberg, Germany,* [2]*Toronto General Hospital, Toronto, ON, Canada,* [3]*Toronto University, Toronto, ON, Canada,* [4]*University of New South Wales, Sydney, NSW, Australia,* [5]*Ingham Institute, Liverpool, NSW, Australia,* [6]*Liverpool Hospital, Liverpool, NSW, Australia*

Chapter Outline

ABBREVIATIONS

CD	cluster of differentiation
dslDNA	double-stranded linear DNA
EIA	enzyme immunoassay
ECM	extracellular matrix
GWAS	genome-wide association studies
HSC	hepatic stellate cell
HAV	hepatitis A

Genomic and Precision Medicine. DOI: http://dx.doi.org/10.1016/B978-0-12-800685-6.00017-5

HBV	hepatitis B
HCV	hepatitis C
HDV	hepatitis D
HEV	hepatitis E
HCC	hepatocellular carcinoma
HGF	hepatocyte growth factor
HLA	human leukocyte antigen
IFN	interferon
ISG	intreferon stimulated gene
NK	natural killer
OR	odds ratio
ORF	open reading frames
SNP	single nucleotide polymorphism
SVR	sustained viral response
TLR	Toll-like receptor
TNF	tumour necrosis factor

INTRODUCTION

Viral hepatitis is one of the major etiologies for liver disease worldwide. The highly diverse hepatitis viruses (named A to E) are the main causative agents of viral hepatitis (Table 17.1). In this chapter, we focus on factors that determine three crucial aspects of viral hepatitis infection: the postexposure outcome of viral infection (fulminant vs acute vs chronic infection); the disease progression toward cirrhosis and hepatocellular carcinoma (HCC) associated with chronic viral hepatitis infection; and the responses to antiviral therapies. Genomic analyses have been used to better understand and manage all three aspects and are outlined below. At the end of the chapter, we outline the potential contributions of genomic medicine that we expect to see in the near future.

THE HEPATITIS VIRUSES

Viral hepatitis is defined as the inflammation of liver caused by viral pathogens. While there are multiple pantropic viruses (e.g., Herpes Simplex Virus, Cytomegalovirus, and Epstein–Barr virus) that can cause hepatitis, there are five nominal hepatitis viruses (hepatitis A–E viruses) that solely or primarily infect and replicate within the hepatocytes of the liver (Table 17.1). These hepatitis viruses (which will be the subject of this chapter) are from five completely separate families and have vastly varying characteristics (Table 17.1). Following exposure to any of these hepatitis viruses, two clinical outcomes are possible: acute or chronic hepatitis infection (Table 17.1).

GENETIC PREDISPOSITION TO VIRAL HEPATITIS

Predisposing factors to viral hepatitis include a number of nongenetic and genetic factors [1–3] (Table 17.2). Genetic factors predispose to persistence

TABLE 17.1 Characteristics of the Five Common Hepatitis Viruses

	Hepatitis A	Hepatitis B	Hepatitis C	Hepatitis D	Hepatitis E
Virological Aspects					
Enveloped	N	Y	Y	Y	N
Size (nm)	27–32	42	40–60	36	32
Type	+ sense ssRNA	dsDNA	+ sense ssRNA	− sense closed circular ssRNA	+ sense ssRNA
Size (kbp)	7.5	3.2	10	1.7	7.5
Open reading frames	1 (polypeptide)	4 (overlapping)	1 (polypeptide)	5 (only 1 utilized, HDAg)	3
Proteins encoded	11	7	7	1	9
Genotypes (human only)	3	10	11	3	4
Clinical Aspects					
Number of new infections per year	~1 million	~10 million	~3 million	~500,000	~20 million
Route of transmission	Fecal-oral	Blood-borne	Blood-borne	Blood-borne	Fecal-oral
Chronicity	Always acute	Children = ~90%, adults = ~10%	~80%	Co-infection = < 5% Super-infection = 80%	Always acute
Number of current chronic infections	N/A	~300 million	~150 million	~10 million	N/A
Deaths each year	~10,000	~600,000	~350,000	~100,000	~50,000
Markers of infection	Serum anti-HAV IgM antibodies	Serum HBsAg or HBV DNA	Serum HCV RNA or anti-HCV IgM antibodies	Serum HDV RNA or HDAg	HEV RNA
Vaccine available?	Y	Y	N	Y-dependent on HBV vaccine	N
Cure available?	N	N-Antivirals only suppress replication, not cure	Y	N	N
Notable features				Satellite virus, dependent on HBsAg for replication	

TABLE 17.2 Viral Hepatitis Genetic Susceptibility Associations

Allele/Polymorphism	Hepatitis C	Hepatitis B	Comment
HLA-DRB1*1302 and HLA-DRB1*0301		Spontaneous elimination of infection	
HLA-DQA1*0501, DQB1*301, HLA-DRB1*1102 and HLA-DRB1*0301		Persistence of infection	
HLA-DRB1*0101, HLA-DRB1*0401, HLA-DRB1*15, HLA-DRB1*1101, HLA-DRB1*0301, HLA-A*2301, HLA-A*1101, HLA-A*03, HLA-B*57 and HLA-Cw*0102	Spontaneous elimination of infection		
HLA-DRB1*0701, HLA-A*01-B*08-Cw*07-DRB1*0301-DBQ1*0201, HLA-Cw*04, HLA-Cw*04-B*53	Persistence of infection		
TNF promoter	Viral replication and clearance	Viral replication and clearance	Polymorphisms at −308 and −238 best characterized
Interleukin-10		Spontaneous elimination of infection	
Vitamin D receptor		Control of viral replication	Expressed on monocytes and lymphocytes
GDNF family receptor alpha 1	At risk of HCC in HCV		
Chemokine (C-X-C motif) ligand 14	At risk of HCC in HCV		Previously known as SCYB14
IFN-λ3/IL28B	Spontaneous clearance Interferon treatment response		To-date the strongest genetic association demonstrated with either spontaneous clearance or treatment responses to HCV Genotype-1 infection

of hepatitis B (HBV) infection given the rate of concordance for surface antigen expression is greater in monozygotic compared to dizygotic twins (Table 17.2). The twin concordance data in hepatitis C (HCV) infection is not as convincing. There are a number of human leukocyte antigen (HLA) alleles are associated with both the persistence and clearance of both HBV and HCV [1,4]. Interestingly and in contrast to HIV progression, homozygosity for HLA class II locus increases the risk of HBV persistence. The HLA class II locus DRB1*1302 is associated with HBV clearance and the DQB1*0301 locus is associated with a self-limiting course of HCV infection. Non-HLA immune-associated genes are also implicated in viral hepatitis with tumour necrosis factor (TNF) promoter polymorphisms resulting in higher TNF secretion being associated with HBV clearance. The killer cell immunoglobulin like receptors (KIR) genes interact with HLA class I molecules and specific KIR heliotypes are associated with HCV clearance [5,6]. Importantly, the majority of genetic predispositions identified in viral hepatitis is linked to viral persistence or clearance and can be directly implicated in the adaptive immune response. The currently documented genetic disease associations with viral hepatitis are of limited use in clinical practice. Presently, genome-wide association studies (GWAS) have been undertaken to characterize hepatitis genetic susceptibility. Therefore, future clinical practice is likely to see panels of genetic susceptibility markers being screened to determined prognosis as well as the likelihood of a treatment response. The overall impact of genomic studies of hepatitis virus is considerable and impacts the stage of infection (acute and chronic), treatment responses, and consequences such as liver cancer (Table 17.3).

ACUTE VIRAL HEPATITIS

Acute viral hepatitis is defined as a self-limiting virus infection that lasts for <6 months. Acute hepatitis is generally difficult to diagnose due to a short clinical course (on the order of weeks) and nonspecific symptoms (if any), including, poor appetite, nausea, vomiting, fever, pain in the upper right part of the abdomen, and jaundice. Confirmation of viral hepatitis infection is generally shown postclearance by blood tests for persistent neutralizing antibodies to viral antigens (Table 17.1). This temporary nature of acute viral hepatitis, lack of infection models, and absence of symptoms in the majority of exposed patients makes research in this area difficult.

During clearance, multiple arms of the immune system are activated, particularly virus-specific CD8+ cytotoxic T-cells [7]. In some patients, fulminant hepatitis can occur in which an overactive antiviral immune response results in extensive multifocal necrosis and acute liver failure. Fulminant hepatitis is the leading cause of death associated with HAV or HEV infection. Following the complete clearance from the liver, there are generally no long-term sequelae associated with any of the hepatitis viruses.

TABLE 17.3 Key Findings Arising from Genomic Studies of Viral Hepatitis

	Hepatitis C	Hepatitis B	Comments
• Acute infection	• IL28B polymorphism associated with acute HCV clearance • Interferon stimulated gene expression (Mx1, ISG15) correlated with viral clearance	• "Stealth virus" • Immune evasion	• Acute HBV or HCV infection has been partially characterized by genomic studies
• Chronic infection	• Th1 immunophenotype perpetuating chronic injury	• T-cell effector function activated and latter B cell related gene expression observed	
• Treatment responses	• IL28B polymorphism associated with acute HCV response to interferon • Interferon stimulated gene induction correlate with treatment responses		• HBV treatment response have been poorly characterized by genomic studies
• Hepatocellular carcinoma	• HCV core protein oncogenic	• HBV x (HBx) protein oncogenic	
• Bio-markers	• MiRNA distinguish stage of disease and HCC • A number of nonspecific inflammatory markers identified	• MiRNA distinguish stage of disease and HCC • A number of nonspecific inflammatory markers identified	
• Future studies	• Predicting HCV IFN treatment responses (especially in nongenotype-1 HCV)	• Predicting HBV IFN and nucleotide/ nucleoside treatment responses	• Determining the pathogenesis of and predicting chronicity, disease progression and development of sequelae such as HCC prognosis

FACTORS AFFECTING THE POSTEXPOSURE CLINICAL OUTCOME OF HEPATITIS INFECTION

As shown in Table 17.1, the rates of acute infection vary amongst the hepatitis viruses. HAV and HEV almost always cause acute hepatitis, while chronic infection is possible with HBV, HCV, and HDV infections. Similarly, the rates of fulminant hepatitis vary between the hepatitis viruses. Clinical factors have been linked to the outcome of infection (clearance, chronicity, or acute liver failure), including: sex of the patient; age of infection; virus genotype; and size of inoculum. These clinical factors have given leads on potential genes that may be involved in controlling (and therefore potentially predicting) the response to hepatitis virus exposure.

Cellular studies have shown additional factors that may play a role, including: the susceptibility of hepatocytes to infection; the level of viral antigen expression; the presentation of viral antigens; and the regulation of immune cell activation, particularly innate immunity.

Postexposure Outcomes in HAV Infection

Studies into HAV infection have been restricted by low numbers of patients, and so poor data exists for host single-nucleotide variants (SNVs) affecting clinical outcome. A WGS study of 10 patients with acute liver failure during HAV infection did not find a single shared mutation [8]. However, a larger study of 30 Argentinean patients with HAV-induced acute liver failure and 102 healthy case controls focusing on the HAV receptor TIM1 showed that the long-form polymorphism of TIM1 (containing a 6-amino-acid 157insMTTTVP insertion) is associated with greater risk of developing severe HAV-induced liver failure, through more efficient binding of HAV to hepatocytes and/or increased activation of NKT cells [9].

Postexposure Outcomes in HBV Infection

After the discovery of Sodium Taurocholate Cotransporting Peptide (NTCP) as the cellular receptor for HBV and HDV [10,11], NTCP SNVs have been found in the human population through GWAS. SNVs in the NTCP gene have been found to predict postexposure outcomes of HBV infection, with the p.Ser267Phe variant of NTCP being associated with increased HBV clearance (odds ratio (OR) = 0.33) and decreased acute-on-chronic liver failure ($p = 0.007$) in a Han Chinese population [12]. Structural modeling suggested that the variant causes alterations in ligand binding, leading to lower binding of HBV to hepatocytes.

The genes that encode HLA molecules are amongst the most recurrent arising from GWAS of viral hepatitis [13,14]. For example, in a multivariate analysis of American Caucasians [15], HLA class I haplotype A*0301 has been shown to be associated with clearance (OR = 0.47), while haplotypes B*08 and

B*44 are associated with HBV persistence (OR = 1.66 and 1.81, respectively). Further, SNPs in and near HLA loci have been associated with HBV chronicity in Saudi [16], Japanese [17,18], Korean [18,19], Chinese [20,21], Thai [22], Europeans, and African Americans [23]. Some of these HLA risk alleles were shown to be associated with significantly decreased HLA surface expression [23], suggesting that the mechanism behind increased susceptibility to chronic infection is a decreased presentation of viral antigens to the immune system. Relatedly, SNPs in HLA loci have also been shown to predict HBV vaccine response in Indonesian [24] (ORs ranging from 0.75 to 2) and Han Chinese cohorts (rs477515, OR = 0.18) [25].

Other immune-associated genes (especially those involved in innate immunity) have also been found to affect postexposure chronicity. SNPs in TLR3 have been associated with HBV clearance in Saudi populations (rs1879026, OR = 0.809) [26] and Han Chinese populations (rs3775290, OR = 0.52) [27]. GWAS of a Han Chinese population by this group showed SNVs in transmembrane protein 2 (TMEM2, rs2297089), complement component 2 (C2, rs9332739), and importantly, interferon (IFN) alpha 2 (IFNA2, rs146352658) and its regulator NLR family member X1 (NLRX1, rs149770693) were also associated with chronic HBV infection [28] with ORs of 2.45, 1.97, 4.08, and 2.34, respectively. These results suggest that IFN-related genes are key determinants in postexposure outcome of infection, which is corroborated by other studies. For example, SNVs in the promoter sequence of IFN-α receptor 1 (IFNAR1) promoter have been associated with chronic HBV infection in Chinese and Caucasian populations [29]. Further in vitro studies showed that risk-conferring SNVs were transcriptionally less active than the protective allele [29].

Postexposure Outcomes in HCV Infection

Like HBV infection, immune-related genes are highly represented in genomic studies on predicting HCV postexposure outcomes. SNPs in HLA-DQB1*0301 and DRB1*1101 alleles predict spontaneous clearance in HCV-exposed patients with ORs up to 3.0 and 2.5, respectively based on meta-analyses of European, American, and Asian studies [1,30]. This was supported in an international multicenter GWAS, in which SNPs in HLA-DQB1*0301 were associated with the resolution of HCV (overall per-allele OR = 1.6) [31].

The IFN response in particular has gained much attention for its role in HCV resolution. SNPs in and near IFNλ3 (also known as IL28B) is associated with HCV clearance (overall per-allele OR = 0.45) [31], whereas SNPs proximal to the IFN-stimulated genes PKR and OAS-1 are associated with chronic HCV infection (OR = 2.75 and 0.43, respectively) [32]. Further, SNVs in the IFNλ4 gene are predictive of clearance of HCV G3 in Caucasian patients (rs12979860, OR = 2.45) [33] and has been shown to affect rate of infection and innate immune response in primary human hepatocytes [34]. Indeed, innate immune responses appear to be key in the initial response against HCV with SNPs in TLR4 shown to be predictive of clearance (rs4986790, rs4986791, OR = 0.40, and 0.30) [35].

Postexposure Outcomes in HDV and HEV Infections

Scant data is available on the effect of host genomics on HDV and HEV infections. One study was conducted on 186 chronic HBV patients, where SNPs in IL10 promoter (rs1800896, present in 17.7% and 37.4% of patients with and without HDV infection) or the vitamin D receptor (rs731236, 86.6%, and 62.7%) were associated with chronic HDV infection [36]. Others have tried and failed to find associations between fulminant hepatitis and genotype, strains, or viral substitutions [37], or between HDV clearance and IL28B genotype [38].

CHRONIC HEPATITIS

The majority of research has focused on chronic infection, as they are the major cause of viral hepatitis-associated morbidity and mortality. Given that HAV and HEV infections are almost always acute and the paucity of HDV studies due to lack of reagents, the literature is skewed toward HBV and HCV and so we will only discuss these for the rest of this chapter.

The natural history of chronic HBV infection can be categorized into five phases [39]: immune tolerance, immune reactive HBeAg-positive, inactive HBV carrier, HBeAg-negative chronic hepatitis, and HBsAg-negative phases. The first phase (immune tolerance) is generally asymptomatic. High levels of HBV serum titers of ~2×10^9 International Units (IU, ≈5 copies HBV DNA [39]) per mL are observed, as a result of low levels of activated HBV-specific CD8$^+$ T-cells [40,41]. When immune tolerance is broken (by poorly defined mechanisms), the HBV infection progresses to the immune reactive HBeAg-positive phase, in which HBV antigen-specific CD8$^+$ cytotoxic T-cells target infected hepatocytes. This causes flares of immune-mediated liver damage and fluctuations in HBV titers [40]. HBV-specific antibodies are produced in inactive HBV carrier phase, leading to HBeAg to HBeAb seroconversion, and elimination of many (but usually not all) HBV-infected hepatocytes. Low HBV titers and limited liver-injury progression are observed. Ineffective HBV suppression can occur in some patients, who then enter the HBeAg-negative chronic hepatitis phase. HBV mutants are common in this phase, driving chronic inflammation with flares of escape mutants followed by spikes of immune-mediated hepatocyte death. A minority of patients' progress to the HBsAg-negative phase, in which HBV titers are at undetectable levels, antibodies to HBsAg (anti-HBs) are detected, and HBV infection is considered cleared. However, even in this phase, reactivation of HBV infection still can occur if the patient is immunocompromised [42].

In contrast to the complicated HBV natural history of infection, chronic HCV infection leads a less well-segregated natural history and only consists of acute and viral persistence phases [7]. Upon exposure, HCV replication spikes to ~10^{12} virions produced per day followed by immune-mediated hepatocyte cell death. Following the acute phase, antibodies against HCV antigens are formed, but the antibodies are nonneutralizing and continual evolution of HCV can lead to a persistent infection in ~80% of cases. Chronic low-level liver turnover occurs over time, causing progression of liver fibrosis toward cirrhosis.

FACTORS AFFECTING LIVER DISEASE PROGRESSION

The most important disease outcomes associated with chronic hepatitis virus infection are liver cirrhosis and HCC, which generally occur late in disease progression on the background of cirrhosis (with the exception of HBV-infected South African black males, who can present with HCC without concomitant cirrhosis). Several key clinical characteristics (summarized in Table 17.2) have been shown to be associated with the disease progression of viral hepatitis, including: patient age; level of immune-mediate liver damage (measured by serum ALT levels); concomitant alcohol consumption or exposure to aflatoxin B1; virus genotype (HBV genotype C and HCV genotype 1), high viral titers (>2000 HBV IU/mL); and the presence viral mutations [43–45].

SNPs in some genes have been shown to predict disease progression in both HBV and HCV infection (Table 17.4). Again, HLA-associated SNPs play a significant role in viral hepatitis disease progression. SNPs in HLA loci have been shown to be predictive of HBV clearance in chronically infected Japanese/Koreans [18,46] and Han Chinese [20,47–49]. However, on the whole, HLA SNPs have been poorly predictive of progression of HCV-mediated liver injury [1]. HLA-associated SNPs are also associated with HBV-HCC in Japanese [46]; Han Chinese [49–51]; and Korean [19,52]; but not in any other East-Asian cohorts [53]. HLA SNPs are associated with risk of progression to HCV-HCC (OR = ~2.8–1.5, with cumulative additional effect) in a Japanese cohort [54]. Indeed, a meta-analysis of eight case-control studies has shown that SNVs in HLA-DRB1*07, -DRB1*12, and -DRB1*15 are associated with increased risk of progression to viral-associated HCC (OR = ~2.88) for Asian populations and in DRB1*07 and DRB1*12 (OR = ~1.65) in multiple populations [55], regardless of etiology.

Further, SNPs in IFNλ4, patatin-like phospholipase domain-containing-3 (PNPLA3) and MHC class I polypeptide-related chain A (MICA) have been reported to predict disease progression in both chronic HBV and HCV infections. In Caucasian patients, IFNλ4 CC genotype (rs12979860) is associated with both HBV- and HCV-associated fibrosis progression (OR = 1.63 and 2.75, respectively) and inflammation (OR = 4.36 and 2.11, respectively) [33].

The rs738409 SNP in PNPLA3 has been associated with significantly worse fibrosis (OR = 1.2) in Japanese chronic HCV patients [56]. This worse prognosis is potentially due to alterations in fat metabolism pathways [57], which HCV uses during its replication cycle [58]. Indeed in chronic HBV infection, rs738409 is associated with steatohepatitis in nonoverweight patients (OR = 2.4 and 13.4, for men and women, respectively), as well as lobular inflammation (OR = 2.2) and intrahepatic iron deposition (OR = 2.8) [59].

MICA is a ligand of the Natural Killer Group 2D Receptor (present on NK and T-cells) and mediates antiviral immune responses. The SNPs rs2596542 and rs2596538 in 5′ flanking region of MICA are linked to low serum levels of MICA and are associated with the progression of HCV-associated HCC

TABLE 17.4 Genomic Alterations Involved in Viral Hepatitis Disease Progression (Cutoff at OR = 2)

Virus	Chr.	Position	Gene	SNP	Risk Allelle
HBV	14	70,245,193	NTCP (SLC10A1)	rs2296651	C
HCV	6	N/A	HLA Class I allele B*08	Haplotype	N/A
HCV	6	N/A	HLA Class I allele B*44	Haplotype	N/A
HBV	6	N/A	HLA Class I allele A*0301	Haplotype	N/A
HBV	6	32,702,467	HLA-DQB1	rs2856718	AA
HBV	6	32,762,235	HLA-DQB2	rs7453920	AA
HBV	6	32,711,222	HLA-DQA2	rs9275572	A
HBV	6	33,087,030	HLA-DPB1	rs9277534	G
HBV	4	186,083,063	TLR3	rs3775290	C
HBV	9	71,690,182	TMEM2	rs2297089	T
HBV	9	21,384,972	IFNA2	rs146352658	T
HBV	11	119,180,140	NLRX1	rs149770693	T
HBV	11	110413469	Ferredoxin 1	rs2724432	C
HBV	2	117640013	DDX18	rs2551677	A
HBV/HCV	6	N/A	HLA-DRB1*07 HLA-DRB1*12 HLA-DRB1*15	Haplotype	N/A
HBV/HCV	19	39248147	IFNλ4	rs12979860	CC
HCV	22	43928847	PNPLA3	rs738409	GC or GG
HCV	17	34320812	MCP-2	rs1133763	C
HCV	3	141744456	RNF7	rs16851720	AA
HCV	2	112013193	MERTK	rs4374383	AG or GG
HCV	6	35534425	TULP1	rs9380516	TT
HCV	12	104028030	GLT8D2	rs2629751	GG
HCV	9	90419249	LOC340515	rs883924	AA

(OR = ~1.4) [60,61] in Japanese populations. The rs2596542 has also been reported to associated with active HBV disease (OR = 1.31) [62] and increase the risk of HBV-associated HCC (OR = 1.2) [63,64].

Serum miRNA are an attractive target for screening of HBV, HCV, and HCC. A total of 523 cases were used to study predominantly serum miRNA expression in HBV and HBV-associated HCC [65]. In this study, serum miRNAs were shown to distinguish HBV and HCV infection (miR-375 and miR-92a) and the presence of HCC (miRNA-25, miRNA-375, and let-7f) [65].

Disease Progression in Chronic HBV Infection

Changes in circulating HBV genomes have been associated with disease progression during chronic HBV infection. HBV quasispecies undergo positive selection and display decreased complexity leading up to HBeAg seroconversion and progression into the inactive HBV carrier phase [66]. This immune pressure selects for escape mutants, which become common in latter phases of HBV infection. The presence of HBV escape mutants with PreS mutations (particularly A1762T and G1764A) predicts HCC risk with adjusted OR of 6.18, shown in a meta-analysis of 43 studies [67].

Immune responses appear to be important in HBV-associated disease progression. Whole transcriptome analysis has shown that the transition between HBV phases are marked by changes in peripheral leukocyte mRNA expression, predominantly composed of highly upregulated immunoglobulin-encoding genes [68]. Somatic copy number variations in T-cell receptor gamma and alpha loci have also been shown to increase the risk of developing HBV-associated HCC (HBV-HCC) in a Korean cohort (low copy number OR = 0.19 and 0.17, respectively) [52]. Further, the rs3775290 SNV in TLR3 has been associated with a decreased risk for progression to HBV-related liver cirrhosis and HCC (OR = 0.32 and 0.49) in Chinese populations [27].

In addition, metabolic genes may also play a role in HBV-associated disease progression. For example, a SNP 16 kb upstream of the bile acid synthesis gene Ferredoxin 1 (rs2724432) was found to have an association with complicated chronic HBV infection (cirrhosis or HCC) with an OR of 2.34 [69]. Further, the rs11866328 SNP in the glutamate receptor ionotropic N-methyl D-aspartate 2A (GRIN2A) gene has been associated with disease progression in two different Chinese populations (OR = 1.65 and 1.73) [70].

Multiple SNPs have been attributed with an altered risk of progression to HBV-HCC. The rs2880301 SNP in PTEN homologue TPTE2 conferred protection against progression to HCC (OR = 0.27) in a Korean cohort with chronic HBV infection, while the rs2551677 SNP in the ATP-dependent RNA Helicase DDX18 conferred increased risk (OR = 3.4) [52]. In various HBV-infected Chinese cohorts, SNPs in KIF1B (rs17401966), lncRNA on Chr8p12 (rs12682266, rs7821974, rs2275959, and rs1573266), and STAT4 (rs7574865) have been reported to confer small increased risk (ORs = ~0.8, ~1.3, and 1.2, respectively) for progression to HCC [51,71–73], though the latter was shown to not be valid in other Asian populations [74].

GWAS have informed the design of predictive gene panels with some success. Recently, a panel of 6 germline SNVs have been found to separate patients into high-, medium-, and low-risk groups and are significantly associated with HBV-HCC survival time (HR = 2.14 and 3.17 for medium- and high-risk groups, respectively) [75]. This panel includes: rs2275959 and rs37821974 upstream of a cancer suppressor gene DLC-1 [76]; SNPs in MHC loci (rs3997872, rs7453920, and rs7768538); and rs1419881 located in the 3′ untranslated region of transcription factor 19, found in previous GWAS of HBV clearance [19].

Disease Progression in Chronic HCV Infection

A more detailed view of virus dynamics during infection using NGS has revealed clinical value in HCV genome changes. Narrowing of the diversity of HCV quasispecies has been shown to predict seroconversion after HCV exposure [77], likely a measure of positive selection HCV quasispecies in the face of active immune pressure [78]. Further, increased neutral evolution is associated rapid fibrosis progression [79], probably due to increased hepatocyte turnover.

Host genomics also plays a role in HCV-associated disease progression, particularly immune-associated genes. SNPs in IFN-associated genes IFN-γ receptor 2 (rs9976971) and IFNλ3 (also known as IL28B; rs8099917) *have been found to be associated with HCV-associated fibrosis progression* in Japanese and Caucasian cohorts (OR = not reported) [56,80]. Further, genetic alterations in immune cell receptors TLR7 (c. 1–120T > G mutation, adjusted OR = 4.187) [81] and CCR5 (and its ligand MCP-2; rs333 and rs1133763, OR = 1.97 and 2.29) have been associated with greater fibrosis [82].

A multicenter GWAS (combined cohort of 2342 HCV-infected patients) was carried out by Patin et al. in search for SNPs that predict progression of HCV-associated fibrosis [56]. SNPs in apoptosis genes RNF7 (rs16851720), MERTK (rs4374383), and TULP1 (rs9380516) were implicated in progression of fibrosis (OR = 0.46, 0.2, and 4.5), suggesting control of cell death is a strong predictor of HCV-associated disease. A multitude of other hits were found in this study including PKD1L1 (rs7800244, OR 0.61), an intronic SNP in GLT8D2 gene (rs2629751, OR = 7.1), and a SNP downstream of the noncoding RNA LOC340515 (rs883924, OR = 2.7) [56]. Studies have also shown increased risk of HCV-associated HCC is dependent on an intronic SNP in DEPDC5 (rs1012068, adjusted OR = 1.96), which is a key component of the cancer-associated mTORC pathway [83].

HEPATITIS VIRUS–ASSOCIATED HCC

Few of the SNPs described above these appear to be somatically mutated in HCC, suggesting they are not driver genes acting in the HCC, but rather alter susceptibility by other means (e.g., alterations in immune surveillance). HCC presents an altered state where the genome has undergone somatic alterations during progression to malignancy. There is a potential role for personalized medicine to find DNA mutations or dysregulated pathways that could inform the use of specific targeted therapies for particular HCCs or predict patient survival. HCC genomics has been addressed in detail in reviews [84] and other chapters of this book. We highlight here several genomic studies specifically focusing on viral hepatitis-associated HCC.

Different etiologies of hepatitis virus-associated HCC appear to affect the mutational landscape of the tumor genome. HBV-associated HCC (HBV-HCC) tends to be associated with more DNA rearrangements [85,86] compared to HCV-HCC. HBV-HCC is also associated with particular transcription profiles that include dysregulation in mitotic cell cycle-, AKT activation-, AXIN1-, developmental-, and DNA imprinting-pathways [87]. Some authors have reported an increase in the number of TP53 R249S mutations in HBV-HCC (39%) compared to non-HBV-HCC (0%) [88], though this is possibly due to the concurrent exposure to Aflatoxin B1 (for which the R249S mutation is a common occurrence) and not HBV itself.

HBV has a double-stranded linear (dsl)DNA form that can randomly integrate into the host cell genome by nonhomologous end joining [89]. HBV DNA integration into oncogenes, such as the mixed-lineage leukemia 4 (MLL4) gene and human telomerase reverse transcriptase (hTERT) gene, have been repeatedly observed in HBV-associated HCCs [90–92]. Further, if HBV DNA integrates upstream of a long interspersed nuclear element (LINE), HBx-LINE1 fusion genes are formed and can produce novel lncRNA. HBx-LINE1 transcripts can be expressed in human HCCs and are associated with a poorer survival (HR = 2.65) [93].

No genes, however, are consistently disrupted by HBV DNA integration [86,90–92,94,95], as opposed to woodchuck hepatitis virus (WHV)-infected woodchucks where WHV DNA integration occurs in the N-Myc gene occurs in ~100% of HCCs [96]. Further, others have shown that HBV integration in genome of the surrounding nontumor liver tissue occurs just as frequently and into the same genes compared to the tumor tissue [86,94], suggesting that integrated HBV DNA in a limited driver of caricinogenesis [97].

Specific genomic mutations are different in HBV-HCC compared to non-HBV-HCC. Inactivation of interferon regulatory factor 2 (IRF2) has been shown to be a rare, but enriched mutation in HBV-HCC, while mutations in β-catenin (CTNNB1) are generally absent [85]. ARID2 (a component of the SWI/SNF chromatin remodeling complex) mutations have been reported to be enriched in HCV-HCC (14%) vs. HBV-HCC (2%) [98]. In these studies, ARID2 mutations tend to cooccur with CTNNB1 mutations (30% in HCV-HCC vs. 10% in HBV-HCC), but exclusive from TP53 (20 vs 40%) and ARID1A mutations [85]. However, larger NGS studies have shown that ARID2 mutations do occur frequently in HBV-HCC (up to 13%) and is not significantly different between etiologies [85,95]. Very few (if any) genomic features distinguish HCV-HCC from any nonviral-associated HCC.

TREATMENT OF HEPATITIS VIRUS INFECTION

The mainstays of viral hepatitis treatment are composed of nucleo(s/t)ide analogues to directly inhibit viral replication, IFN treatment to activate host antiviral immunity or a combination of the two.

For HBV infection, the gold standard is anti-HBs antibody seroconversion, signifying complete suppression of the virus infection. However, this occurs rarely with current therapies. Therefore, the majority of studies measure as an endpoint progression to the inactive HBV carrier phase (via HBeAg-seroconversion), wherein the risk of fibrosis progression is lower compared to the HBeAg-positive phase. Virological measurements, such as decrease in HBV titers, are also used as a surrogate for treatment response. As the stable HBV cccDNA template is not targeted by current therapies, these virological changes can be only temporary with viral titers rebounding after therapy cessation.

In HCV infection, complete viral clearance is observed in the vast majority of patients treated with recently developed nucleo(s/t)ide inhibitors (e.g., Declatasvir, Sofosbuvir, and Simeprevir). However, access to these highly-effective agents is poor for the majority of HCV-infected patients due to prohibitive price. Therefore, many patients are still treated with pegylated IFN therapy, which has variable response rates from one-third (Genotype 1-infected) to two-thirds (nongenotype 1-infected) of patients [99].

Treatment outcome has been shown to be dependent on multiple clinical features aside from viral genotype, such as viral titer, fibrosis level, and age (Table 17.2). In addition, recent GWAS have found genetic predictors of treatment outcome, which have highlighted particular host molecular pathways that affect therapeutic response (Table 17.5). Some of these results have led to informing clinically relevant decisions and segregation of patients into likely and unlikely responders.

Treatment of Chronic HBV Infection

Very few SNPs in either the host or HBV genome have been found to predict HBV response. Multiple SNPs in HLA loci (rs9276370 and rs7756516) have

TABLE 17.5 Genomic Alterations Involved in Patient Responses to Antiviral Treatment (Cutoff at OR = 2)

Virus	Chr.	Position	Gene	SNP	Risk Allele
HCV	19	39252525	IL28B	rs8099917	CT or TT
		3213196		rs1127354	CC
HCV	20	3213247	ITPA	rs7270101	AA

been associated with increased response to direct acting antiviral therapies and IFN treatment in a small cohort of Taiwanese–Han-Chinese patients, though this has yet to be verified in larger populations [48]. While there was a positive relationship found with SNPs in the IL28b gene (rs8099917) in a single early study, subsequent trials have found no correlation between IL28b genotype and therapeutic outcome in chronic HBV infection [100,101]. A small study has shown an intrahepatic transcriptomic profile (41 genes) that predicts combined response (HBeAg negativity, and HBV-DNA levels ≤2000 IU/mL) to Adefovir and IFN [102]. The majority of these were genes associated with MHC Class II presentation, chemokine activity, and IFN-stimulated genes.

Treatment of Chronic HCV Infection

The quintessential success story of personalized medicine in viral hepatitis is the use of IL28b genotype (rs8099917) to predict HCV therapeutic response. This single SNP strongly predicts response (OR = 20–30) to combination pegylated-IFN/Ribavirin therapy in Japanese [103,104], Taiwanese [104], Americans with European or African descent [105], Australian [106], and white European [107] populations. The predictive strength of IL28B genotype is strongest in patients with G1 where IFN treatment response is generally poorer.

IL28B genotype is associated with immune pressure against HCV, shown by its association with decreased positive pressure in HCV quasispecies (which is also associated with treatment response [108]). SNPs in the promoter of the IFN-stimulated gene MxA has also been reported to predict HCV therapeutic response, but to a much lower level compared to IL28b genotype (OR = ~0.49) [32,109].

Finally, Peg-IFN/ribavirin therapy generates frequent side effects, including drug-induced thrombocytopenia and anemia. SNPs in the inosine triphosphate pyrophosphatase (ITPA) gene (rs1127354 and rs7270101) have been associated with ITPA deficiency and protection against Ribavirin-induced thrombocytopenia (ORs = 0.2 to 0.5) in Japanese [110,111], American [112,113], and Egyptian [114] populations.

FUTURE IMPACT OF GENOMICS AND PERSONALIZED MEDICINE

Apart from some exceptions (most notably IL28B genotype), each of these SNPs or alleles by themselves have a small impact, which makes them difficult to be clinically actionable. Further, only few studies to date have been conducted with the patient number with enough statistical power to detect the true risk conferred by SNVs; for example, some have suggested that an odds ratio in the range of 1.2–2 would require GWAS on the order of thousands of patients rather than the hundreds as past studies have shown [115]. While meta-analyses can increase statistical power, some of the SNVs presented here may be shown as false positives in future. As DNA sequencing technologies become increasingly cheap, we will have an unprecedented amount of data at our disposal and potentially a flood of false-positive hits. The upcoming problems lie in proper statistical power, transparent bioinformatics analysis, and high-quality clinical phenotypes to associate with.

HCC sequencing may be possible in future to target specific pathways based on genomic mutations. At the moment, sequencing of every tumor is too expensive and too inaccurate to be clinically useful. Further, as only a single chemotherapeutic is used for HCC (Sorafenib), there is no actionable choice and therefore no direct clinical incentive to differentiate HCC subtypes. However, the clear heterogeneity in the transcriptomes and genomes of viral hepatitis-associated HCCs, development of novel therapies to target specific pathways requires characterization and differentiation of the various HCC subtypes. This is likely to be the niche for personalized medicine in the treatment of viral-associated HCC.

With HAV and HEV causing no ongoing sequelae after clearance, future studies are likely to remain focused on HBV, HCV, and HDV infections. Further, while novel HCV antivirals are extremely effective in all genotypes, potential therapeutic resistance to these antivirals in the future may require personalized genomic medicine to predict responsive patient groups. Indeed, lack of public access to these expensive agents makes genomic medicine in predicting response to IFN-based therapies still pertinent to clinical needs. Relatedly, virological response to antiviral therapy occurs in ~25% of HDV-infected patients [116–118], highlighting the need for targeted, personalized therapies in HDV infection.

Finally, in order to improve preventative measures against HCC and decompensated liver failure for patients currently with chronic hepatitis virus infections, there is a need to segregate populations into risk groups to define candidates for enhanced screening. While current markers are individually weak, few have combined in a panel of genomic predictors.

CONCLUSION

In conclusion, SNPs in virus receptors and immune-associated genes form the core of predicting the clinical outcomes to exposure, the risk of disease progression and the responses to treatment for viral hepatitis. However, except for IL28B, few discovered SNVs have a strong enough correlation to be clinically relevant, despite several large-scale GWAS. Future studies are likely to focus on integration with other predictors such as transcriptomic or epigenomic data, with the hope of stratifying patients with more accuracy.

REFERENCES

[1] Yee LJ. Host genetic determinants in hepatitis C virus infection. Genes Immun 2004;5(4):237–45.

[2] Wasley A, Alter MJ. Epidemiology of hepatitis C: geographic differences and temporal trends. Semin Liver Dis 2000;20(1):1–16.

[3] Thomas DL. Hepatitis C epidemiology. Curr Top Microbiol Immunol 2000;242:25–41.

[4] Wang FS. Current status and prospects of studies on human genetic alleles associated with hepatitis B virus infection. World J Gastroenterol 2003;9(4):641–4.

[5] Williams AP, Bateman AR, Khakoo SI. Hanging in the balance. KIR and their role in disease. Mol Interv 2005;5(4):226–40.

[6] Martin MP, Carrington M. Immunogenetics of viral infections. Curr Opin Immunol 2005;17(5):510–6.

[7] Rehermann B, Nascimbeni M. Immunology of hepatitis B virus and hepatitis C virus infection. Nat Rev Immunol 2005;5(3):215–29.

[8] Long D, Fix OK, Deng X, Seielstad M, Lauring AS. Whole genome sequencing to identify host genetic risk factors for severe outcomes of hepatitis a virus infection. J Med Virol 2014;86(10):1661–8.

[9] Kim HY, Eyheramonho MB, Pichavant M, Gonzalez Cambaceres C, Matangkasombut P, Cervio G, et al. A polymorphism in TIM1 is associated with susceptibility to severe hepatitis A virus infection in humans. J Clin Invest 2011;121(3):1111–8.

[10] Ni Y, Lempp FA, Mehrle S, Nkongolo S, Kaufman C, Falth M, et al. Hepatitis B and D viruses exploit sodium taurocholate co-transporting polypeptide for species-specific entry into hepatocytes. Gastroenterology 2014;146(4):1070–83.

[11] Yan H, Zhong G, Xu G, He W, Jing Z, Gao Z, et al. Sodium taurocholate cotransporting polypeptide is a functional receptor for human hepatitis B and D virus. Elife 2012; 1:e00049.

[12] Peng L, Zhao Q, Li Q, Li M, Li C, Xu T, et al. The p.Ser267Phe variant in SLC10A1 is associated with resistance to chronic hepatitis B. Hepatology 2015;61(4):1251–60.

[13] Jiang X, Ma Y, Cui W, Li MD. Association of variants in HLA-DP on chromosome 6 with chronic hepatitis B virus infection and related phenotypes. Amino Acids 2014;46(8):1819–26.

[14] Tong H, Bock CT, Velavan TP. Genetic insights on host and hepatitis B virus in liver diseases. Mutat Res Rev Mutat Res 2014;762:65–75.

[15] Thio CL, Thomas DL, Karacki P, Gao X, Marti D, Kaslow RA, et al. Comprehensive analysis of class I and class II HLA antigens and chronic hepatitis B virus infection. J Virol 2003;77(22):12083–7.

[16] Al-Qahtani AA, Al-Anazi MR, Abdo AA, Sanai FM, Al-Hamoudi W, Alswat KA, et al. Association between HLA variations and chronic hepatitis B virus infection in Saudi Arabian patients. PLoS ONE 2014;9(1):e80445.

[17] Mbarek H, Ochi H, Urabe Y, Kumar V, Kubo M, Hosono N, et al. A genome-wide association study of chronic hepatitis B identified novel risk locus in a Japanese population. Hum Mol Genet 2011;20(19):3884–92.

[18] Nishida N, Sawai H, Matsuura K, Sugiyama M, Ahn SH, Park JY, et al. Genome-wide association study confirming association of HLA-DP with protection against chronic hepatitis B and viral clearance in Japanese and Korean. PLoS ONE 2012;7(6):e39175.

[19] Kim YJ, Kim HY, Lee JH, Yu SJ, Yoon JH, Lee HS, et al. A genome-wide association study identified new variants associated with the risk of chronic hepatitis B. Hum Mol Genet 2013;22(20):4233–8.

[20] Hu Z, Liu Y, Zhai X, Dai J, Jin G, Wang L, et al. New loci associated with chronic hepatitis B virus infection in Han Chinese. Nat Genet 2013;45(12):1499–503.

[21] Jiang K, Ma XP, Yu H, Cao G, Ding DL, Chen H, et al. Genetic variants in five novel loci including CFB and CD40 predispose to chronic hepatitis B. Hepatology 2015;62(1):118–28.

[22] Posuwan N, Payungporn S, Tangkijvanich P, Ogawa S, Murakami S, Iijima S, et al. Genetic association of human leukocyte antigens with chronicity or resolution of hepatitis B infection in thai population. PLoS ONE 2014;9(1):e86007.

[23] Thomas R, Thio CL, Apps R, Qi Y, Gao X, Marti D, et al. A novel variant marking HLA-DP expression levels predicts recovery from hepatitis B virus infection. J Virol 2012;86(12):6979–85.

[24] Png E, Thalamuthu A, Ong RT, Snippe H, Boland GJ, Seielstad M. A genome-wide association study of hepatitis B vaccine response in an Indonesian population reveals multiple independent risk variants in the HLA region. Hum Mol Genet 2011;20(19):3893–8.

[25] Pan L, Zhang L, Zhang W, Wu X, Li Y, Yan B, et al. A genome-wide association study identifies polymorphisms in the HLA-DR region associated with non-response to hepatitis B vaccination in Chinese Han populations. Hum Mol Genet 2014;23(8):2210–9.

[26] Al-Qahtani A, Al-Ahdal M, Abdo A, Sanai F, Al-Anazi M, Khalaf N, et al. Toll-like receptor 3 polymorphism and its association with hepatitis B virus infection in Saudi Arabian patients. J Med Virol 2012;84(9):1353–9.

[27] Huang X, Li H, Wang J, Huang C, Lu Y, Qin X, et al. Genetic polymorphisms in Toll-like receptor 3 gene are associated with the risk of hepatitis B virus-related liver diseases in a Chinese population. Gene 2015;269(2):218–24.

[28] Zhao Q, Peng L, Huang W, Li Q, Pei Y, Yuan P, et al. Rare inborn errors associated with chronic hepatitis B virus infection. Hepatology 2012;56(5):1661–70.

[29] Zhou J, Lu L, Yuen MF, Lam TW, Chung CP, Lam CL, et al. Polymorphisms of type I interferon receptor 1 promoter and their effects on chronic hepatitis B virus infection. J Hepatol 2007;46(2):198–205.

[30] Hong X, Yu RB, Sun NX, Wang B, Xu YC, Wu GL. Human leukocyte antigen class II DQB1*0301, DRB1*1101 alleles and spontaneous clearance of hepatitis C virus infection: a meta-analysis. World J Gastroent 2005;11(46):7302–7.

[31] Duggal P, Thio CL, Wojcik GL, Goedert JJ, Mangia A, Latanich R, et al. Genome-wide association study of spontaneous resolution of hepatitis C virus infection: data from multiple cohorts. Ann Intern Med 2013;158(4):235–45.

[32] Knapp S, Yee LJ, Frodsham AJ, Hennig BJ, Hellier S, Zhang L, et al. Polymorphisms in interferon-induced genes and the outcome of hepatitis C virus infection: roles of MxA, OAS-1 and PKR. Genes Immun 2003;4(6):411–9.

[33] Eslam M, Hashem AM, Leung R, Romero-Gomez M, Berg T, Dore GJ, et al. Interferon-lambda rs12979860 genotype and liver fibrosis in viral and non-viral chronic liver disease. Nat Commun 2015;6:6422.

[34] Sheahan T, Imanaka N, Marukian S, Dorner M, Liu P, Ploss A, et al. Interferon lambda alleles predict innate antiviral immune responses and hepatitis C virus permissiveness. Cell Host Microbe 2014;15(2):190–202.

[35] Al-Qahtani AA, Al-Anazi MR, Al-Zoghaibi F, Abdo AA, Sanai FM, Khan MQ, et al. The association of toll-like receptor 4 polymorphism with hepatitis C virus infection in Saudi Arabian patients. Biomed Res Int 2014;2014:357062.

[36] Karatayli SC, Ulger ZE, Ergul AA, Keskin O, Karatayli E, Albayrak R, et al. Tumour necrosis factor-alpha, interleukin-10, interferon-gamma and vitamin D receptor gene polymorphisms in patients with chronic hepatitis delta. J Viral Hepat 2014;21(4):297–304.

[37] Smith DB, Simmonds P. Hepatitis E virus and fulminant hepatitis—a virus or host-specific pathology? Liver Int 2015;35(4):1334–40.

[38] Visco-Comandini U, Lapa D, Taibi C, Angeletti C, Capobianchi MR, Garbuglia AR. No impact of interleukin-28B polymorphisms on spontaneous or drug-induced hepatitis delta virus clearance. Dig Liver Dis 2014;46(4):348–52.

[39] European Association For The Study Of The Liver E. EASL clinical practice guidelines: Management of chronic hepatitis B virus infection. J Hepatol 2012;57(1):167–85.

[40] Wang HY, Chien MH, Huang HP, Chang HC, Wu CC, Chen PJ, et al. Distinct hepatitis B virus dynamics in the immunotolerant and early immunoclearance phases. J Virol 2010;84(7):3454–63.

[41] Kennedy PT, Sandalova E, Jo J, Gill U, Ushiro-Lumb I, Tan AT, et al. Preserved T-cell function in children and young adults with immune-tolerant chronic hepatitis B. Gastroenterology 2012;143(3):637–45.

[42] Hoofnagle JH. Reactivation of hepatitis B. Hepatology 2009;49(5 Suppl):S156–65.

[43] Yang HI, Yeh SH, Chen PJ, Iloeje UH, Jen CL, Su J, et al. Associations between hepatitis B virus genotype and mutants and the risk of hepatocellular carcinoma. J Natl Cancer Inst 2008;100(16):1134–43.

[44] Lai M, Hyatt BJ, Nasser I, Curry M, Afdhal NH. The clinical significance of persistently normal ALT in chronic hepatitis B infection. J Hepatol 2007;47(6):760–7.

[45] Tsang PS, Trinh H, Garcia RT, Phan JT, Ha NB, Nguyen H, et al. Significant prevalence of histologic disease in patients with chronic hepatitis B and mildly elevated serum alanine aminotransferase levels. Clin Gastroenterol Hepatol 2008;6(5):569–74.

[46] Kamatani Y, Wattanapokayakit S, Ochi H, Kawaguchi T, Takahashi A, Hosono N, et al. A genome-wide association study identifies variants in the HLA-DP locus associated with chronic hepatitis B in Asians. Nat Genet 2009;41(5):591–5.

[47] Yan Z, Tan S, Dan Y, Sun X, Deng G, Wang Y. Relationship between HLA-DP gene polymorphisms and clearance of chronic hepatitis B virus infections: case-control study and meta-analysis. Infect Genet Evol 2012;12(6):1222–8.

[48] Chang SW, Fann CS, Su WH, Wang YC, Weng CC, Yu CJ, et al. A genome-wide association study on chronic HBV infection and its clinical progression in male Han-Taiwanese. PLoS ONE 2014;9(6):e99724.

[49] Hu L, Zhai X, Liu J, Chu M, Pan S, Jiang J, et al. Genetic variants in human leukocyte antigen/DP-DQ influence both hepatitis B virus clearance and hepatocellular carcinoma development. Hepatology 2012;55(5):1426–31.

[50] Li S, Qian J, Yang Y, Zhao W, Dai J, Bei JX, et al. GWAS identifies novel susceptibility loci on 6p21.32 and 21q21.3 for hepatocellular carcinoma in chronic hepatitis B virus carriers. PLoS Genet 2012;8(7):e1002791.

[51] Jiang DK, Sun J, Cao G, Liu Y, Lin D, Gao YZ, et al. Genetic variants in STAT4 and HLA-DQ genes confer risk of hepatitis B virus-related hepatocellular carcinoma. Nat Genet 2013;45(1):72–5.

[52] Clifford RJ, Zhang J, Meerzaman DM, Lyu MS, Hu Y, Cultraro CM, et al. Genetic variations at loci involved in the immune response are risk factors for hepatocellular carcinoma. Hepatology 2010;52(6):2034–43.

[53] Sawai H, Nishida N, Mbarek H, Matsuda K, Mawatari Y, Yamaoka M, et al. No association for Chinese HBV-related hepatocellular carcinoma susceptibility SNP in other East Asian populations. BMC Med Genet 2012;13:47.

[54] Urabe Y, Ochi H, Kato N, Kumar V, Takahashi A, Muroyama R, et al. A genome-wide association study of HCV-induced liver cirrhosis in the Japanese population identifies novel susceptibility loci at the MHC region. J Hepatol 2013;58(5):875–82.

[55] Lin ZH, Xin YN, Dong QJ, Wang Q, Jiang XJ, Zhan SH, et al. Association between HLA-DRB1 alleles polymorphism and hepatocellular carcinoma: a meta-analysis. BMC Gastroenterol 2010;10:145.

[56] Patin E, Kutalik Z, Guergnon J, Bibert S, Nalpas B, Jouanguy E, et al. Genome-wide association study identifies variants associated with progression of liver fibrosis from HCV infection. Gastroenterology 2012;143(5) 1244–1252 e1241–e1212.

[57] Romeo S, Kozlitina J, Xing C, Pertsemlidis A, Cox D, Pennacchio LA, et al. Genetic variation in PNPLA3 confers susceptibility to nonalcoholic fatty liver disease. Nat Genet 2008;40(12):1461–5.

[58] Syed GH, Amako Y, Siddiqui A. Hepatitis C virus hijacks host lipid metabolism. Trends Endocrinol Metab 2010;21(1):33–40.

[59] Brouwer WP, van der Meer AJ, Boonstra A, Pas SD, de Knegt RJ, de Man RA, et al. The impact of PNPLA3 (rs738409 C > G) polymorphisms on liver histology and long-term clinical outcome in chronic hepatitis B patients. Liver Int 2015;35(2):438–47.

[60] Kumar V, Kato N, Urabe Y, Takahashi A, Muroyama R, Hosono N, et al. Genome-wide association study identifies a susceptibility locus for HCV-induced hepatocellular carcinoma. Nat Genet 2011;43(5):455–8.

[61] Lo PH, Urabe Y, Kumar V, Tanikawa C, Koike K, Kato N, et al. Identification of a functional variant in the MICA promoter which regulates MICA expression and increases HCV-related hepatocellular carcinoma risk. PLoS ONE 2013;8(4):e61279.

[62] Al-Qahtani AA, Al-Anazi M, Abdo AA, Sanai FM, Al-Hamoudi W, Alswat KA, et al. Genetic variation at −1878 (rs2596542) in MICA gene region is associated with chronic hepatitis B virus infection in Saudi Arabian patients. Exp Mol Pathol 2013;95(3):255–8.

[63] Kumar V, Yi Lo PH, Sawai H, Kato N, Takahashi A, Deng Z, et al. Soluble MICA and a MICA variation as possible prognostic biomarkers for HBV-induced hepatocellular carcinoma. PLoS ONE 2012;7(9):e44743.

[64] Tong HV, Toan NL, Song LH, Bock CT, Kremsner PG, Velavan TP. Hepatitis B virus-induced hepatocellular carcinoma: functional roles of MICA variants. J Viral Hepat 2013;20(10):687–98.

[65] Li LM, Hu ZB, Zhou ZX, Chen X, Liu FY, Zhang JF, et al. Serum microRNA profiles serve as novel biomarkers for HBV infection and diagnosis of HBV-positive hepatocarcinoma. Cancer Res 2010;70(23):9798–807.

[66] Cheng Y, Guindon S, Rodrigo A, Wee LY, Inoue M, Thompson AJ, et al. Cumulative viral evolutionary changes in chronic hepatitis B virus infection precedes hepatitis B e antigen seroconversion. Gut 2013;62(9):1347–55.

[67] Liu S, Zhang H, Gu C, Yin J, He Y, Xie J, et al. Associations between hepatitis B virus mutations and the risk of hepatocellular carcinoma: a meta-analysis. J Natl Cancer Inst 2009;101(15):1066–82.

[68] Vanwolleghem T, Hou J, van Oord G, Andeweg AC, Osterhaus AD, Pas SD, et al. Re-evaluation of hepatitis B virus clinical phases by systems biology identifies unappreciated roles for the innate immune response and B cells. Hepatology 2015;62(1):87–100.

[69] Al-Qahtani A, Khalak HG, Alkuraya FS, Al-hamoudi W, Alswat K, Al Balwi MA, et al. Genome-wide association study of chronic hepatitis B virus infection reveals a novel candidate risk allele on 11q22.3. J Med Genet 2013;50(11):725–32.

[70] Liu L, Li J, Yao J, Yu J, Zhang J, Ning Q, et al. A genome-wide association study with DNA pooling identifies the variant rs11866328 in the GRIN2A gene that affects disease progression of chronic HBV infection. Viral Immunol 2011;24(5):397–402.

[71] Zhang H, Zhai Y, Hu Z, Wu C, Qian J, Jia W, et al. Genome-wide association study identifies 1p36.22 as a new susceptibility locus for hepatocellular carcinoma in chronic hepatitis B virus carriers. Nat Genet 2010;42(9):755–8.

[72] Chan KY, Wong CM, Kwan JS, Lee JM, Cheung KW, Yuen MF, et al. Genome-wide association study of hepatocellular carcinoma in Southern Chinese patients with chronic hepatitis B virus infection. PLoS ONE 2011;6(12):e28798.

[73] Huang M, Pan Y, Liu J, Qi F, Wen J, Xie K, et al. A genetic variant at KIF1B predicts clinical outcome of HBV-related hepatocellular carcinoma in Chinese. Cancer Epidemiol 2014;38(5):608–12.

[74] Clark A, Gerlach F, Tong H, Hoan NX, Song le H, Toan NL, et al. A trivial role of STAT4 variant in chronic hepatitis B induced hepatocellular carcinoma. Infect Genet Evol 2013;18:257–61.

[75] Li C, Bi X, Huang Y, Zhao J, Li Z, Zhou J, et al. Variants identified by hepatocellular carcinoma and chronic hepatitis B virus infection susceptibility GWAS associated with survival in HBV-related hepatocellular carcinoma. PLoS ONE 2014;9(7):e101586.

[76] Zhang LH, Qin LX, Ma ZC, Ye SL, Liu YK, Ye QH, et al. Allelic imbalance regions on chromosomes 8p, 17p and 19p related to metastasis of hepatocellular carcinoma: comparison between matched primary and metastatic lesions in 22 patients by genome-wide microsatellite analysis. J Cancer Res Clin Oncol 2003;129(5):279–86.

[77] Farci P, Shimoda A, Coiana A, Diaz G, Peddis G, Melpolder JC, et al. The outcome of acute hepatitis C predicted by the evolution of the viral quasispecies. Science 2000;288(5464):339–44.

[78] Humphreys I, Fleming V, Fabris P, Parker J, Schulenberg B, Brown A, et al. Full-length characterization of hepatitis C virus subtype 3a reveals novel hypervariable regions under positive selection during acute infection. J Virol 2009;83(22):11456–66.

[79] Wang XH, Netski DM, Astemborski J, Mehta SH, Torbenson MS, Thomas DL, et al. Progression of fibrosis during chronic hepatitis C is associated with rapid virus evolution. J Virol 2007;81(12):6513–22.

[80] Nalpas B, Lavialle-Meziani R, Plancoulaine S, Jouanguy E, Nalpas A, Munteanu M, et al. Interferon gamma receptor 2 gene variants are associated with liver fibrosis in patients with chronic hepatitis C infection. Gut 2010;59(8):1120–6.

[81] Schott E, Witt H, Neumann K, Taube S, Oh DY, Schreier E, et al. A Toll-like receptor 7 single nucleotide polymorphism protects from advanced inflammation and fibrosis in male patients with chronic HCV-infection. J Hepatol 2007;47(2):203–11.

[82] Hellier S, Frodsham AJ, Hennig BJ, Klenerman P, Knapp S, Ramaley P, et al. Association of genetic variants of the chemokine receptor CCR5 and its ligands, RANTES and MCP-2, with outcome of HCV infection. Hepatology 2003;38(6):1468–76.

[83] Miki D, Ochi H, Hayes CN, Abe H, Yoshima T, Aikata H, et al. Variation in the DEPDC5 locus is associated with progression to hepatocellular carcinoma in chronic hepatitis C virus carriers. Nat Genet 2011;43(8):797–800.

[84] Li S, Mao M. Next generation sequencing reveals genetic landscape of hepatocellular carcinomas. Cancer Lett 2013;340(2):247–53.

[85] Guichard C, Amaddeo G, Imbeaud S, Ladeiro Y, Pelletier L, Maad IB, et al. Integrated analysis of somatic mutations and focal copy-number changes identifies key genes and pathways in hepatocellular carcinoma. Nat Genet 2012;44(6):694–8.

[86] Jiang Z, Jhunjhunwala S, Liu J, Haverty PM, Kennemer MI, Guan Y, et al. The effects of hepatitis B virus integration into the genomes of hepatocellular carcinoma patients. Genome Res 2012;22(4):593–601.

[87] Boyault S, Rickman DS, de Reynies A, Balabaud C, Rebouissou S, Jeannot E, et al. Transcriptome classification of HCC is related to gene alterations and to new therapeutic targets. Hepatology 2007;45(1):42–52.

[88] Amaddeo G, Cao Q, Ladeiro Y, Imbeaud S, Nault JC, Jaoui D, et al. Integration of tumour and viral genomic characterizations in HBV-related hepatocellular carcinomas. Gut 2015;64(5):820–9.

[89] Bill CA, Summers J. Genomic DNA double-strand breaks are targets for hepadnaviral DNA integration. Proc Natl Acad Sci U S A 2004;101(30):11135–40.

[90] Brechot C. Pathogenesis of hepatitis B virus-related hepatocellular carcinoma: old and new paradigms. Gastroenterology 2004;127(5 Suppl 1):S56–61.

[91] Paterlini-Brechot P, Saigo K, Murakami Y, Chami M, Gozuacik D, Mugnier C, et al. Hepatitis B virus-related insertional mutagenesis occurs frequently in human liver cancers and recurrently targets human telomerase gene. Oncogene 2003;22(25):3911–6.

[92] Sung WK, Zheng H, Li S, Chen R, Liu X, Li Y, et al. Genome-wide survey of recurrent HBV integration in hepatocellular carcinoma. Nat Genet 2012;44(7):765–9.

[93] Lau CC, Sun T, Ching AK, He M, Li JW, Wong AM, et al. Viral-human chimeric transcript predisposes risk to liver cancer development and progression. Cancer Cell 2014;25(3):335–49.

[94] Jiang S, Yang Z, Li W, Li X, Wang Y, Zhang J, et al. Re-evaluation of the Carcinogenic Significance of Hepatitis B Virus Integration in Hepatocarcinogenesis. PLoS ONE 2012;7(9):e40363.

[95] Huang J, Deng Q, Wang Q, Li KY, Dai JH, Li N, et al. Exome sequencing of hepatitis B virus-associated hepatocellular carcinoma. Nat Genet 2012;44(10):1117–21.

[96] Jacob JR, Sterczer A, Toshkov IA, Yeager AE, Korba BE, Cote PJ, et al. Integration of woodchuck hepatitis and N-myc rearrangement determine size and histologic grade of hepatic tumors. Hepatology 2004;39(4):1008–16.

[97] Tu T, Budzinska MA, Shackel NA, Jilbert AR. Conceptual models for the initiation of hepatitis B virus-associated hepatocellular carcinoma. Liver Int 2015;35(7):1786–800.

[98] Li M, Zhao H, Zhang X, Wood LD, Anders RA, Choti MA, et al. Inactivating mutations of the chromatin remodeling gene ARID2 in hepatocellular carcinoma. Nat Genet 2011;43(9):828–9.

[99] Martinot-Peignoux M, Marcellin P, Pouteau M, Castelnau C, Boyer N, Poliquin M, et al. Pretreatment serum hepatitis C virus RNA levels and hepatitis C virus genotype are the main and independent prognostic factors of sustained response to interferon alfa therapy in chronic hepatitis C. Hepatology 1995;22(4 Pt 1):1050–6.

[100] Galmozzi E, Vigano M, Lampertico P. Systematic review with meta-analysis: do interferon lambda 3 polymorphisms predict the outcome of interferon-therapy in hepatitis B infection? Aliment Pharmacol Ther 2014;39(6):569–78.

[101] Zhang Q, Lapalus M, Asselah T, Laouenan C, Moucari R, Martinot-Peignoux M, et al. IFNL3 (IL28B) polymorphism does not predict long-term response to interferon therapy in HBeAg-positive chronic hepatitis B patients. J Viral Hepat 2014;21(7):525–32.

[102] Jansen L, de Niet A, Makowska Z, Dill MT, van Dort KA, Terpstra V, et al. An intrahepatic transcriptional signature of enhanced immune activity predicts response to peginterferon in chronic hepatitis B. Liver Int 2015;35(7):1824–32.

[103] Tanaka Y, Nishida N, Sugiyama M, Kuroski M, Matsuura K, Sakamoto N, et al. Genome-wide association of IL28B with response to pegylated interferon-alpha and ribavirin therapy for chronic hepatitis C. Nat Genet 2009;41(10):1105–9.

[104] Ochi H, Maekawa T, Abe H, Hayashida Y, Nakano R, Imamura M, et al. IL-28B predicts response to chronic hepatitis C therapy—fine-mapping and replication study in Asian populations. J Gen Virol 2011;92(Pt 5):1071–81.

[105] Ge D, Fellay J, Thompson AJ, Simon JS, Shianna KV, Urban TJ, et al. Genetic variation in IL28B predicts hepatitis C treatment-induced viral clearance. Nature 2009;461(7262):399–401.

[106] Suppiah V, Moldovan M, Ahlenstiel G, Berg T, Weltman M, Abate ML, et al. IL28B is associated with response to chronic hepatitis C interferon-alpha and ribavirin therapy. Nat Genet 2009;41(10):1100–4.

[107] Rauch A, Kutalik Z, Descombes P, Cai T, Di Iulio J, Mueller T, et al. Genetic variation in IL28B is associated with chronic hepatitis C and treatment failure: a genome-wide association study. Gastroenterology 2010;138(4) 1338–1345, 1345 e1331–1337.

[108] Aparicio E, Franco S, Parera M, Andres C, Tural C, Clotet B, et al. Complexity and catalytic efficiency of hepatitis C virus (HCV) NS3 and NS4A protease quasispecies influence responsiveness to treatment with pegylated interferon plus ribavirin in HCV/HIV-coinfected patients. J Virol 2011;85(12):5961–9.

[109] Hijikata M, Mishiro S, Miyamoto C, Furuichi Y, Hashimoto M, Ohta Y. Genetic polymorphism of the MxA gene promoter and interferon responsiveness of hepatitis C patients: revisited by analyzing two SNP sites (-123 and -88) in vivo and in vitro. Intervirology 2001;44(6):379–82.

[110] Tanaka Y, Kurosaki M, Nishida N, Sugiyama M, Matsuura K, Sakamoto N, et al. Genome-wide association study identified ITPA/DDRGK1 variants reflecting thrombocytopenia in pegylated interferon and ribavirin therapy for chronic hepatitis C. Hum Mol Genet 2011;20(17):3507–16.

[111] Ochi H, Maekawa T, Abe H, Hayashida Y, Nakano R, Kubo M, et al. ITPA polymorphism affects ribavirin-induced anemia and outcomes of therapy—a genome-wide study of Japanese HCV virus patients. Gastroenterology 2010;139(4):1190–7.

[112] Fellay J, Thompson AJ, Ge D, Gumbs CE, Urban TJ, Shianna KV, et al. ITPA gene variants protect against anaemia in patients treated for chronic hepatitis C. Nature 2010;464(7287):405–8.

[113] Thompson AJ, Fellay J, Patel K, Tillmann HL, Naggie S, Ge D, et al. Variants in the ITPA gene protect against ribavirin-induced hemolytic anemia and decrease the need for ribavirin dose reduction. Gastroenterology 2010;139(4):1181–9.

[114] Ahmed WH, Furusyo N, Zaky S, Eldin AS, Aboalam H, Ogawa E, et al. Pre-treatment role of inosine triphosphate pyrophosphatase polymorphism for predicting anemia in Egyptian hepatitis C virus patients. World J Gastroenterol 2013;19(9):1387–95.

[115] Thursz M. Pros and cons of genetic association studies in hepatitis B. Hepatology 2004;40(2):284–6.

[116] Gunsar F, Akarca US, Ersoz G, Kobak AC, Karasu Z, Yuce G, et al. Two-year interferon therapy with or without ribavirin in chronic delta hepatitis. Antiviral Ther 2005;10(6):721–6.

[117] Niro GA, Ciancio A, Gaeta GB, Smedile A, Marrone A, Olivero A, et al. Pegylated interferon alpha-2b as monotherapy or in combination with ribavirin in chronic hepatitis delta. Hepatology 2006;44(3):713–20.

[118] Wedemeyer H, Yurdaydin C, Dalekos GN, Erhardt A, Cakaloglu Y, Degertekin H, et al. Peginterferon plus adefovir versus either drug alone for hepatitis delta. N Engl J Med 2011;364(4):322–31.

Index

Note: Page numbers followed by "*f*" and "*t*" refer to figures and tables, respectively.

Printed in the United States
By Bookmasters